Phytolith Systematics

Emerging Issues

ADVANCES IN ARCHAEOLOGICAL AND MUSEUM SCIENCE

Volume 1: PHYTOLITH SYSTEMATICS: Emerging Issues
Edited by George Rapp, Jr. and Susan C. Mulholland

Phytolith Systematics

Emerging Issues

Edited by

George Rapp, Jr. and

Susan C. Mulholland

University of Minnesota

Duluth, Minnesota

Published in cooperation with the
Society for Archaeological Sciences

Springer Science+Business Media, LLC

Library of Congress Cataloging-in-Publication Data

Phytolith systematics : emerging issues / edited by George Rapp, Jr., and Susan C. Mulholland.
p. cm. -- (Advances in archaeological and museum science ; v. 1)
Includes bibliographical references and index.
ISBN 978-1-4899-1157-5
1. Archaeological geology. 2. Plant remains (Archaeology)
I. Rapp, George Robert, 1930- . II. Mulholland, Susan C.
III. Series.
CC77.5.P47 1992
930.1--dc20 92-9588
CIP

ISBN 978-1-4899-1157-5 ISBN 978-1-4899-1155-1 (eBook)
DOI 10.1007/978-1-4899-1155-1

Originally published by Plenum Press, New York in 1992
Softcover reprint of the hardcover 1st edition 1992

This volume is dedicated to

MALCOLM H. WIENER

a philanthropist and archaeologist
with a keen vision of the role
of natural science in archaeology.

CONTRIBUTORS

Steven R. Bozarth, Research Assistant of Geography and the Palynology Laboratory, University of Kansas.

Vaughn M. Bryant, Jr., Professor and Head of Anthropology and Director of the Palynology Laboratory, Texas A&M University.

Linda Scott Cummings, Director of the Paleoresearch Laboratories; Faculty Affiliate, Colorado State University.

Elizabeth H. Dinan, Graduate Student of Archaeology, University of Missouri Columbia.

Martin J. Hodson, Senior Lecturer in Environmental Biology, School of Biological and Molecular Sciences, Oxford Polytechnic.

John G. Jones, Adjunct Assistant Professor of Anthropology and Associate Director of the Pollen Laboratory, Texas A&M University; Doctoral Research Fellow, Smithsonian Tropical Research Institute.

Lawrence Kaplan, Professor of Biology, University of Massachusetts Boston.

Elizabeth J. Lawlor, Graduate Student of Anthropology, University of California Riverside.

Susan C. Mulholland, Adjunct Assistant Professor of Anthropology and Associate Director of the Archaeometry Laboratory, University of Minnesota Duluth; Adjunct Assistant Professor, Center for Ancient Studies, University of Minnesota Twin Cities.

Amy L. Ollendorf, PhD candidate, Center for Ancient Studies and Limnological Research Center, University of Minnesota, Twin Cities.

Deborah M. Pearsall, Assistant Professor of Archaeology, University of Missouri Columbia.

Alix H. Powers, SERC Post-doctoral Fellow of Archaeology, University of Cambridge.

George Rapp, Jr., Professor of Geology and Archaeology and Director of the Archaeometry Laboratory, University of Minnesota Duluth; Professor, Center for Ancient Studies, University of Minnesota Twin Cities; Research Professor of Archaeology, Boston University.

Arlene Miller Rosen, Research Fellow, Environmental Sciences and Energy Research, Weizmann Institute of Science.

Irwin Rovner, Associate Professor of Anthropology, North Carolina State University.

John C. Russ, Visiting Associate Professor of Materials Science and Engineering, North Carolina State University.

Allan G. Sangster, Professor of Natural Sciences, Glendon College, York University.

Mary B. Smith, Biology Technician, University of Massachusetts Boston.

Lesley A. Sneddon, Regional Ecologist, Nature Conservancy Boston.

Page C. Twiss, Professor of Geology, Kansas State University.

SERIES FOREWORD

This volume is the first in the *Advances in Archaeological and Museum Science* series sponsored by the Society for Archaeological Sciences. The purpose of this biennial series is to provide summaries of advances in closely defined topics in archaeometry, archaeological science, environmental archaeology, preservation technology and museum conservation.

The Society for Archaeological Sciences (SAS) exists to encourage interdisciplinary collaboration between archaeologists and colleagues in the natural and physical sciences. SAS members are drawn from many disciplinary fields. However, they all share a common belief that physical science techniques and methods constitute an essential component of archaeological field and laboratory studies.

The General Editors wish to express their appreciation to Renee S. Kra and Frances D. Moskovitz of *Radiocarbon* for their special expertise and assistance in the production of this volume. We also appreciate the contribution of the two reviewers for their excellent comments and suggestions. The General Editor responsible for undertaking the development of this volume was R. E. Taylor.

PREFACE

The decision to undertake this volume is based on our firm belief that phytolith studies can make a major contribution to the solution of archaeological problems related to human use of plants and the environmental context of human evolution and cultural change. Archaeological applications of phytolith systematics remain handicapped by the paucity of data on many important plant groups as well as the availability of adequate keys and/or comparative collections. We hope this volume will add materially to the corpus of available systematics for some of these important plant groups. The annotated bibliography in particular should provide a major "jump-start" for new researchers in the field.

As a further aid to readers not familiar with botanical nomenclature, both old and new forms of selected family names have been listed throughout the text. The names of these families (Poaceae/Gramineae, Fabaceae/Leguminosae, Asteraceae/Compositae, Brassicaceae/Cruciferae, Apiaceae/Umbelliferae) were judged to be in approximately equal use in the recent literature. Additional families for which alternative names have been proposed are most widely known by the old form (*i.e.*, Palmae), which is used in this text. No attempt was made to standardize generic or specific epithets to a single reference.

The first chapters recruited for this volume were based on papers presented at the Third Phytolith Research Workshop on January 23–24, 1988, in Columbia, Missouri. We thank Deborah Pearsall and Dolores Piperno for arranging this conference. Many thanks go to the anonymous reviewers who reviewed individual chapters and the book as a whole. The comments and suggestions were generally very helpful. Dwight Brown provided particularly insightful comments and criticisms.

For the last decade all of the books published under the editorship or authorship of the senior editor would not have the grammatical and typographical quality were it not for the broad talents and sharp eyes of College of Science and Engineering editor Judith Holz. James K. Huber assisted with the indices and translation of Spanish articles; Kathy Skelton and Avis Hedin checked the citations in the annotated bibliography. We also want to thank Avis Hedin, Cindy Ristow, Tessa Schwanekamp and Michael Tinsley for word processing the manuscript. A special acknowledgment is made to Susan Kesti and Mary Menzel of the University of Minnesota, Duluth, Library staff for countless hours of tracking down and obtaining the widely scattered phytolith literature.

George Rapp, Jr.
Susan C. Mulholland

CONTENTS

FIGURES

TABLES

1

PHYTOLITH SYSTEMATICS: AN INTRODUCTION

SUSAN C. MULHOLLAND and GEORGE RAPP, JR.

ABSTRACT

Phytoliths, microscopic mineral deposits in plants, have the potential for encoding significant archaeological and paleoenvironmental information. Although common throughout the plant kingdom and preserved under varied environmental conditions, only recently have phytoliths been examined in the systematic manner that yields data useful for interpretation. The most crucial area remains systematics—the description of shape and distribution data within the plant kingdom. Calcium phytoliths have been observed from the beginnings of microscopy, yet crystal and amorphous shapes are still described only in the most general of terms. Opal phytoliths have received much recent attention, but many plant taxa are yet to be carefully examined. A wide variety of approaches has been taken towards phytolith classification, often based on specific applications of the data. This volume presents some of the current research on various topics within phytolith systematics.

BACKGROUND

Phytoliths are botanical microfossils that can provide significant paleoenvironmental and archaeobotanical information (Rovner 1983b, 1988). Phytoliths are here defined as mineral deposits that form in and between plant cells. Cysts from one-celled organisms such as diatoms and algae have been studied separately and thus are not considered here. Many plants absorb various chemical elements in solution from groundwater, which are then deposited in controlled locations within the plant. Calcium deposits usually are crystalline, particularly calcium oxalate but also calcium phosphate and calcium carbonate (Lanning 1961). However, some calcium deposits are amorphous (Arnott 1976: 73). Opaline silica deposits are generally amorphous in nature (Jones & Handreck 1967) but crystalline phases have been reported in some material (Wilding & Drees 1974). Much recent systematic research has focused on opal phytoliths, although as

Phytolith Systematics, Edited by G. Rapp, Jr. and
S.C. Mulholland, Plenum Press, New York, 1992

indicated by papers in this volume (Cummings, Jones & Bryant), calcium phytoliths also possess the necessary attributes for microfossil applications.

Distribution of opal phytoliths in angiosperms, gymnosperms and pteridophytes has been reviewed by Piperno (1988: 21, 23–37). Opal phytolith production is widespread in both monocotyledons and dicotyledons. Families that are well known as consistent accumulators of identifiable silica bodies include Poaceae or Gramineae (grass), Cyperaceae (sedge), Ulmaceae (elm), Fabaceae or Leguminosae (bean), Cucurbitaceae (squash) and Asteraceae or Compositae (sunflower). However, opal deposition is not universal because many families yield little or no identifiable phytoliths. This does not discount all members of these families. Even families that generally lack silica may have species that produce distinctive phytoliths. For example, the Pinaceae generally lack phytoliths (Piperno 1988: 21) but a distinctive type is reported for *Pseudotsuga menzeisii* (Brydon, Dore & Clark 1963; Norgren 1973).

Distribution of calcium phytoliths is also widespread. Calcium oxalate is reported in numerous families (McNair 1932) and occurs in every plant part (Franceschi & Horner 1980). Some families appear to be consistent accumulators. The Cactaceae are noted for calcium crystals (Franceschi & Horner 1980). Scurfield, Michell and Silva (1973) report crystals from stems of woody perennials; Chattaway (1953, 1955, 1956) reports on similar occurrences in many tropical woody species. Calcium carbonate is reported for the Urticaceae (nettle), Moraceae (mulberry), Acanthaceae (acanthus) and Cannabinaceae (hemp) (Simkiss & Wilbur 1989: 125). Although most deposits are in cell walls, distinctive structures called cystoliths do form in some plants.

Phytoliths tend to be liberated and deposited through decay-in-place mechanisms (Dimbleby 1978: 129). Deposition normally occurs through surface or shallow subsurface decomposition of plants; thus phytoliths are incorporated directly into soils and sediments. Fire and strong wind erosion, however, can and do expose phytoliths to wind transport. Transport over considerable distances is indicated by recovery of phytoliths in atmospheric dust (Folger, Burckle & Heezen 1967; Twiss, Suess & Smith 1969). The question of water transport has not been addressed in detail, although ocean currents are thought to be responsible for some anomalous phytolith concentrations near the Cape Verde Islands (Melia 1984: 354). Phytoliths have also been reported in tap water (Baker 1959b: 82).

Phytoliths can be stable components of sediment horizons (Rovner 1988: 155). Although susceptible to solution under conditions of extreme pH, under normal conditions phytoliths can survive for long periods of time. Paleocene sedimentary rocks approximately 60 million years old have yielded recognizable opal phytoliths (Jones 1964). More recent sediments that contain phytoliths include periglacial sediments (Fredlund, Johnson & Dort 1985) and soils from a wide range of modern habitats: swamps, prairies, grasslands and forests. Phytoliths have also been found in ocean cores (Bukry 1987) and lake cores

(Piperno 1985b). Calcium phytoliths have not been reported from many sedimentary contexts but have been recovered from recent coprolites (Bryant 1974).

The combination of fairly widespread production in plants and good preservation in many sediments gives phytoliths a great potential value for archaeological and paleobotanical work. As with any fossil system, however, there are advantages and disadvantages to phytolith analysis. A small number of scientists is investigating many facets of phytolith production and deposition in order to study ancient environments and early agriculture.

The degree to which plant taxa can be identified by phytoliths is one major area of research. Identification of phytoliths from sediments is based directly on a detailed knowledge of modern plant phytolith production. Various studies indicate that phytolith production is high in many families of monocots and dicots (Franceschi & Horner 1980; Piperno 1988). In microfossil applications, phytoliths need to be in some way distinctive as well as durable in sediments. Studies for archaeological and paleoecological goals (Pearsall 1979; Piperno 1985a) thus differ in the attributes examined from those for botanical taxonomy or compositional analysis (Metcalfe 1960; Lanning 1961). Morphology is emphasized over distribution patterns within tissue or elemental composition.

The amount of morphological variation differs from taxon to taxon but the specificity of identification in some taxa is greater than that offered by other microfossils. For example, grass pollen grains are largely unidentifiable below family level. Opal phytoliths, although subject to factors of multiplicity (production of many forms in one taxon) and redundancy (occurrence of a form in many taxa) (Rovner 1971), are abundant and morphologically diverse in the grasses (Mulholland 1989). Sedges are another family where detailed analysis of phytolith shapes may yield significant information not available from other microfossils (Ollendorf, Mulholland & Rapp 1987).

Taphonomic issues are also crucial for accurate interpretation of fossil material. The different methods of dispersal associated with each type of microfossil affect distribution in sediments. Pollen grains, as part of the reproductive process, are produced only in flower parts and are transferred from flower to flower by either wind or insects. Wind-pollinated plants produce numerous grains that can be transported great distances, yielding information on regional vegetation. Insect-pollinated plants produce far fewer grains and so are usually under-represented in the regional pollen record. Phytoliths, in contrast, occur within many plant parts, and are released by destruction of the plant tissue. Normal decomposition in the humus layer introduces phytoliths directly into a soil, producing a record of local vegetation. Fire and strong erosional processes introduce phytoliths into the atmosphere and hydrosphere where long-distance transport is possible.

The various physical and chemical processes active in soils and sediments affect fossils in different ways. Opaline silica is susceptible to dissolution under strongly alkaline conditions (Iler 1979: 41, 65). Strongly alkaline sediments (such as shell middens) therefore would not be expected to contain many identifiable phytoliths. However, some factors appear to ameliorate the dissolution process. Carbon coatings on the surfaces of phytoliths can shield against high pH conditions, as indicated by recovery of phytoliths in sediments with pH of 8.2–9.6 (Lewis 1981). Higher aluminum content and lower water content within the opaline silica deposit also retard dissolution (Bartoli & Wilding 1980). Calcium oxalate phytoliths are subject to dissolution in acidic environments. Under the proper conditions, calcium crystals can be preserved (Andrejko, Cohen & Raymond 1983).

HISTORY

Phytoliths have been recognized from the beginnings of microscopy. Loeuwenhoek observed calcium phytoliths in 1675 (Arnott 1976: 57). Silica phytoliths are reported somewhat later; the first study appeared in 1835 (Rovner 1983a). In 1836, C. G. Ehrenberg began systematic descriptions of phytoliths from plants and sediments (Baker 1960). Ehrenberg considered these bodies to be skeletons from micro-organisms living inside plants. In 1866, however, he recognized that some forms were mineralized plant tissue rather than independent organisms. Also in 1866, Ruprecht defined the term "phytolith" as microscopic opaline bodies deposited in plants (Baker 1959a). After the period of initial discovery, phytolith research was sporadic. Descriptive work by various microscopists continued for the next several decades. For example, phytoliths from dustfalls were classified using Ehrenberg's system (Baker 1960).

Research relating phytoliths to plant anatomy and physiology was undertaken around 1900 (Piperno 1988: 4–5). Netolitzky (1929) published a review of early botanical research on opal phytoliths; see McNair (1932) for distribution of calcium phytoliths (Franceschi & Horner 1980: 381). Rovner (1983a) reports a steadily increasing amount of work between the two World Wars. The relation of phytoliths to plant physiology and applications to agronomy continued to be investigated during this time. A great deal of information on phytolith occurrence, morphology and variation is reported in these works, focusing on distribution of minerals in plant families. Phytoliths also were used in conjunction with other epidermal characteristics for taxonomic purposes (Prat 1936). Although significant taxonomic correlations were noted in epidermal microstructure, this methodology was essentially abandoned with the onset of chromosomal research.

The immediate post-World War II period brought a great increase in opal phytolith research. Soil scientists applied phytolith analysis to paleoenvironmental reconstruction as well as paleosol identification and dating. Environmental reconstruction focused mainly on distinguishing between prairie and forest soils (Witty & Knox 1964). Most studies analyzed modern soils as references in order to characterize various environments by phytolith content. Although grass phytoliths were present in all modern sediments, the quantity was much lower in forest soils. These data were then applied to paleosols and soils in various geographic areas (Gross 1973; Jones & Hay 1975).

Phytoliths were associated with their plant progenitors only infrequently in pedological studies; sediment samples were the primary concern. Systematic description of opal phytoliths in plants is more recent, although foreshadowed by work in the early 1900s. Botanists have documented phytoliths (as well as other anatomical structures) in scanning electron microscope studies of epidermal tissue (Hayward & Parry 1980; Terrell & Wergen 1981). Leaf and inflorescence tissue have been most intensively studied, although stem and root tissue have also been found to contain silica. Other botanical studies have focused on deposition processes (Sangster & Parry 1981) or the function of silica in the plant (Kaufman *et al* 1981).

Calcium phytoliths have not been studied extensively from a paleoecological viewpoint. Although fairly widely distributed in plants (Franceschi & Horner 1980), few instances of recovery from sediments are reported. This situation may reflect poor preservation in many sedimentary contexts. Coprolites, representing a special depositional context, have yielded calcium phytoliths (Bryant 1974; Cummings 1989). The scarcity of reports from other contexts, however, may reflect poor recovery methods rather than actual absence of material. Silica phytoliths were not reported from many contexts until techniques were developed specifically to recover opaline silica. Many such techniques incorporate acid washes and so would not recover calcium phytoliths.

Two types of research are currently under way in phytolith studies: systematics of modern plant phytoliths and interpretation of phytoliths recovered from sediments (including archaeological contexts). A great deal of information on opal phytolith production and distribution has been generated in the last two decades as a result of archaeological and paleoecological studies. Numerous species from Ecuador (Pearsall 1979), Hawaii (Pearsall & Trimble 1984), Panama (Piperno 1984, 1985a) and the central United States (Brown 1984) have been examined. In addition, botanical studies have reported distribution of silica in a wide range of plant taxa. Rediscovery of the German literature especially brought attention to numerous studies of the occurrence of silica in various families. Calcium phytoliths, particularly calcium oxalate, have been reported in the botanical literature consistently. Several reviews (Arnott 1976; Franceschi & Horner 1980) summarize occurrence in plant families and parts. A fairly large body of data can be applied to interpretation of phytoliths recovered from sediments. However, numerous taxa have not been examined, and detailed morphological studies are needed for many others.

Paleoecological studies have applied opal phytolith analysis to a wide variety of samples. Paleoenvironmental reconstruction has been made from glacial age sediments (Twiss, Suess & Smith 1969; Fredlund, Johnson & Dort 1985) as well as more recent sediments (Jones & Beavers 1964; Wilding & Drees 1973). Phytoliths have been used to identify paleosol horizons (Dormaar & Lutwick 1969) and past vegetation (Gross 1973). Cores of ocean sediments have also yielded information on paleoenvironment, including wind direction (Melia 1984; Bukry 1987).

Numerous archaeological research questions have been addressed in phytolith research. Identification of cultigens includes maize (Pearsall 1978; Piperno 1984), rice (Fujiwara, Jones & Brockwell 1985) and various Old World cereals (Helbaek 1961; Rosen 1987). Coprolites have yielded phytoliths attributed to cactus or agave (Bryant 1974). Studies of farming practice include identification of field surfaces (Pearsall & Trimble 1984), mulching practices (Turner & Harrison 1981) and irrigation (Rosen 1987). Phytoliths from archaeological sediments have also been used to attempt environmental reconstruction. Carbone (1977) interpreted past environments, mostly forests, by comparison to modern A horizons; Lewis (1981) and MacDonald (1974) documented changing types of prairie from Paleoindian to recent times.

SYSTEMATICS

Most calcium phytoliths are crystalline, usually forming one of five common shapes (Franceschi & Horner 1980: 381). Raphides are needle-like individual crystals that aggregate into bundles within a single cell. The styloid is similar in shape to individual raphides but is thicker and occurs singly. Prismatic calcium phytoliths form various blocky geometric shapes. The druse is an aggregate of many crystals into a rough spherical shape. Crystal sand refers to a group of small crystals occurring together in a single cell. Combinations and variations of the five common shapes can also occur.

The five common crystal shapes have been recognized since at least the early 1900s (Haberlandt 1914). Franceschi and Horner (1980: 383) report that shape and location characteristics have been used in classification. Their summary tables (pp 384–393, 397–401) illustrate the variety of plant parts and species that produce calcium phytoliths. More detailed descriptions of phytolith shape, however, will be needed for identification of plant origin from fossil material.

The chemical composition of calcium crystals can also be used in classifications. Calcium oxalate occurs in monohydrate (whewellite) and dihydrate (weddellite) forms. Either one or the other is reported for individual species (Franceschi & Horner 1980: 397–401). Other calcium compounds reported include carbonates and phosphates (Lanning 1961).

Several approaches have been taken to opal phytolith systematics. One is to treat each phytolith type as a separate entity and assign Latin names following the Linnean system. Taxa are described solely on the basis of morphology; groupings are made following the rules for parataxa (Bukry 1979). Although it is recognized that phytoliths formed in plants, no effort is made to identify the original plant. This approach was originated by Ehrenberg (1854). At first he considered opal phytoliths to be siliceous skeletons of micro-organisms; as such, the Linnean taxonomy was appropriate (similar to diatoms). However, he kept this taxonomic scheme even when later acknowledging that opal phytoliths are actually silicified cells or sections of plants. Ehrenberg eventually described and classified over 90 phytolith "species."

Micropaleontologists still generally follow this approach. Dumitrica (1973) formalized Ehrenberg's system as parataxa, describing three Gramineae (Poaceae) and one Equisetale paragenera. This system is prevalent in reports from ocean cores, although more recent work recognizes other classification approaches (discussed below). Bukry (1979) feels the descriptive approach of Twiss, Suess and Smith (1969) is equally as useful as the Linnean taxonomic approach. Locker and Martini (1986) feel neither approach is very satisfactory, although they ultimately follow the parataxonomic approach. Melia (1984) merely identifies bodies as opal phytoliths, sidestepping the entire issue of classification.

The rationale behind the parataxonomic approach is twofold. First, the International Code of Botanical Nomenclature allows the assignment of Linnean names to skeletal parts of plants and animals. This procedure is often useful in paleontology where origin of the fossils is unknown. For example, the conodonts is one group classified into parataxa. However, phytoliths are known to originate from plants. Extant taxa produce phytoliths that can often be identified to family or lower level.

The second reason underlying the parataxonomic approach rests on the degree to which identification of the original plant is thought possible. Dumitrica (1973: 940) quotes Smithson (1958: 154) as saying that phytoliths can at best be identified only to tribe level, not species. Bukry (1979) states that the "fundamental forms" are widely distributed in plants and, furthermore, a single leaf can produce different types of phytoliths. These observations are true to some degree, particularly for the Poaceae (Smithson 1958 addresses only this family). Multiplicity and redundancy potentially play a large role in grass opal phytolith production. However, even in the Poaceae, more variability in morphology and more consistency in production of phytoliths are present than is apparent from superficial studies. Identification of original plants from fossil material requires detailed morphological studies on local modern reference material. Brown (1984) elaborates on the classification of Twiss, Suess and Smith (1969), illustrating the great diversity within basic grass phytolith shapes. Pearsall (1979) and Piperno (1984) specifically address the identification of maize by cross-shaped bodies. Piperno (1985a) further identifies silica in many tropical plant families, including dicotyledons. Her work indicates that opal phytoliths are often distinctive at the genus or species level (Piperno 1988: 248–253).

Botanists have approached opal phytolith systematics based on articulated tissue segments. Since orientation and location information, as well as morphology, are available with in situ phytoliths, a greater degree of identification can be achieved than is possible with disaggregated phytoliths alone. Combined with other anatomical characteristics, a fine degree of identification normally can be made. The phytolith types are named after morphological, locational and orientation characteristics (eg, saddles, spheres, horizontally elongated with sinuous edges).

The botanical approach is illustrated in Metcalfe (1960). Phytoliths (or silica bodies) are one of a number of anatomical characteristics recorded for the Poaceae. Location and orientation of phytoliths are described, as well as morphology. For example, phytoliths over veins are distinguished from those between veins. One subfamily, the Oryzoideae, is identified by orientation of dumbbell-shaped phytoliths perpendicular to leaf length (as opposed to parallel). Sedges and palms are treated in much the same manner (Tomlinson 1969; Metcalfe 1971). The extensive research in the early 1900s was also oriented toward the botanical aspect of phytoliths; context within the plant was stressed.

A third approach is to base classification solely on the morphology of the phytolith. Classifications that incorporate in situ orientation and location characteristics are not applicable to disaggregated phytoliths. Most sedimentary contexts do not yield articulated fragments. In compensation, disaggregated phytoliths allow a greater amount of three-dimensional morphology to be described because tissue surrounding in situ phytoliths tends to obscure morphological details.

Two methods of describing phytolith types have been applied in the morphological approach. In one, only morphological descriptions and names are used. Phytolith types are based on geometric shapes, usually in outline. This method has the advantage of being easily applicable and understandable. The other method relates phytolith types to plant anatomy more than does the geometric method. Phytolith types are named after the plant elements that are silicified. For example, long cylinders may be sclereids or tracheids. This method requires a greater knowledge of plant anatomy. Morphology is still the only characteristic used but with this method it is used to identify origin in plant tissue. The advantages are obvious—the more information recoverable from phytoliths, the more valuable interpretation will be.

Twiss, Suess and Smith (1969) illustrated the geometric method of morphological classification for disaggregated phytoliths. Their classification groups grass phytoliths into 4 classes with a total of 26 types. The morphological names are based on the appearance in outline: saddles, dumbbells, etc. This scheme has been widely applied (Lewis 1981; Fredlund, Johnson & Dort 1985). Brown (1984) expanded on Twiss's classification and added trichomes (a class based on plant anatomy) to the list of geometric classes. Piperno (1988) illustrates (and advocates) the anatomical method. All primary divisions within her scheme are based on cell type; secondary divisions are based on morphology. Interpretation of fossil material is enhanced by incorporating anatomical information. For example, identification of sclereids indicates presence of woody plants even when species was not determined (Piperno 1988: 155).

CURRENT STATE OF PHYTOLITH SYSTEMATICS

Classification schemes are constructed to bring some order to large bodies of data. For phytoliths one can construct classifications based on cladistics, genetics, plant anatomy, morphometry or other variations resulting from the innate nature of phytolith formation in plants. For archaeological research, useful classifications must address the state of the phytoliths as found in archaeological and paleoenvironmental contexts. The overwhelming majority of fossil phytoliths from these sedimentary contexts are single morphs, disarticulated from the structural network in the original host plant. Hence, classification schemes designed by those seeking to apply phytolith studies to the solution of archaeological problems use a morphometric or morphometric/anatomical context rather than the cladistic context of botany.

The papers collected in this volume address various aspects of phytolith systematics. A short history and a delineation of general issues in phytolith classification are presented in this introductory chapter. A variety of approaches to systematics have been employed historically; many are reviewed or developed further in this volume. A more detailed review of the classifications developed in Europe is presented by Powers, providing additional historical background (for a general review of phytolith analysis worldwide, see Piperno 1988: 1–10). Pearsall and Dinan propose a classification system for opal phytoliths based on more than a decade of research. This classification provides a descriptive framework for all phytoliths based on photographic types and morphology.

Many chapters address specific plant taxa or parts. Classifications of grass silica-bodies (Mulholland & Rapp) and sedge cones (Ollendorf) partly review and partly reorganize the groups based on new data. Twiss discusses the correlation between grass silica-body shapes and C_3/C_4 pathways, with implications for paleoclimatic interpretation. Identification of food plants is the subject of several papers. Rosen discusses wheat and barley, as well as weed grasses and date palms. Kaplan presents data on a wide variety of cultigens. Cummings illustrates and describes numerous calcium and silica phytoliths from wild and domestic plants. Bozarth has worked with dicots, mostly wild, from the central United States. Jones and Bryant have tackled the classification of calcium phytoliths of the cacti. Sangster and Hodson provide a timely reminder that underground portions of plants can also be silica accumulators.

The final chapter and appendix address the future and past of phytolith systematics. Rovner and Russ present a new approach to classification, drawing on automated techniques developed for stereology and morphometry. Principles and techniques of classification will provide a rich field for research for some time to come. The annotated bibliography is provided as a starting point for beginning phytolith analysts. Although the bibliography is not exhaustive (particularly in regard to calcium phytoliths), an attempt has been made to include as many citations as possible that contain systematic information. The citations range from articles dealing exclusively with phytolith systematics, morphology and classification to those that contain a few phytolith descriptions or illustrations as part of a larger study. The brief annotations act as a guide to the depth and type of information available.

Much work remains to be done in phytolith systematics in order to provide increasingly sophisticated interpretations of archaeological material. The kind of study Piperno (1988) has done in Panama needs to be duplicated in other areas. Detailed morphological descriptions of many groups have yet to be made. Russ and Rovner (1989) point to the general lack of quantitative data on the extent of variation in size and shape parameters for phytolith forms within a taxon. However, phytoliths are already being applied to a wide variety of archaeological and paleoecological situations. A second volume derived from the Third Phytolith Research Workshop is devoted to applications of phytolith analysis (Pearsall and Piperno, in prep); additional research is published in a wide variety of journals.

A considerable amount of phytolith information is currently available. In addition to specific studies on morphology, much systematics is "buried" in botanical literature. Studies of plant anatomy may contain information on phytolith location. Plant taxonomies potentially contain morphological data from studies based on articulated segments. Even studies of compositional analysis can indicate whether specific taxa contain silica, although whether distinct bodies are present is often not specified. These data need to be collated and supplemented by additional, regional studies as well as thorough studies of major plant taxa. This volume constitutes part of the continuing effort to increase our knowledge of phytolith systematics.

REFERENCES

Andrejko, MJ, Cohen, AD and Raymond, R, Jr 1983 Origin of mineral matter in peat. *In* Raymond, R, Jr and Andrejko, MJ, eds, *Mineral Matter in Peat.* Los Alamos, New Mexico, Los Alamos National Laboratory: 3–24.

Arnott, HJ 1976 Calcification in higher plants. *In* Watabe, N and Wilbur, KM, eds, *The Mechanisms of Mineralization in the Invertebrates and Plants.* Columbia, South Carolina, University of South Carolina Press: 55–78.

Baker, G 1959a Fossil opal-phytoliths and phytolith nomenclature. *Australian Journal of Science* 21: 305–306.

_____1959b Opal phytoliths in some Victorian soils and "red rain" residues. *Australian Journal of Botany* 7: 64–87.

_____1960 Phytolitharien. *Australian Journal of Science* 22: 392–393.

Bartoli, F and Wilding, LP 1980 Dissolution of biogenic opal as a function of its physical and chemical properties. *Soil Science Society of America, Journal* 44: 873–878.

Brown, DA 1984 Prospects and limits of a phytolith key for grasses in the central United States. *Journal of Archaeological Science* 11: 345–368.

Bryant, VM, Jr 1974 The role of coprolite analysis in archaeology. *Bulletin of the Texas Archeological Society* 45: 1–28.

Brydon, JE, Dore, WG and Clark, JS 1963 Silicified plant asterosclerids preserved in soil. *Soil Science Society of America, Proceedings* 27: 476–477.

Bukry, D 1979 Comments on opal phytoliths and stratigraphy of Neogene silico-flagellates and coccoliths at Deep Sea Drilling Project Site 397 off northwest Africa. *In* Shamback, JD, ed, *Initial Reports of the Deep Sea Drilling Project* 49: 977–1009. Washington, DC, US Government Printing Office.

_____1987 North Atlantic Quaternary silicoflagellates, Deep Sea Drilling Project Leg 94. *In* Orlfsky, S, ed, *Initial Reports of the Deep Sea Drilling Project* 94(2): 779–783. Washington, DC, US Government Printing Office.

Carbone, VA (ms) 1977 Environment and prehistory in the Shenandoah Valley. PhD dissertation, The Catholic University of America: 227 p.

Chattaway, MM 1953 The occurrence of heartwood crystals in certain timbers. *Australian Journal of Botany* 1: 27–38.

_____1955 Crystals in woody tissues: Part I. *Tropical Woods* 102: 55–74.

_____1956 Crystals in woody tissues: Part II. *Tropical Woods* 104: 100–124.

Cummings, L Scott (ms) 1989 Coprolites from medieval Christian Nubia: An interpretation of diet and nutritional stress. PhD dissertation, University of Colorado: 204 p.

Dimbleby, GW 1978 *Plants and Archaeology.* Atlantic Heights, New Jersey, Humanities Press, Inc: 190 p.

Dormaar, JF and Lutwick, LE 1969 Infrared spectra of humic acids and opal phytoliths as indicators of paleosols. *Canadian Journal of Soil Science* 49: 29–37.

Dumitrica, P 1973 Phytolitharia. *In* Kaneps, AG, ed, *Initial Reports of the Deep Sea Drilling Project* 13(2): 940–943. Washington, DC, US Government Printing Office.

Ehrenberg, CG 1854 *Mikrogeologie,* 2 volumes. Leipzig, Leopold Voss: 374 p.

Folger, DW, Burckle, LH and Heezen, BC 1967 Opal phytoliths in a North Atlantic dust fall. *Science* 155: 1243–1244.

Franceschi, VR and Horner, HT, Jr 1980 Calcium oxalate crystals in plants. *Botanical Review* 46: 361–427.

Fredlund, GG, Johnson, WC and Dort, W, Jr 1985 A preliminary analysis of opal phytoliths from the Eustis Ash Pit, Frontier County, Nebraska. Institute for Tertiary-Quaternary Studies, *TER-QUA Symposium Series* 1: 147–162.

Fujiwara, H, Jones, R and Brockwell, S 1985 Plant opals (phytoliths) in Kakadu archaeological sites: A preliminary report.

In Jones, R, ed, *Archaeological Research in Kakadu National Park*. Australian National Parks and Wildlife, Special Publication 13, Canberra, Australian National University: 155–164.

Gross, ER (ms) 1973 Buried soils of the drainageways in the driftless area of the upper Mississippi Valley. PhD dissertation, University of Minnesota: 195 p.

Haberlandt, G 1914 *Physiological Plant Anatomy*. London, Macmillan and Co: 777 p.

Hayward, DM and Parry, DW 1980 Scanning electron microscopy of silica deposits in the culms, floral bracts, and awns of barley (*Hordeum sativum* Jess.). *Annals of Botany* 46: 541–548.

Helbaek, H 1961 Studying the diet of ancient man. *Archaeology* 14: 95–101.

Iler, RK 1979 *The Chemistry of Silica*. New York, John Wiley & Sons: 866 p.

Jones, LHP and Handrek, KA 1967 Silica in soils, plants, and animals. *Advances in Agronomy* 19: 107–149.

Jones, RL 1964 Note on occurrence of opal phytoliths in some Cenozoic sedimentary rocks. *Journal of Paleontology* 38: 773–775.

Jones, RL and Beavers, AH 1964 Variation of opal phytolith content among some great soil groups in Illinois. *Soil Science Society of America, Proceedings* 28: 711–712.

Jones, RL and Hay, WW 1975 Bioliths. *In* Giesking, JE, ed, *Soil Components. II*. New York, Springer-Verlag: 481–496.

Kaufman, PB, Dayanandan, P, Takeoka, Y, Bigelow, WC, Jones, JD and Iler, R 1981 Silica in shoots of higher plants. *In* Simpson, TL and Volcani, BE, eds, *Silicon and Siliceous Structures in Biological Systems*. New York, Springer-Verlag: 409–449.

Lanning, FC 1961 Calcite in *Lesquerella ovalifolia* trichomes. *Science* 133: 380.

Lewis, RO 1981 Use of opal phytoliths in paleoenvironmental reconstruction. *Journal of Ethnobiology* 1: 175–181.

Locker, S and Martini, E 1986 Phytoliths from the southwest Pacific, site 591. *In* Blakeslee, JH, ed, *Initial Reports of the Deep Sea Drilling Project* 90(2): 1079–1084. Washington, DC, US Government Printing Office.

MacDonald, LL (ms) 1974 Opal phytoliths as indicators of plant succession in North-Central Wyoming. MS thesis, University of Wyoming: 71 p.

McNair, JB 1932 The interrelation between substances in plants: Essential oils and resins, cyanogen and oxalate. *American Journal of Botany* 19: 255–271.

Melia, MB 1984 The distribution and relationships between palynomorphs in aerosols and deep-sea sediments off the coast of northwest Africa. *Marine Geology* 58: 345–371.

Metcalfe, CR 1960 *Anatomy of the Monocotyledons. I. Gramineae*. Oxford, Clarendon Press: 731 p.

_____1971 *Anatomy of the Monocotyledons. II. Cyperaceae*. Oxford, Clarendon Press: 597 p.

Mulholland, SC 1989 Phytolith shape frequencies in North Dakota grasses: A comparison to general patterns. *Journal of Archaeological Science* 16: 489–511.

Netolitzky, F 1929 Die kieselkorper. *In* Linsbauer, K, ed, *Hanbuch der Pflanzenanatomie* 3(1a): 1–19. Berlin, Gebruder Borntraeger.

Norgren, JA (ms) 1973 Distribution, form and significance of plant opal in Oregon soils. PhD dissertation, Oregon State University: 165 p.

Ollendorf, AL, Mulholland, SC and Rapp, G, Jr 1987 Phytoliths from some Israeli sedges. *Israel Journal of Botany* 36: 125–132.

Pearsall, DM 1978 Phytolith analysis of archaeological soils: Evidence for maize cultivation in Formative Ecuador. *Science* 199: 177–178.

_____(ms) 1979 The application of ethnobotanical techniques to the problem of subsistence in the Ecuadorian Formative. PhD dissertation, University of Illinois: 267 p.

Pearsall, DM and Trimble, MK 1984 Identifying past agricultural activity through soil phytolith analysis: A case study from the

Hawaiian Islands. *Journal of Archaeological Science* 11: 119–133.

Piperno, DR 1984 A comparison and differentiation of phytoliths from maize and wild grasses: Use of morphological criteria. *American Antiquity* 49: 361–383.

_____1985a Phytolith analysis and tropical paleo-ecology: Production and taxonomic significance of siliceous forms in New World plant domesticates and wild species. *Review of Palaeobotany and Palynology* 45: 185–228.

_____1985b Phytolith analysis of geological sediments from Panama. *Antiquity* 59: 13–19.

_____1988 *Phytolith Analysis: An Archaeological and Geological Perspective.* New York, Academic Press: 280 p.

Prat, H 1936 La systematique des Graminees. *Annales des Sciences Naturelles, Botanique, Series* 10, 18: 165–258.

Rosen, AM 1987 Phytolith studies at Shiqmim. *In* Levy, TE, ed, Shiqmim I: Studies concerning Chalcolithic societies in the Northern Negev Desert, Israel (1982–1984), *British Archaeological Reports International Series* 356: 243–249.

Rovner, I 1971 Potential of opal phytoliths for use in paleoecological reconstruction. *Quaternary Research* 1: 343–359.

_____(ms) 1983a Multi-disciplinary sense and nonsense: Is a science of phytoliths really necessary? Paper presented at the 149th AAAS Annual Meeting, Detroit, Michigan.

_____1983b Plant opal phytolith analysis: Major advances in archaeobotanical research. *In* Schiffer, M, ed, *Advances in Archaeological Method and Theory* 6: 225–266. New York, Academic Press.

_____1988 Macro- and micro-ecological reconstruction using plant opal phytolith data from archaeological sediments. *Geoarchaeology* 3: 155–163.

Russ, JC and Rovner, I 1989 Stereological identification of opal phytolith populations from wild and cultivated *Zea*. *American Antiquity* 54: 784–792.

Sangster, AG and Parry, DW 1981 Ultrastructure of silica deposits in higher plants. *In* Simpson, TL and Volcani, BE, eds, *Silicon and Siliceous Structures in Biological Systems.* New York, Springer-Verlag: 383–407.

Scurfield, G, Michell, AJ and Silva, SR 1973 Crystals in woody stems. *Botanical Journal of the Linnean Society of London* 66: 277–289.

Simkiss, K and Wilbur, KM 1989 Plant mineralization—Ions, silicification, and the transpiration stream. *In* Simkiss, K and Wilbur, KM, *Biomineralization: Cell Biology and Mineral Deposition.* New York, Academic Press: 106–130.

Smithson, F 1958 Grass opal in British soils. *Journal of Soil Science* 9: 148–155.

Terrell, EE and Wergen, WP 1981 Epidermal features and silica deposition in lemmas and awns of *Zizania* (Gramineae). *American Journal of Botany* 68: 697–707.

Tomlinson, PB 1969 *Anatomy of the Monocotyledons. II. Palmae.* Oxford, Clarendon Press: 453 p.

Turner, BL, II and Harrison, PD 1981 Prehistoric raised-field agriculture in the Maya lowlands. *Science* 213: 399–405.

Twiss, PC, Suess, E and Smith, RM 1969 Morphological classification of grass phytoliths. *Soil Science Society of America, Proceedings* 33: 109–115.

Wilding, LP and Drees, LR 1973 Scanning electron microscopy of opaque opaline forms isolated from forest soils in Ohio. *Soil Science Society of America, Proceedings* 37: 647–650.

_____1974 Contributions of forest opal and associated crystalline phases to fine silt and clay fractions of soils. *Clays and Clay Minerals (Clay Mineral Society, Proceedings of the Conference)* 22: 295–306.

Witty, JE and Knox, EG 1964 Grass opal in some chestnut and forested soils in north central Oregon. *Soil Science Society of America, Proceedings* 28: 685–688.

2

GREAT EXPECTATIONS: A SHORT HISTORICAL REVIEW OF EUROPEAN PHYTOLITH SYSTEMATICS

ALIX H. POWERS

"Much of my unassisted self. . . I struggled through the alphabet as if it had been a bramble-bush; getting considerably worried and scratched by every letter."
Great Expectations, Charles Dickens

ABSTRACT

This paper is a review of some of the fundamental European studies of phytoliths in the nineteenth and twentieth centuries. From the earliest observations by German workers during the last century, the identification and classification of phytoliths derived from botanical and sedimentary sources has continued slowly but with persistence. The expansion in the study and classification of phytoliths derived from British archaeological sites reflects a change in the status of environmental archaeology in this country in the last ten years. Excavators are becoming increasingly aware of the advantages that may be gained from the study of all microfossils, including phytoliths, from archaeological deposits. As the classification of phytoliths and their sources is continually refined and updated, the expectations of the quality and amount of information that can be derived from these microfossils also increases.

INTRODUCTION

Studies of phytolith morphology and the production of rigorous classification schemes that may be systematically applied to archaeological and non-archaeological material have occupied the time of numerous analysts. An accurate and precise system of classification is as fundamental for the interpretation of ancient phytolith collections as the individual characters in an alphabet, which when used in correct and ordered combinations can impart considerably more information than the sum of the parts. The search for an appropriate, robust, competent "alphabet" of phytolith types has "worried and scratched" at analysts for many years.

Phytolith Systematics, Edited by G. Rapp, Jr. and
S.C. Mulholland, Plenum Press, New York, 1992

Much of the early work on the recognition of the existence of opal phytoliths and subsequent identification of (possible) sources was performed in Europe. Some of the very earliest accounts originate from Germany and England.

It is the intention of this paper to consider the contributions of European workers in the discipline, from the nineteenth century to the present day; it is not to be a definitive collection, but an overview of the papers published on the subject. It is also intended to provide some insight into the manner by which the study of archaeologically derived phytoliths has affected and been affected by the expectations of the excavator. Over time, the questions asked of the recovered phytoliths have changed from a "simple" identification of the source plant to more complex questions of paleoenvironmental and paleoeconomic reconstruction.

DARWIN AND FRIENDS: NOT QUITE THE BEGINNING . . .

In 1845, one Charles Darwin Esquire, late of Her Majesty's Ship *Beagle*, reported to the Geological Society of London on the many and varied accounts of dust fall on vessels in transit in the Atlantic Ocean (Darwin 1846). Dust fall on vessels sailing the African side of the Atlantic Ocean was not an unusual phenomenon. A number of early reports describe heavy falls of dust whose origins are variously attributed to deserts (Alexander 1822; Berghaus 1842) or volcanic activity (Clarke in Berghaus 1842; Horner 1847). The significance of maritime dust, or of its composition, started some years earlier with the work of Christian Gottfried Ehrenberg, a German doctor with a penchant for adventure.

INFUSORIA AND *PHYTOLITHARIA*

Christian Ehrenberg was a most remarkable man. In 1820, turning his back on the medical life, for which he had gained his medical degree some two years previously, he became a naturalist accompanying General von Minutoli on an "antiquarian journey into Egypt" (Duncan 1877: 57).

In the course of this expedition Ehrenberg and fellow naturalist Hemprich were to investigate the Libyan desert (Duncan 1877). This may have helped to spark Ehrenberg's interest in the red desert dusts that swirl across oceans and continents alike. In the six subsequent years Ehrenberg made many trips in Europe and Africa, losing an alarming number of companions (nine in total), but gaining considerable insight into the fauna and flora of the area. In 1830, after expeditions to Asia, Ehrenberg settled down to devoting his time to expeditions of another nature, those of the microscopic world. In 1830 he published a memoir of the identification, classification and geographical distribution of "*infusoria*" (microscopic plant and animal life). This initial piece of work was to have a profound effect on scientists of the period. Cuvier, a contemporary of the period, wrote: "This discovery entirely changes our ideas, and especially upsets many systems; it is one of those which constitute epochs in the sciences" (Cuvier in Duncan 1877: 58).

Ehrenberg's classification of what he described as *"phytolitharia"* (silicified plant material) merely constituted a floral element to add to that of micro-organisms such as Foraminifera, Protozoa and Coelenterata. A paper in 1841 described some 71 different types of micro-organisms recovered from North and South American samples (Ehrenberg 1841), and one in 1845 concerned the contents of volcanic ash from the Ascension Islands (Ehrenberg 1845). It was at this time that Ehrenberg was approached by Darwin to examine the dust that collected on the deck of the *Beagle*.

Darwin's description of the onboard appearance of the dust while the *Beagle* lay at anchor in the Cape Verde Archipelago is both detailed and vivid:

> On the 16th January (1833), when the Beagle was ten miles off the N.W. end of St. Jago, some very fine dust was found adhering to the under side of the horizontal wind vane at the mast head, it appeared to have been filtered by the gauze from the air . . . (and) the dust probably came from the coast of Africa. The atmosphere was so hazy that the visible horizon was only one mile distant. During our stay of three weeks at St. Jago . . . the atmosphere was often hazy, and very fine dust was almost constantly falling, so that the astronomical instruments were roughened and a little injured. The dust collected in the Beagle was excessively fine grained and of reddish brown colour; it does not effervesce with acids . . . (Darwin 1846: 27).

The dust collected by the ever-observant Darwin was examined by Ehrenberg, who recorded 67 different types of *infusoria*, of which some 34 were *phytolitharia* (Darwin 1846). The very persistent appearance of the fine dust covering ship and instruments and generally making a thorough nuisance of itself meant that it was observed and recorded.

In an era when the exploration of the microscopic world was just beginning and information was at a premium, Ehrenberg worked on the identification of microscopic life forms from sedimentary sources. He studied a wide range of source materials, including "mud" from mouths and deltas of North European rivers (Ehrenberg 1845); clay (Ehrenberg 1851b); and a gypsum bed that contained 44 different forms of "beautiful siliceous *infusoria*," of which 6 were *phytolitharia* (Ehrenberg 1851a). Of particular interest is a brief note on Ehrenberg's study of the "Black Earth" of Central Russia. It was hypothesized at the time that the soil derived from water-lain material. On examination it turned out to contain 22 forms of non-aqueous *phytolitharia*, and Ehrenberg concluded that it was "a soil formed of the debris of ancient forests" (Ehrenberg 1851c: 113). This early exhibition of paleoenvironmental reconstruction, highlighting a significant ecological change from forest cover to open land, is an early precursor of modern phytolith studies. Such a dramatic ecological change, whether a result of climatic or human activity, or both, is of primary importance to the archaeologist and environmental historian. These precise aspects constitute part of the fundamental work that is currently under way in Sheffield on environmental and economic reconstruction of the prehistoric landscape of the Outer Hebrides, northwest Scotland (see below).

In 1854, Ehrenberg published his seminal textbook on the microfossils recovered from fluvial and alluvial sediments, entitled *Mikrogeologie*. Started some 14 years previously, the book contains 41 detailed copper plates of the microfossils he had noted and classified. His classification of microscopic fauna and flora included four "paragenera" of *phytolitharia*, an artificial taxonomy based on phytoliths derived from the Poaceae (Gramineae) and Equisetaceae families.

In many ways Ehrenberg may be considered the "father of phytolith studies." Although he was not the first to examine and publish work on phytoliths (that honor goes to another German scholar by the name of Struve, who submitted a dissertation on silica in plants to Berlin University in 1835), without much of Ehrenberg's pioneering work, the phytolith workers of the second half of this century would have needed to examine the fundamental principles of phytolith form and distribution from their rudiments. Ehrenberg did much to lay the foundations of the hunt for the botanical sources of the phytoliths he recorded. He continued his work on sedimentary material right up to his death in 1876 (Duncan 1877).

THE START OF BOTANICAL INVESTIGATIONS: GUNTZ AND GROB

Work on microscopic fossils continued in Germany, with the emphasis moving from sedimentary investigations by Ehrenberg (also Ruprecht 1866) to observations of phytolith sources. At the end of the nineteenth century, work began in earnest on the elucidation and classification of phytoliths and phytolith-producing plants.

Guntz in 1886 (cited in Grob 1896) studied 130 species of grasses in an attempt to determine if there was any connection between the cell anatomy of leaves and the climatic area in which the species grew. He elucidated four different climatic groups on the basis of their leaf structure: bamboo, savannah, meadow and steppe grasses. Since phytoliths form within cells, and create duplicates of cells, any work on the differentiation of plants by their cell structure is important. Eighty years later, Twiss, Suess and Smith (1969) produced a classification of phytoliths derived from specific ecozones, eg, prairie grasses.

In a substantial paper entitled "Beitrage zur Anatomie der Epidermis der Gramineenblatter" (Contribution on the anatomy of the epidermis of the grass leaf) Grob (1896) provides a comprehensive review of early papers on the anatomy of the Poaceae (Gramineae) family published from 1860 to the 1890s. A wide variety of grasses were studied, from wild species such as *Phragmites*, *Elymus* and *Dactylis*, to familiar domesticates. In particular, two papers published in 1875 by Hohnel (cited in Grob 1896) studied *Panicum miliaceum* (common millet); *Sorghum vulgare* (sorghum); *Avena sativa* (oat); *Triticum spelta* (spelt wheat); *Hordeum vulgare* (six-rowed barley) and *Secale cereale* (rye), with detailed discussion of the morphology of epidermal cells, silica cells, prickle hairs and bristle hairs. These same divisions of silica-accumulating cells can be seen in the much later but more familiar work of Metcalfe (1960), whose detailed description of the anatomy of grasses includes a section on characteristic silica cells in the epidermis.

Grob (1896) himself describes several categories of long and short cells in epidermal layers, including "Kieselkurzzellen" (silica short cells), "Korkkurzzellen" (cork short cells) and four different types of hair cells (see Table 2.1). The short silica cells are divided into those associated with phloem cells (tissues that translocate nutrients) and what Grob describes as "assimilation tissue" (possibly xylem tissues that transport water within the plant). By far the most numerous silicified cells are those associated with the phloem tissues; of these, five types are common to many species. Grob considers two tribes of grasses (see Table 2.2), listing the percentage of the genera studied that showed occurrences of the common silica cell types I–V. A further five types are restricted to one or two of the species that Grob studied, and are given no specific name.

Grob (1896) also has a section dealing with the "topography" of the leaf and the various types and forms of silica cells in, for example, the leaf margin or around the central vein. There is a detailed description of the differences between species.

This early work by Hohnel, Guntz and Grob on the morphology and location of silica cells in a number of grass plants is fundamental to many later phytolith studies, including those of the author. These three particular papers dealt with what may be considered the four most important issues in studies of phytolith-producing plants:

(a) What types of grass cells become silicified? (eg, hair cells, short cells)
(b) Are individual silicified cell types universal or restricted to one or two species?
(c) Docs the part of the plant ("topography") have any bearing on silicification; and how does this compare between species?
(d) Is there any connection between cell anatomy and the climate/area in which the species are grown?

The pertinence of these questions is still apparent in modern studies of phytoliths and phytolith-producing species, particularly in archaeological investigations.

TABLE 2.1. Grob's Classification of Epidermal Cell Morphology*

(i) Long cells
(ii) "Bladder" cells
(iii) Short cells
  (a) Silica short cells
    (1) **Silica short cells associated with phloem tissue**

Principal Type (common to many species)		Secondary Type (Seen in only 1 species)
I	Cross, dumbbell & knot cells	VI
II	Saddle cells	VII
III	Rice cells	VIII
IV	Circular & elliptical cells	IX
V	Rod & lamina cells	X

    (2) **Silica short cells associated with assimilation tissue (xylem?)**
      i. Cross cells
      ii. *Olyra* cells (found in the species *Olyra*)
      iii. Prickle cells
      iv. Silica cells associated with parenchyma cells in *Pharus* & *Leptaspis*

  (b) Cork short cells
  (c) "In between" cells (intracellular)

*After Grob 1896.

TABLE 2.2. Genera Studied by Grob (1896)

Panicoideen	Poaeoideen
1. Maydeae	7. Phalarideae
2. Andropogoneae	8. Agrostideae
3. Zoysieae	9. Aveneae
4. Trigestineae	10. Clorideae
5. Paniceae	11. Festuceae
6. Oryzeae	12. Hordeae
	13. Bambuseae

THE GERMAN SCHOOL

The work of the early German workers Guntz and Grob was extended by colleagues over the next few years. Numerous papers were published that dealt with the silicification of cereals and other members of the Poaceae (Gramineae) family (see Table 2.3). Botanists concentrated on extending their knowledge and understanding of cell silicification in members of the grass family. Piperno (1988) draws our attention to the domination of this family in studies of this period, with a query as to why this should be the case. Although it is impossible to divine the precise intentions of the phytolith analysts of the late nineteenth and early twentieth centuries, it is conceivable to hazard a guess. The Poaceae (Gramineae) is a large plant family, easy to find in a wide variety of locations. There has been considerable interest in members of this plant family (specifically the domesticated cereals) for a very long time. Is it therefore really surprising that there should be an emphasis on the microstructure of these plants? Finally, it should not be forgotten that academic studies tend to follow trends. Workers and their research topics tend to follow and build upon previous published studies, starting with the familiar and branching into new ground.

FROM DUST TO DIRT . . .

The botanical investigations of the German school, although dominant in terms of numbers and scope, were not the exclusive holders of the "phytolith flame" in the latter half of the nineteenth century and the start of the twentieth. The publications of Ehrenberg (as reported in English journals) and Darwin in the mid-1800s did encourage some small amount of phytolith work in Britain. Gregory's (1855) study was based on the analysis of sixty soil samples from a variety of locations worldwide, including North and South America, the West Indies, New Zealand, India, Britain and Northern Europe. Gregory worked extensively with Ehrenberg's *Mikrogeologie* for the identification of (chiefly) diatoms and phytolitharia, making no attempt to classify his unknown specimens.

During the period of the World War II and directly following, microscopic examination of soils took place in Russia. Smithson (1958) cites the work of Tyurin (1937), Usov (1943), Parfenova (1956), Yarilova (1956) and Parfenova and Yarilova (1956) in connection with the study of silica and phytolith accumulation in Russian soils. This perpetuation of interest in the phytolith content of soils, although neither as "high profile" nor as extensive as comparable botanical studies, was the forerunner of the very significant work of Welsh soil scientists and ultimately, therefore, of analysis of archaeological sediments.

TABLE 2.3. German Botanical Papers Published in the Late 19th and Early 20th Centuries

Year	Author	Species studied	
1899	Formanek	*Oryza sativa*	*Panicum miliaceum*
		Hordeum	*Avena sativa*
		Lolium temulentum	*Bromus secalinus*
		Avena fatua	*Setaria viridus*
		Panicum crus galli	*Triticum repens*
1905	Neubauer	*Triticum sativum*	*Triticum repens*
		Secale cereale	*Bromus secalinus*
		Zea mays	*Alopecurus pratensis*
		Lolium temulentum	*Hordeum sativum*
		Avena sativa	*Avena fatua*
		Oryza sativa	*Panicum miliaceum*
		Panicum crus galli	*Setaria glauca*
		Setaria viridis	
1908a	Mobius	General	
1908b	Mobius	*Callisia repens*	
1914	Netolitzky	*Panicum miliaceum*	*Setaria italica*
		Digitaria sanguinalis	
		Echinochloa crus galli	
1914	Frohnmeyer	*Zea mays*	*Tripsacum dactyloides*
		Coix lacrymae	*Saccharum officinarum*
		Sorghum halepense	*Andropogon schimperi*
		Panicum plicatum	*Panicum sanguinale*
		Oryza sativa	*Phalaris arundinacea*
		Alopecurus pratensis	*Phleum pratense*
		Phleum boehmeri	*Avena sativa*
		Trisetum flavescens	*Festuca arundinacea*
		Poa pratensis	*Uniola latifolia*
		Arundo donax	*Arundo phragmites*
		Elymus sibiricus	*Secale cereale*
		Lolium perenne	*Hordeum distichum*
		Bambusa nana	*Bambusa arundinacea*
1929	Netolitzky	General	
1930	Frey-Wyssling	General chemistry	

WELSH PHYTOLITH STUDIES OF "... INTERESTING OBJECTS ..."

The mid-1950s saw the re-emergence of phytolith studies in western Europe. Following the dramatic change in circumstances for German scholars in the 1940s, phytolith studies recommenced in the Principality of Wales, at the University College of North Wales, Bangor.

After publications on the micromineralogy of soils in North Wales (Smithson 1953, 1954), in the latter of which Smithson describes the appearance of silicified sponge spicules, there appeared the first major investigation of opaline silica deposits in British soils (Smithson 1956). Smithson made no attempt to identify his siliceous material at this stage, and indeed no precise details of morphology are given, other than a basic division of forms into complete small organisms (eg, diatoms) and fragments of larger organisms (eg, phytoliths) (Smithson 1956).

Smithson cites the work of Netolitzky (1929) as having stated that phytoliths are present in grasses and sedges. He also made a prophetic statement: "Little work appears to have been carried out on the occurrence in soils of these interesting objects and it is possible that they might, with careful study, provide information about former vegetation and soil conditions of the sites where they occur" (Smithson 1956: 128).

In the following five years, Smithson produced four more papers, including one on the discrimination of sponge spicules from phytoliths found in soils (Smithson 1959); methodology (Parry & Smithson 1958a; Smithson 1961); and a detailed paper on grass phytoliths containing approximately 150 scale drawings of the morphologies he found (Smithson 1958). The paper dealt with the problem of identifying the source of the phytolith material that was recovered from three geographic locations whose major floral components consisted of ten British grasses (see Table 2.4 for details). Smithson also examined the phytoliths recovered from the dung of animals grazed on the location sites studied. This concerted approach to sediment, botanical and secondary organic deposit analysis is one that has been successfully used in archaeological studies in Britain (see below). Sedimentary samples from archaeological horizons have been compared with groups of phytoliths recovered from botanical specimens and modern fecal remains (Powers, Gilbertson & Jones 1986; Powers & Gilbertson 1987; Powers, Padmore & Gilbertson 1989).

Work in the 1950s and 1960s dealt with specific groups of silicified cells and their form and distribution, with information on three-dimensional structure of phytoliths within grasses (eg, Parry & Smithson 1958b, c, 1964, 1966). A summary of their classification of phytoliths in 1964 and 1966 can be seen in Table 2.5. Many of the studies of cell silicification and specific plant anatomy were aided by the publication of two books concerned with the Poaceae (Gramineae) family. The first was a book by Hubbard, one-time Deputy Director of the Royal Botanic Gardens, Kew. First published in 1954, Hubbard's *Grasses*

TABLE 2.4. Sources of Soil and Grass Species Studied by Smithson*

Site Location	Grasses studied
A. Rough enclosed pasture for cattle and sheep, about 300 ft. above sea level, near Menai Bridge, Anglesey.	Festuceae *Cynosurus cristatus* *Poa annua* *Dactylis glomerata* Aveneae *Anthoxanthum odoratum* *Holcus lanatus* Agrosteae *Agrostis tenuis*
B. Rough open pasture for sheep 1,700 to 1,800 ft. above sea level, near Aber, Caernarvonshire.	Festuceae *Festuca ovina* Nardeae *Nardus stricta*
C. Rough enclosed pasture for ponies, very wet, about 350 ft. above sea level, near Pentir, Caernarvonshire.	Danthonieae *Molinia caerulea* *Sieglingia decumbens*

*After Smithson 1958, Table 1.

TABLE 2.5. Taxonomy of Silicified Grass Cells*

1964	**1966**
(1) Membrane silicification—unconnected particles that adhere to cell membranes	
(2) Intercellular silicification—forming between cells, where plant has excess silica available	
(3) Intracellular silicification (a) Costal rods & dumb-bells (b) Hat-shaped intercostal silica bodies (c) Intercostal long cells (d) Bulliform cells (e) Hairs & hair bases –Macrohairs –Microhairs –Prickle hairs –Papillae (f) Cork cells (g) Stomata cells (h) Sclerenchyma cells	Intracellular silicification (a) Costal rods (b) Hats (c) Long cells –scutiform –round (d) Hairs –Macrohairs –Microhairs –Prickle hairs –Papillae (e) Other epidermal cells eg, stomata cells (f) Sub-epidermal cells –sclerenchyma –xylem

*After Parry and Smithson 1964, 1966

was a definitive text on the structure, identification and distribution of British grasses. Hubbard's nomenclature for the tribe, sub-tribe and genus subdivisions of the great grass family was adopted by most workers.

Of even more significance for phytolith work was the production of Metcalfe's *Anatomy of the Monocotyledons. I. Gramineae* in 1960. The text lists the detailed structure of grass families, from the general to the specific. In a section dealing with "diagnostic microscopical characters," Metcalfe (1960: xxxvii) lists some 20 types of silica bodies located in the epidermis, together with details of various forms of hairs, stomata and long cells (see Table 2.6). It is soon apparent from the most casual of observations that Metcalfe uses the same divisions isolated some years earlier by, for example, Grob and Guntz.

OF DESCRIPTION, ELECTRON AND DISEASE

Workers at the University College of North Wales, Bangor, continued to study phytoliths and the process of plant cell silicification from a botanical standpoint. From the late 1960s to the 1980s a succession of papers was published in journals such as *Annals of Botany* on the topic of cell silicification. A selection of these papers has been listed in Table 2.7, together with details of the species studied and the type of analysis employed. The species studied continued generally to be those of the grass family, but the type of analyses employed changed distinctly. The earlier papers primarily concerned the observations of silicified cells by the use of standard light microscopy techniques. Blackman (1971) is a particularly good example of this type of "descriptive" paper as it contains diagrammatic results of observations of phytoliths in situ in the leaf and culm sheaths and leaf and culm blades of 26 species of grasses. Particular attention is drawn to differences in morphology of silica bodies deriving from costal and intercostal zones and from abaxial and adaxial leaf blade surfaces (Blackman 1971).

Much of the listed work after Blackman 1971 (see Table 2.7) concentrated on the investigation of silicification by the use of electron-probe microanalyzer and various types of electron microscopes. These studies were less concerned with the form of individual phytoliths than with the concentration and distribution of silica within specific plant parts. The apparent interest in *Phalaris canariensis* (canary grass) exhibited by Welsh phytolith workers in the early 1980s (see Table 2.7) stems from investigations that revealed the implication of silicified cells in instances of esophageal cancer clusters (O'Neill *et al* 1980; Newman & Mackay 1983; Sangster *et al* 1983). Similarly, the carcinogeneity of *Setaria italica* (foxtail millet) and *Pteridium aquilinum* (bracken) have been investigated (Hodson & Parry 1982; Hodson, Sangster & Parry 1982; O'Neill *et al* 1982; Parry & Hodson 1982; Parry, Hodson & Newman 1985).

TABLE 2.6. Types of Diagnostic Silica Bodies Found in Grass Leaf Epidermis*

(a)	Tall and narrow
(b)	Cubical
(c)	Cuboid
(d)	Round
(e)	Elliptical
(f)	Crescent-shaped
(g)	Oblong
(h)	Fitting into concavity in a cork cell
(i)	Saddle-shaped
(j)	Acutely angled
(k)	Elongated-smooth
(l)	Elongated-sinuous
(m)	Cross-shaped
(n)	Intermediate between cross and dumb-bell
(o)	Dumbbell shaped
	(i) with narrow middle portions
	(ii) with wide middle portions
(p)	Shortly dumb-bell shaped
(q)	Dumb-bell shaped but varying in appearance with the focus
(r)	Nodular
(s)	Oryza type
(t)	Crenate-vertical

*From Metcalfe 1960: xlii.

The extensive botanical studies by a number of Welsh analysts provided a valuable source-base of information on the processes of cell wall silicification and the formation of phytoliths in grass species. Without these studies on which to build, many of the archaeological and "applied" botanical studies of ancient and modern phytolith sources would have been hindered by a lack of basic information. It would be difficult to interpret data and to assess their potential for useful information if basic biological processes responsible for the formation of the "evidence" were not clearly understood.

NON-BANGOR BOTANISTS

There were a number (admittedly small) of botanical phytolith studies from British institutions outside Bangor. Studies include observations on cell wall silicification in grass species (eg, Dinsdale, Gordon & George 1979; Perry,

TABLE 2.7. Some Papers Published on Cell Silicification by Workers at the University of North Wales in the 1960s to 1980s

Year	Author	Species studied	Part of plant studied	Analytical techniques
1968	Blackman & Parry	*Secale cereale*	General	LM
1969	Sangster & Parry	*Oryza sativa* *Cynodon dactylon* *Sieglingia decumbens*	Bulliform cell	PC, LM
1970a	Sangster	*Oryza sativa* *Cynodon dactylon* *Sieglingia decumbens*	Leaves	PC, LM
1970b	Sangster	*Sieglingia decumbens*	Leaves	PC, LM
1971	Blackman	26 range grass species	General	LM
1973	Hayward & Parry	*Hordeum sativum*	General	LM, EPM
1973	Soni & Parry	*Oryza sativa*	Inflorescence bracts	EPM
1975	Parry & Kelso	*Molinia caerulea* *Sorghum bicolor*	Roots	EPM, SEM
1975	Hayward & Parry	*Hordeum sativum*	Leaves	SEM
1976	Sangster & Parry	*Sorghum bicolor*	Roots	PC, EPM, SEM
1977	Parry & Kelso	*Saccharum officinarum*	Roots	SEM, EPM, CORA
1979	Montgomery & Parry	*Molinia caerulea*	Roots	TEM, CORA
1980	Hayward & Parry	*Hordeum sativum*	Culms, bracts & awns	EPM, SEM
1981	Bennett & Parry	*Hordeum sativum* *Avena sativa* *Secale cereale* *Triticum aestivum*	Caryopses	EPM, SEM
1981	Wadham & Parry	*Oryza sativa*	General	PC, MO
1982	Bennett	*Hordeum sativum* *Avena sativa* *Triticum aestivum*	Roots	EPM
1982	Bennett	*Hordeum sativum*	Leaves	PC, CORA, TEM
1983	Sangster, Hodson & Parry	*Phalaris canariensis*	Inflorescence bracts	PC, LM, SEM, EPM
1984	Hodson, Sangster & Parry	*Phalaris canariensis*	Lemma	PC, CORA, LM, STEM
1985	Hodson, Sangster & Parry	*Phalaris canariensis*	Glumes	PC, TEM, LM, STEM
1986	Hodson & Parry	*Phalaris canariensis*	Roots, culm, leaves	PC, LM, SEM, EPM

Key: CORA = Corinth analytical microscope (a form of analytical transmission electron microscopy); EPM = electron-probe microanalysis; LM = light microscopy; MO = macro observation; PC = plant culture; SEM = scanning electron microscopy; STEM = scanning transmission electron microscopy; TEM = transmission electron microscopy

Williams & Fry 1987); effects of silicon on plants grown in aqueous solution (eg, Barber & Shone 1966; Adatia & Besford 1986); effects of silica bodies on the reduction of insect attack (eg, Moore 1984a, 1984b); and further studies on the macrohairs of canary grass (Perry, Mann & Williams 1984; Perry *et al* 1984).

All the sedimentary and botanical phytolith work carried out in Europe in the past 150 years provided a bank of information that could be utilized and exploited by the archaeological world. In the search for new sources of archaeological information to complement already established analyses such as palynology, phytolith analysis has come to the fore. Ancient, archaeological deposits are frequently rich in phytoliths, and in some preservational circumstances such as calcareous sand dunes, when carbon-based microfossils (eg, pollen) do not survive, phytoliths may be invaluable in paleoenvironmental and paleoeconomic reconstruction.

THE ARCHAEOLOGICAL WORLD

The archaeological application of phytolith studies to British and European sites has sadly "lagged" behind those of our American colleagues. Work over the past 10 years has been concentrated at three main institutions: the Institute of Archaeology in London, the University of East Anglia and the University of Sheffield.

Workers in London have studied phytoliths from ungulate teeth (Armitage 1975; Scaife unpublished), but these were present in such low numbers that only cautious attempts at identification were made, using the classifications of Metcalfe (1960), Parry and Smithson (1964), Twiss, Suess and Smith (1969) and Rovner (1971). Insufficient phytoliths were recovered to create any new morphotypes or classifications. Macphail (1981) studied phytoliths as part of a full geophysical and botanical investigation of the "dark earth" that is often found on Roman and Medieval urban sites. He also used the classification of Twiss, Suess and Smith and Rovner to categorize his finds (Macphail 1981).

Peter Murphy at the University of East Anglia discovered sheets, or "aggregations," of phytoliths from mineralized coprolites from the Roman levels of Culver Street, Colchester (Murphy, personal communication, 1986). Phytoliths were also recovered from a "pillow-stain" in a grave in the Early Bronze Age ring-ditch of Bowthorpe, Norwich. Murphy listed some 28 types of phytoliths that he recovered from the Bowthorpe and associated comparative samples (Murphy 1986).

These initial studies of British archaeological samples showed that phytoliths could be successfully used for the interpretation of a number of different sample types. They also provided the impetus for the much more extensive studies of prehistoric midden sediments from the Outer Hebrides.

SHEFFIELD—THE QUANTITATIVE "BLACK-BOX" ANALOG APPROACH

Much of the work on phytoliths in Britain and America in recent years has concentrated almost exclusively on the identification of the *source* of phytolith material.

When the opportunity arose for the investigation of phytoliths from archaeological sites in the Outer Hebrides (a chain of small islands on the north Atlantic seaboard of Scotland), the decision was made to take a "black-box" approach to the problem. Instead of trying to identify the species or families from which individual phytoliths originated, samples were assessed in terms of the composition of the whole phytolith group or "suite" recovered. A threefold strategy was evolved: to quantitatively and qualitatively identify suites of phytoliths in whole assemblages, and to compare these ancient suites with those of modern analogs from sediments associated with (possible) source habitats/sedimentary sources—both "natural" and anthropogenic in origin. A simple robust classification scheme was created (Powers, Gilbertson & Jones 1986; Powers, Padmore & Gilbertson 1989) with a limited number of categories (or "morphotypes"). This classification has since undergone a number of metamorphoses, with the addition of new types and the subdivision of some of the broader categories (Powers, in press). However, the main premise of the classification remains the same: based on the visual appearance of the phytolith rather than an attempt to tie form to a particular anatomical structure (exceptions to this being hairs and stomata).

This approach has been used to identify the sources of organic material recovered from the Late Bronze Age/Early Iron Age midden sites on the coastal sand dunes and "machair" (coastal calcareous dune pasture) of the Outer Hebrides. In preservational conditions where pollen is rarely recovered from these archaeological sites, phytolith studies proved to be one of the chief tools in paleoenvironmental and paleoeconomic reconstruction analysis. Investigations are based primarily on the quantitative comparison of organic horizons within the ancient deposits to modern analog studies of botanical specimens and organic "elements" such as thatching, bedding and byring material, in a manner similar to the work of Smithson (1958).

Data, based on the proportions of the various morphotypes within phytolith suites, are subject to statistical analysis in order to form some idea of the similarities and differences between groups of samples (Powers, Padmore & Gilbertson 1989; Powers & Padmore, in press). Although the resulting information will not enable the analyst to tell the excavator what species were present in the precise manner that, for example, a palynologist can, the quality of information available from this type of data set is very high if the correct questions are asked.

Questions addressed to this type of phytolith data have to be based on a full and frank exchange of ideas between the (idealistic?) excavator and the (realistic?) analyst, in an attempt to satisfy expectations on both sides. Large, general questions, such as "What is the source of the organic material on this site?" must be honed down and refined into a series of more precise, less general questions to which the analyst may reasonably be expected to provide some sort of answer; for example, "Is it possible to differentiate between cattle and sheep feces on the basis of the phytoliths they contain?" The work at Sheffield has indicated that it is possible to make such a differentiation between sheep and cattle feces from animals grazed on the machair (see Powers, Padmore & Gilbertson 1989), and this is one small piece towards the solution of the large and complex archaeological "jigsaw" of the interpretation of the prehistoric middens.

Discussions between excavator and analyst *before* an excavation can often inspire the excavator to sample particular contexts or sites that (s)he would not previously have considered for phytolith analysis. The use of "black-box" systematics for phytolith analysis ensures that the scope of the technique is always expanding to include new types of contexts, sites and geographical locations.

SHEFFIELD—THE "GRAY-BOX" APPROACH

As a direct opposite to the "black-box," no preconceptions approach, the gray-box approach is an attempt to compare archaeological and modern cereal remains. The aim is to identify the various "elements" of cereal plants (*ie,* root, culm, leaf, inflorescence) on the basis of their (varying) phytolith suites, and to see whether it is possible to identify particular species or genera of common domesticated cereals (Harvey & Powers, in preparation). Investigations of this nature have been performed on a small scale by Miller Rosen (1980, 1983, 1987) and the initial results look very promising.

CONCLUSIONS

It is ironic that the continent in which phytolith studies can truly be thought of to have originated has for many years lagged behind the advances of American workers. The patient, thorough work of individuals such as Ehrenberg, who seems to have inspired much interest and academic investigation of the microscopic world, and of Grob, Guntz and their fellow German botanists created the foundations upon which much of our modern studies are based.

Despite the disruptions of a World War, the "phytolith flame" continued to be kept alight, seemingly by the singular efforts of Smithson and Parry in Wales, to survive into the 1980s and 1990s. It is now undergoing a revival in the archaeological world, with excavators seeing the potentials for phytolith analysis in conditions where pollen analysis is neither practicable nor suitable.

One of the major problems in the examination of phytoliths is that there is no single clear classification scheme that everyone employs. As a relatively "new" (with apologies to the Germans) science, phytolith analysis is in a state of flux, with continual exchange, improvements and refinements occurring almost weekly.

As our source-base of botanical information improves, so will our potential for the elucidation of individual phytoliths that are exclusive to particular families or species of plants. This must not, however, be the sole, driving force behind archaeological investigations. Species are not everything. The potential for the use of suite composition for the identification of ecological and economic information is large and should not be underrated.

ACKNOWLEDGMENTS

The author wishes to thank Archaeological Operations and Conservation (formally the Central Excavation Unit) of the Scottish Development Department, who funded the original phytolith work in 1985–1987; and the Science and Engineering Research Council for funding current studies (1988-1991). Thanks also to Dave Gilbertson, Jennifer Hillam and Cathy Groves for their editorial advice and assistance.

REFERENCES

Adatia, MH and Besford, RT 1986 The effects of silicon on cucumber plants grown in recirculating nutrient solution. *Annals of Botany* 58: 343–351.

Alexander, J 1822 Blowing sand. *Edinburgh (New) Philosophical Journal* 7: 404–405.

Armitage, PL 1975 The extraction and identification of opal phytoliths from the teeth of ungulates. *Journal of Archaeological Science* 2: 187–197.

Barber, DA and Shone, MGT 1966 The absorption of silica from aqueous solution by plants. *Journal of Experimental Botany* 17: 569–578.

Bennett, DM 1982a An ultrastructural study on the development of silicified tissues in the leaf tip of barley (*Hordeum sativum* Jess). *Annals of Botany* 50: 229–237.

_____1982b Silicon deposition in the roots of *Hordeum sativum* Jess, *Avena sativa* L.and *Triticum aestivum* L. *Annals of Botany* 50: 239–245.

Bennett, DM and Parry, DW 1981 Electron-probe microanalysis studies of silicon in the epicarp hairs of the caryopses of *Hordeum sativum* Jess, *Avena sativa* L., *Secale cereale* L. and *Triticum aestivum* L. *Annals of Botany* 48: 645–654.

Berghaus, 1842 On falls of dust on vessels traversing the Atlantic. *Edinburgh (New) Philosophical Journal* 32: 134–136.

Blackman, E 1971 Opaline silica bodies in the range grasses of southern Alberta. *Canadian Journal of Botany* 49: 769–781.

Blackman, E and Parry, DW 1968 Opaline silica deposition in rye (*Secale cereale* L.). *Annals of Botany* 32: 199–206.

Darwin, C 1846 An account of the fine dust which often falls on vessels in the Atlantic Ocean. *Quarterly Journal of the Geological Society of London* 2: 26–30.

Dinsdale, D, Gordon, AH and George, S 1979 Silica in the mesophyll cell walls of Italian rye grass (*Lolium multiflorum* Lam. cv. RvP). *Annals of Botany* 44: 73–77.

Duncan, PM 1877 Anniversary address of the president. *Quarterly Journal of the Geological Society of London* 13: 56–60.

Ehrenberg, CG 1841 Uber Vertreitung und Einflufs des mikroskopischen Lebens in Sud- und Nordamerika. *Monatsberichte der Koniglich Preussischen Akademie der Wissenschaften zu Berlin*: 139–144.

_____1845 On the muddy deposits at the mouths and deltas of various rivers in Northern Europe, and the infusorial animalcules found in those deposits. *Quarterly Journal of the Geological Society of London* 1: 251–257.

_____1846 Uber die Vulkanischen Phytolitharien der insel Ascension. *Monatsberichte der Koniglich Preussischen Akademie der Wissenschaften zu Berlin*: 191–202.

_____1848 On organic substances, recognizable by the microscope, found in the meteoric ashes which fell on the 1st of May 1812 in the island of Barbados and changed day into night. *Quarterly Journal of the Geological Society of London* 4(2): 19–23.

_____1851a On a gypsum bed containing Infusoria. *Quarterly Journal of the Geological Society of London* 7(2): 33–34.

_____1851b On the tchornoi zem of Russia. *Quarterly Journal of the Geological Society of London* 7(2): 112–113.

_____1851c On an extensive rock-formation of siliceous polycystina from the Nicobar Islands. *Quarterly Journal of the Geological Society of London* 7(2): 118.

_____1854 *Mikrogeologie*. Leipzig, Leopold Voss: 374 p.

Formanek, J 1899 Ueber kie Erkennung der in den Nahrungs-und Futtermitteln vorkommenden Spelzen. *Zeitschrift fur Untersuchung der Nahrungs-und Genussmittel, so wie der Gebrauchsgegenstande* 11: 833–842.

Frey-Wyssling, A 1930 Vergleich zwischen der Ausscheidung von Kieselsaure und Kalziumsalzen in der Pflanze. *Berichte der Deutschen Botanischen Gesellschaft* 48: 184–191.

Frohnmeyer, M 1914 Die Entstehung und Ausbildung der Kieselzellen bei Gramineen. *Bibliotheca Botanica* 21: 1–41.

Gregory, W 1855 On the presence of diatomaceae, phytolitharia and sponge spicules in soils which support vegetation. *Proceedings of the Botanical Society of Edinburgh*: 69–72.

Grob, A 1896 Beitrage zur Anatomie der Epidermis der Gramineenblatter. *Bibliotheca Botanica* 36: 1–63.

Hayward, DM and Parry, DW 1973 Electron-probe microanalysis studies of silica distribution in barley (*Hordeum sativum* L.). *Annals of Botany* 37: 579–591.

_____1975 Scanning electron microscopy of silica deposits in the leaves of barley (*Hordeum sativum* L.). *Annals of Botany* 39: 1003–1009.

_____1980 Scanning electron microscopy of silica deposits in culms, floral bracts and awns of barley (*Hordeum sativum* Jess). *Annals of Botany* 46: 541–548.

Hodson, MJ and Parry, DW 1982 The ultrastructure and analytical microscopy of silicon deposition in the aleurone layer of the caryopsis of *Setaria italica* (L.) Beauv. *Annals of Botany* 50: 221–228.

_____1986 Silicon deposition in the roots, culm and leaf of *Phalaris canariensis* L. *Annals of Botany* 58: 167–177.

Hodson, MJ, Sangster, AG and Parry, DW 1982 Silicon deposition in the inflorescence bristles and macrohairs of *Setaria italica* (L.) Beauv. *Annals of Botany* 50: 843–850.

_____1984 An ultrastructural study on the development of silicified tissues in the lemma of *Phalaris canariensis* L. *Proceedings of the Royal Society of London Series* B22: 413–425.

_____1985 An ultrastructural study on the development phases and silicification of the glumes of *Phalaris canariensis* L. *Annals of Botany* 55: 649–665.

Horner, L 1847 Anniversary address of the president. *Quarterly Journal of the Geological Society of London* 3: lii–liii.

Hubbard, CE 1968 *Grasses*. Second edition, Harmondsworth, Penguin Books: 463 p.

Macphail, RI 1981 Soil and botanical studies of the "dark earth." *In* Dimbleby, D and Jones, M, eds, The environment of man, *British Archaeological Reports* 87: 309–329.

Metcalfe, CR 1960 *Anatomy of the Monocotyledons. I. Gramineae.* Oxford, Clarendon Press: 731 p.

Mobius, M 1908a Uber die Festlegung der Kalksalze und Kiesel-korper in den Pflanzenzellen. *Berichte der Deutschen Botanischen Gesellschaft* 26A: 29–37.

_____1908b Uber ein eigentumliches Vorkommen von Kieselkorpern in der Epidermis und den Bau des Blattes von *Callisia repens*. *Wiesner-Festschrift* ?: 81–91.

Moore, D 1984a The role of silica in protecting Italian rye grass (*Lolium multiflorum*) from attack by dipterous stem-boring larvae (*Oscinella frit* and other related species). *Annals of Applied Biology* 104: 161–166.

_____1984b The distribution of opaline silica bodies in the leaf sheaths of two perennial rye-grass cultivars differing in their susceptibility to attack by dipterous stem borers. *Grass and Forage Science* 39: 205–208.

Montgomery, DJ and Parry, DW 1979 The ultrastructure and analytical microscopy of silicon deposition in the intercellular spaces of the roots of *Molinia caerulea* (L.) Moench. *Annals of Botany* 44: 79–84.

Murphy, P 1986 Botanical evidence. *In* Lawson, AJ, ed, Barrow excavations in Norfolk 1950–82, *East Anglian Archaeological Reports* No 29: 43–45.

Netolitzky, F 1914 Die Hirse aus antiken Funden. *Sitzbuch der Keiserliche Akademie Wissenschaft der Mathematisch-Naturwissenschaft* 123(6): 725–759.

_____1929 Die Kieselkorper. *In* Linsbauer, K, ed, *Handbuch der Pflanzenanatomie* 3/1a: 1–19. Berlin: Gebruder Borntraeger.

Neubauer, H 1905 Mikrophotographien der fur die Nahrungs-und Futtermitteluntersuchung wichtigsten Gramineenspelzen. *Landwirtschaftliche Jahrbucher* 34: 973–984.

Newman, RH and Mackay, AL 1983 Silica spicules in canary grass. *Annals of Botany* 52: 927–929.

O'Neill, CH, Clarke, G, Hodges, G, Jordan, P, Newman, R, Pan, Q-Q, Ge, M, Liu, F-S, Chang, Y-M and Toulson, E 1982 Silica fragments from millet bran in mucosa surrounding oesophageal tumours in patients in Northern China. *The Lancet* 1: 1202–1206.

O'Neill, CH, Hodges, GM, Riddle, PN, Jordan, PW, Newman, RH and Flood, RJ 1980 A fine fibrous silica contaminant of flour in the high oesophageal cancer area of North-east Iran. *International Journal of Cancer* 26: 517–628.

Parfenova, EI 1956 Investigations on the minerals of podzolic soils in relation to their genesis. *Kora Vyvetrivaniya* 2: 31–44.

Parfenova, EI and Yarilova, EA 1956 The formation of secondary minerals in connexion with the migration of elements. *Pochvovedenie* 4: 38–42.

Parry, DW and Hodson, MJ 1982 Silica distribution in the caryopsis and inflorescence bracts of foxtail millet [*Setaria italica* (L.) Beauv.] and its possible significance in carcinogenesis. *Annals of Botany* 49: 531–540.

Parry, DW, Hodson, MJ and Newman, RH 1985 The distribution of silicon deposits in the fronds of *Pteridium aquilinum* L. *Annals of Botany* 55: 77–83.

Parry, DW and Kelso, M 1975 The distribution of silicon deposits in the roots of *Molinia caerulea* (L.) Moench. and *Sorghum bicolor* (L.) Moench. *Annals of Botany* 39: 995–1001.

_____1977 The ultrastructure and analytical microscopy of silicon deposits in the roots of *Saccharum officinarum* (L.). *Annals of Botany* 41: 855–862.

Parry, DW and Smithson, F 1958a Techniques for studying opaline silica in grass leaves. *Annals of Botany* 22: 543–549.

_____1958b Silicification of branched cells in the leaves of *Nardus stricta* L. *Nature* 182: 1460–1461.

_____1958c Silicification of bulliform cells in grasses. *Nature* 181: 1549–1550.

_____1964 Types of opaline silica depositions in the leaves of British grasses. *Annals of Botany* 28: 169–185.

_____1966 Opaline silica in the inflorescences of some British grasses and cereals. *Annals of Botany* 30: 525–538.

Perry, CC, Mann, S and Williams, RJP 1984 Structural and analytical studies of the silicified macrohairs from the lemma of the grass *Phalaris canariensis* L. *Proceedings of the Royal Society of London Series* B222: 427–438.

Perry, CC, Mann, S, Williams, RJP, Watt, F, Grime, GW and Takacs, J 1984 A scanning proton microprobe study of macrohairs from the lemma of the grass *Phalaris canariensis* L. *Proceedings of the Royal Society of London Series* B222: 439–445.

Perry, CC, Williams, RJP and Fry, SC 1987 Cell wall biosynthesis during silicification of grass hairs. *Journal of Plant Physiology* 126: 437–448.

Piperno, DR 1988 *Phytolith Analysis: An Archaeological and Geological Perspective.* San Diego, Academic Press: 280 p.

Powers, AH and Gilbertson, DD 1987 Phytolith analysis of the archaeological sites at Baleshare and Hornish Point, Outer Hebrides. University of Sheffield Heritage Landscape Programme Contract Report No. 2 for the Central Excavation Unit, Scottish Development Department, Edinburgh.

Powers, AH, Gilbertson, DD and Jones, GEM 1986 The prospects for environmental discrimination in coastal sand dunes on the basis of studies of phytolith type and abundance. University of Sheffield Heritage Landscape Programme Contract Report No. 1 for the Central Excavation Unit of the Scottish Development Department, Edinburgh.

Powers, AH and Padmore, J in press. Advances in the use of quantitative methods and statistical analysis in the study of opal phytoliths. *In* Jordi, J, ed, *Opal Phytolith and Calcium Oxalate Research and its Applications for Palaeoecology and Archaeology.* Praxis. Barcelona, Monografies d'Arqueologia Aplicada.

Powers, AH, Padmore, J and Gilbertson, DD 1989 Studies of late prehistoric and modern opal phytoliths from coastal sand dunes and machair in north west Britain. *Journal of Archaeological Science* 16: 27–45.

Rosen (Miller), AV (ms) 1980 Phytoliths as potential indicators of farming techniques. Paper presented at the 45th Annual Meeting of the Society for American Archaeology, Philadelphia, Pennsylvania.

Rosen (Miller), AV (ms) 1983 Phytoliths of cereals from two Negev Chalcolithic sites: Identification and edaphic implications. Paper presented at the 48th Annual Meeting of the Society for American Archaeology, Pittsburgh, Pennsylvania.

Rosen (Miller), AV 1987 Phytolith studies at Shiqmim. *In* Levy, TE, ed, Shiqmim I: Studies concerning Chalcolithic societies in the Northern Negev Desert, Israel (1982–1984), *British Archaeological Reports International Series* 356: 243–249.

Rovner, I 1971 Potential of opal phytoliths for use in palaeoecological reconstruction. *Quaternary Research* 1: 343–359.

Ruprecht, F 1866 Geobotanical investigations on chernozem. USSR, Academy of Science.

Sangster, AG 1970a Intracellular silica deposition in immature leaves in three species of the Gramineae. *Annals of Botany* 34: 245–257.

_____1970b Intracellular silica deposition in mature and senescent leaves of *Sieglingia decumbens* (L.) Bernh. *Annals of Botany* 34: 557–570.

Sangster, AG, Hodson, MJ and Parry, DW 1983 Silicon deposition and anatomical studies in the inflorescence bracts of four *Phalaris* species with their possible relevance to carcinogenesis. *New Phytologist* 93: 105–122.

Sangster, AG, Hodson, MJ, Parry, DW and Rees, JA 1983 A developmental study of silicification in the trichomes and associated epidermal structures of the inflorescence bracts of the grass *Phalaris canariensis* L. *Annals of Botany* 52: 171–187.

Sangster, AG and Parry, DW 1969 Some factors in relation to bulliform cell silicification in the grass leaf. *Annals of Botany* 33: 315–323.

_____1976 Endodermal silicon and their linear distribution in developing roots of *Sorghum bicolor* (L.) Moench. *Annals of Botany* 40: 361–371.

Scaife, RG (ms) Plant biogenic (phytolith) remains impacted in equine incisors.

Smithson, F 1953 The micro-mineralogy of North Wales soils. *Journal of Soil Science* 4: 194–210.

_____1954 The rocks and soils of North Wales. *Transactions of the Anglesey Antiquarian Society* 1–16.

_____1956 Silica particles in some British soils. *Journal of Soil Science* 7: 122–129.

_____1958 Grass opal in British soils. *Journal of Soil Science* 9: 148–154.

_____1959 Opal sponge spicules in soils. *Journal of Soil Science* 10: 105–109.

_____1961 The microscopy of the silt fraction. *Journal of Soil Science* 12: 145–157.

Soni, SL and Parry, DW 1973 Electron probe microanalysis of silicon deposition in the inflorescence bracts of the rice plant (*Oryza sativa*). *American Journal of Botany* 60: 111–116.

Struve, GA (ms) 1835 De silicia in plantis nonnullis. PhD dissertation, University of Berlin.

Twiss, PC, Suess, E and Smith, RN 1969 Morphological classification of grass phytoliths. *Soil Science Society of America, Proceedings* 33: 109–115.

Tyurin, IV 1937 On the biological accumulation of silica in soils. *Problems of Soviet Soil Science* 4: 3–23.

Usov, NI 1943 Biological accumulations of silica in soils. *Pochvovedenie* 9–10: 30–36.

Wadham, MD and Parry, DW 1981 The silicon content of *Oryza sativa* L. and its effect on the grazing behaviour of *Agriolimax reticulatus* Muller. *Annals of Botany* 48: 399–402.

Yarilova, EA 1956 Mineralogical investigation of a sub-alpine chernozem on andesite basalt. *Kora Vyvetrivaniya* 2: 45–60.

3

DEVELOPING A PHYTOLITH CLASSIFICATION SYSTEM

DEBORAH M. PEARSALL and ELIZABETH H. DINAN

ABSTRACT

The phytolith classification developed at the Paleoethnobotany Laboratory of the University of Missouri-Columbia is a hierarchically arranged photo identification aid of numbered phytolith types. Currently eleven major categories are defined based on cellular origin; expansion to include additional categories is easily accomplished. Lower order divisions are based on morphological details. Category 10, epidermal quadrilaterals, is described in detail as an example of the system. Classification issues are discussed, focusing on recording of unknown and redundant forms and methods of examination.

INTRODUCTION

In this paper we report on our ongoing efforts to develop a phytolith classification system for the Paleoethnobotany Laboratory of the University of Missouri-Columbia. Our system is still evolving; it has proven to be flexible enough to change as our knowledge of phytoliths expands and as new types are encountered. Our purpose in describing how we developed this system is two-fold. First, we feel it is important for all analysts working in applied (sediment) phytolith analysis to document identifications by reference to standard types. Such types must be accurately described. Second, because study of comparative plant specimens may lag behind analysis of archaeological or geological samples, unknowns may be common in samples. These must also be described for eventual identification and named or numbered so that they can be tallied in samples. We present our classification system as one example of how these goals can be accomplished.

Our system, a hierarchically arranged photo identification aid of numbered phytolith types, has been under development since 1985. At that time, we converted an earlier system, built up by numbering types in the order they were encountered in samples, to the hierarchically organized system we describe below. The photo identification aid is composed of more than 300 distinctive phytolith

Phytolith Systematics, Edited by G. Rapp, Jr. and S.C. Mulholland, Plenum Press, New York, 1992

types that have been observed in archaeological samples, modern surface sediment or comparative plant specimens and described in detail, photographed and numbered. This system is not a key to a specific region or to a specific group of plant taxa, although it contains detailed descriptions of many diagnostic phytolith types. Rather, it is a descriptive device, which allows any phytolith, known or unknown, encountered in a sediment or plant sample to be classified with like-phytoliths.

Why have we been working on such a classification system? Pearsall's early work in phytolith analysis was focused almost exclusively on identification of maize (Pearsall 1978, 1979, 1982; reviewed in Pearsall 1989) and on reconstructing vegetation formations in grassland settings (Pearsall 1981; Pearsall & Trimble 1983, 1984). Identification and counting of grass short cell phytoliths formed the basis for this work. A modified version of the Twiss, Suess and Smith (1969) grass short cell classification system was developed and used successfully. As the research focus of the lab broadened to include vegetation reconstruction in tropical forest settings and identification of utilized plants other than maize, many more types of phytoliths besides grass short cells had to be counted. We quickly encountered problems in recording the occurrence of these "non-grass" phytoliths. How types were classified varied among laboratory assistants and between projects. Although there are published photos of silica bodies produced by plants other than grasses (*ie*, Piperno 1985, 1988; Bozarth 1986, 1987, among others), in any given sediment or plant sample one still encounters unknown (or at least unpublished) phytolith types. In order to tally such types consistently until they can be identified, as well as insure that identified types are classified correctly, we established standard types by photographing each new phytolith, describing it, and assigning it a reference number based on its similarity to other types.

We are working actively to refine and expand our classification system and have as a long-term goal the development of a series of regional phytolith keys to plant taxa that can be utilized by any researcher working in the region covered. We begin our discussion by describing how the system was developed. We then give an example of its application and discuss scanning procedures. Finally, we discuss the approach taken in developing our system in relation to general issues of phytolith classification.

DEVELOPMENT OF THE SYSTEM

As mentioned above, our first attempt at classifying the diversity of phytolith types observed in sediment samples was to assign each new type a number, based on the order in which it was observed during scanning. As the number of types grew into the second hundred, this system became increasingly difficult to use.

Scanners would forget that a type had already been numbered and assign a second number; some kinds of phytoliths were "lumped" into broad types, others "split" into very finely divided types.

Influenced by the publication of Brown's (1984) hierarchical key to grass phytoliths, we decided to transform our randomly ordered system into a hierarchical classification system. Unlike Brown's system, it is designed to accommodate all phytoliths, not just grasses. Also, as discussed above, it is not a key to plant taxa, but an aid for describing and classifying phytoliths. To accomplish the transformation to a hierarchical system, we first divided all known types into broad categories, numbered 10, 20, 30, 40, 50, 60, 70, 80, and 90. Most of these divisions are based on the cellular origin of the phytoliths. Phytoliths are produced in different kinds of plant tissues, including the epidermis and mesophyll. In addition, within each tissue area, different types of cells accumulate silica. Within the epidermis, for example, there may be long cells, short cells, bulliform cells and dermal appendages (hairs, prickles, hair bases). This first hierarchical level of the system is designed to accommodate the various cell types that become silicified. It is easy to add new major types of silica bodies as they are recognized. For instance, the system now includes categories 100 (cystoliths) and 110 (sclereids), as well as 21-25, new categories of epidermal non-quadrilateral cells. If the anatomical origin of the phytolith is unknown, it is given a temporary number in category 60 (anatomical origin unknown).

Current groupings in the first hierarchical level of the system can be described as follows (see Fig 3.1 for some representative types):

10 Epidermal quadrilaterals: large cells of the epidermis, quadrilateral, square to elongated, often very two-dimensional in appearance, with variable surface and edge characteristics.

20 Epidermal non-quadrilaterals: large cells of the epidermis, irregular in shape, usually two-dimensional in appearance.

30 Short cells: Poaceae (Gramineae), produced in the leaf vein area.

40 Dermal appendages: prickles, hairs, edge spines, including base cells.

50 Bulliform cells: Poaceae (Gramineae), located in the mesophyll or epidermis, usually three-dimensional in appearance, highly silicified, shape variable, including rectilinear, rounded, and keystone.

60 Anatomical origin unknown: a category for plant silica of unknown origin in the plant.

70 Honeycombed material: amorphous silicified material, usually of irregular shape, lightly silicified, breaking easily.

80 Stegmata and other spherical or spheroidal bodies: mesophyll cells of spherical-spheroidal shape.

90 Other biogenic silica: diatoms, sponge spicules and the like.

100 Cystoliths: silicified outgrowths of cell walls, occurring in the epidermis and sub-epidermis, variable shape, usually with nodular surfaces.

110 Sclereids: sclerenchyma cells, varied in form, but often elongated and angled.

Second order divisions within the broad categories described above are given Roman numerals. This division reflects natural classes of phytoliths, usually based on shape. In category 10, for example, categories I and II divide epidermal quadrilaterals by relative length and width. In category 30, phytoliths are generally divided by the presence of spikes and/or lobes. The second order level in category 40 divides forms into segmented hairs, non-segmented hairs and hair bases. In category 50, angularity is used to divide forms. In several groups, shape is not used for second order divisions. In two categories, 20 and 80, surface characteristics are used, while in category 70 it is relative density of the silica making up the amorphous bodies.

Third order divisions are given capital letters. At this level, divisions generally reflect edge characteristics: whether the edge is smooth, sinuous, serrated, and so on, whether edges are straight or curved. However, in categories 30 and 40, the capital letters tend to reflect further refinements in shape and, in category 70, they reflect general shape.

Fourth order divisions are given lower-case letters. At this level, divisions reflect a variety of characteristics particular to each broad category of phytolith types. In category 10, for example, surface decoration is reflected; in category 50, further refinements in edge characteristics; in the majority of the other categories, further refinements in shape.

There are also fifth, sixth, seventh, eighth, and ninth order divisions, which are only included in some of the major categories. Each discrete phytolith type has its own numerical designation based on shape, size, surface and edge characteristics and related to all other like-types in the system. Because the kinds and numbers of like-types vary among the major categories, some phytoliths have much longer numbers than other types. This will be illustrated in more detail below, but two examples will demonstrate the form final numbers may take. The designation 40IAb200Aa consists of eight hierarchical levels. This phytolith is in category 40, dermal appendages. "I" means "non-segmented"; "A," "edge spine" (the type of non-segmented dermal appendage); "b," "outer edge curved"; "2," "very long outer edge"; "00," "acute tip"; "A," "smooth surface"; and "a," "smooth edges." The number for phytolith type 80VIII reflects only two hierar-

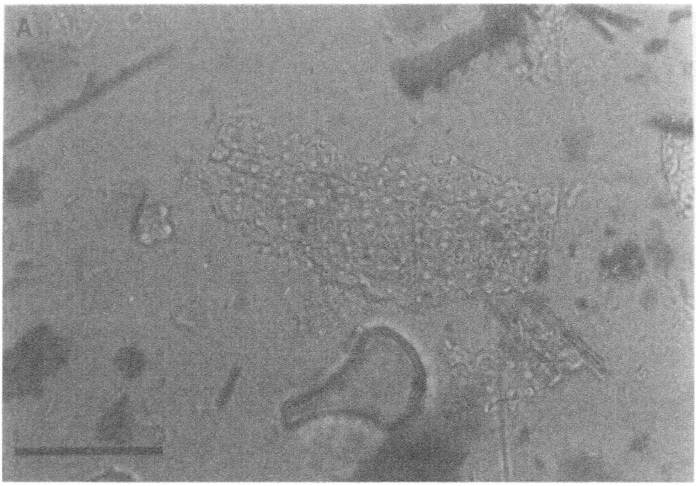

A. Epidermal quadrilaterals (Category 10) and bulliform cell (Category 50) from the leaf of *Oryza sativa* (Poaceae or Gramineae).

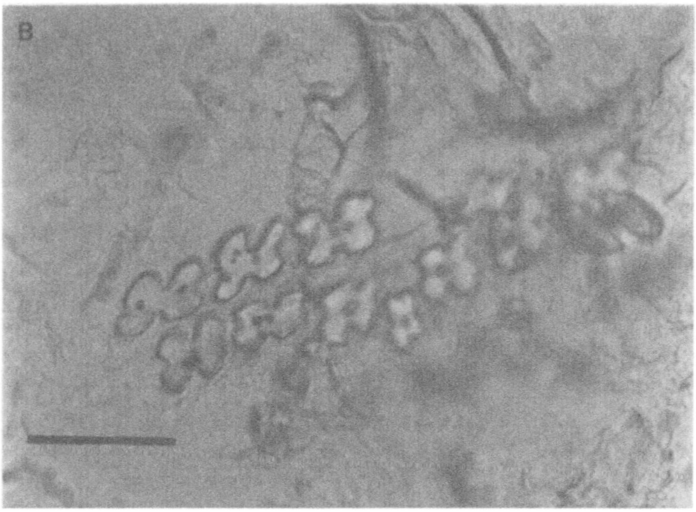

B. Short cells (Category 30) from *Zea mays* (Poaceae or Gramineae) leaf.

Fig 3.1A–K. Photographs of some representative phytoliths in the first hierarchical level of the Missouri classification system. Bar is 50 microns.

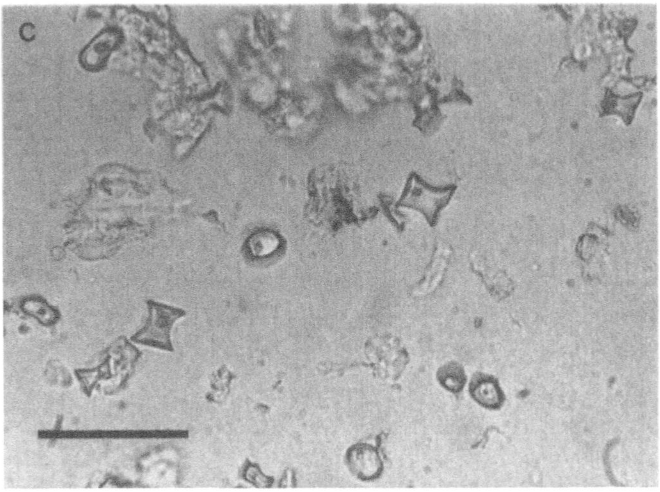

C. Short cells (Category 30) from *Schizostachym glaucifolium* (Poaceae or Gramineae) leaf.

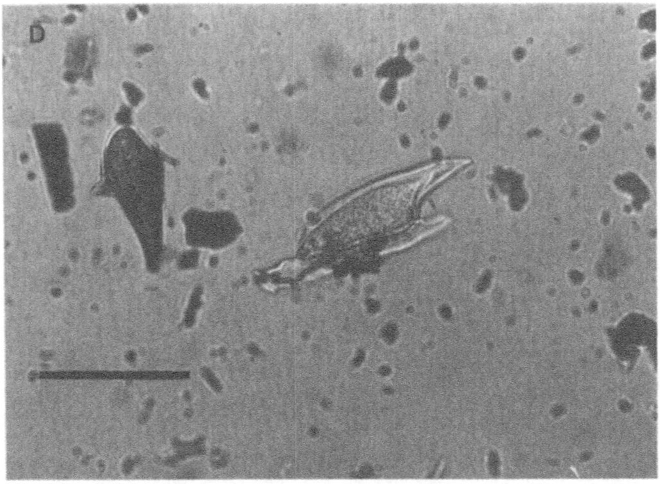

D. Edge spine, a type of dermal appendage (Category 40) from *Saccharum spontaneum* (Poaceae or Gramineae) leaf.

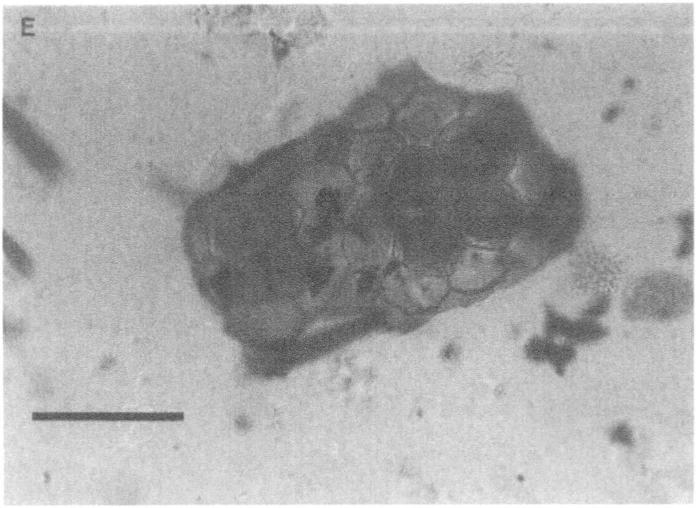

E. Hair cell base (Category 40) from *Cordia polyantha* (Boraginaceae) leaf.

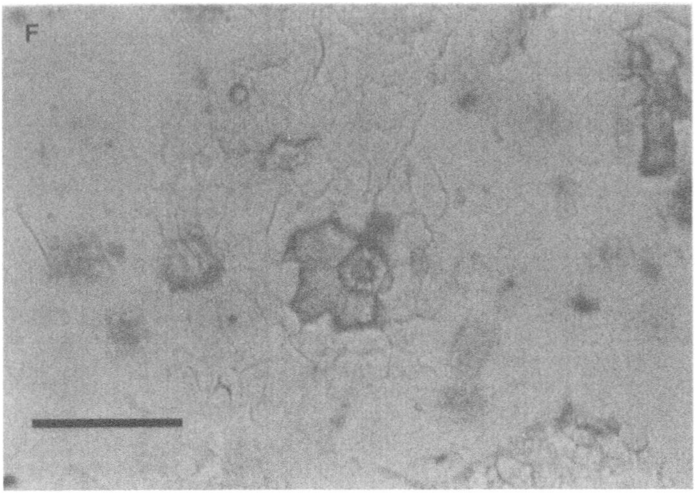

F. Hair cell base (Category 40) with surrounding non-quadrilateral epidermal cells (Category 20) from a modern soil in Yap.

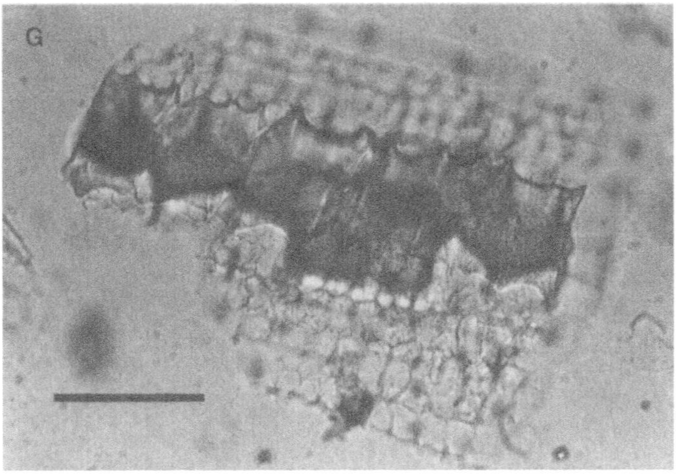

G. Bulliform cells (Category 50) and associated epidermal quadrilaterals (Category 10) from a modern soil in Yap.

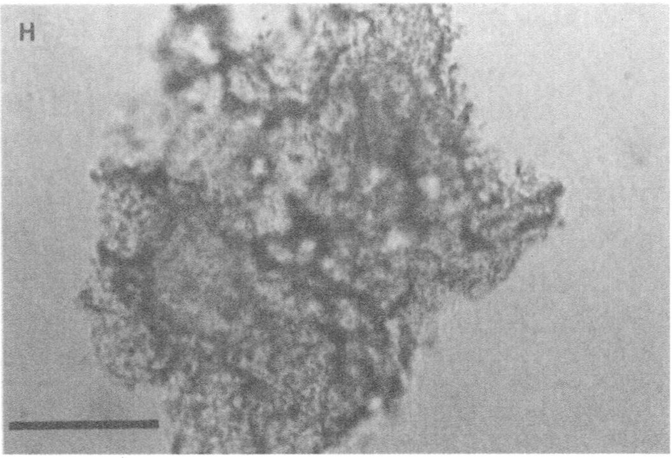

H. Honeycombed material (Category 70) from an archaeological soil sample.

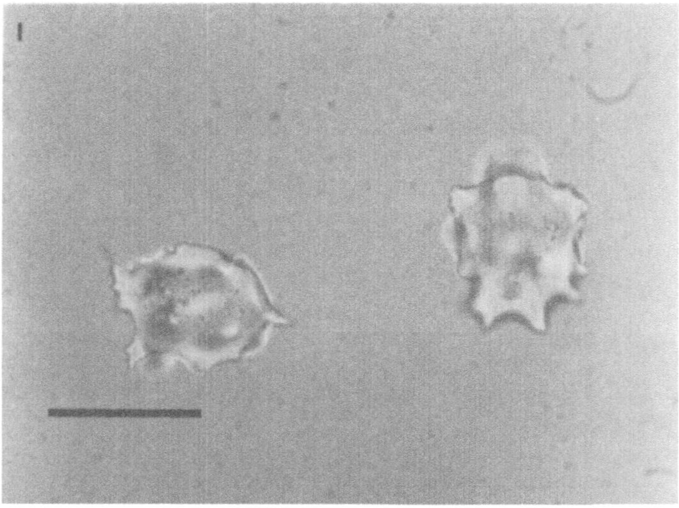

I. Spherical (Category 80) seed epidermal phytoliths from *Maranta arundinacea* (Marantaceae).

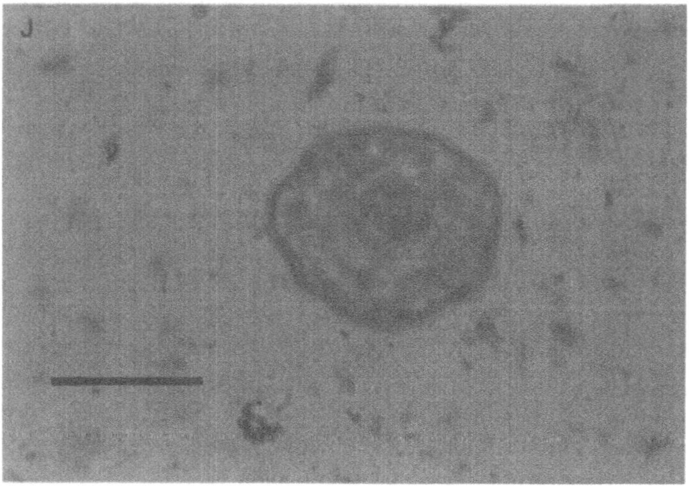

J. Spherical (Category 80) faceted phytolith from *Lagenaria siceraria* (Cucurbitaceae) fruit.

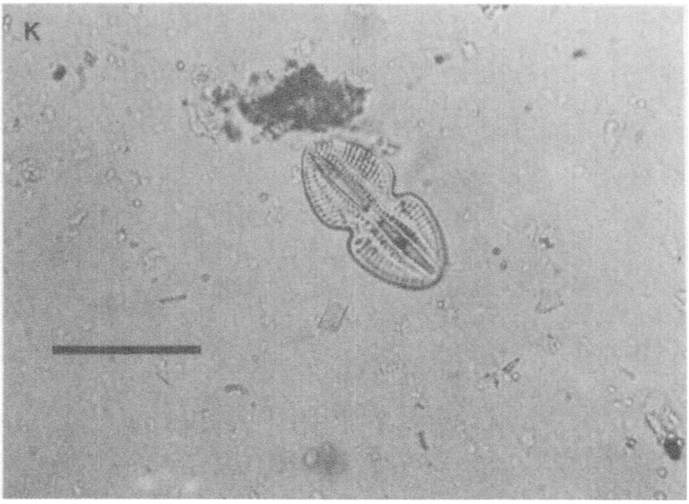

K. Diatom, an example of other biogenic silica (Category 90).

chical levels. Category 80 designates stegmata and other spherical and spheroidal bodies; "VIII" means "smooth surface." No characteristics have thus far been found that further divide smooth surface sphere-like bodies. Additions to the system occur as they are encountered during scanning, or as new types are published by other researchers.

Before giving an example of one segment of the classification system, there are a number of things we have learned in setting up our system, which we would like to share with other analysts. An important first step in setting up a classification system or identification aid is to develop and use a dictionary of terms for describing phytoliths. We always use terms in their precise (dictionary) meaning. A common mistake, for instance, is to describe a phytolith as "round" in shape, when "spherical" is the correct way to describe a three-dimensional object. Our dictionary contains precise descriptions of what is meant by "elongated," "serrated," "sinuous," "rugulose" and so on. Descriptions begin with the standard meaning of the term used, then elaborate points that are important for distinguishing among closely related forms. Terminology used in description of pollen grains is often a useful starting point for phytolith description. We try to avoid arbitrary names for types, which carry no meaning for another researcher. Everyone at the Missouri lab knows what a "cooling tower" is, but this lab designator is useless unless accompanied by a drawing or photograph. We are in the process of "retiring" all such names.

Developing a classification system or key is an ongoing process. One needs to use the system while it is growing and changing. For this reason, it is vital that all changes and additions be documented and dated. In this way research done using the system at different stages of development can be compared. We recommend keeping a logbook of all changes affecting previously numbered types and dating and initialing all photographs and description cards.

All the types in the Missouri classification system are represented by photos. It is useful to take new photos of types periodically; this is one way to insure that classification is consistent and to catch problem areas in the system. File photographs should be taken at the same magnification and printed at the same scale. Source of the photo (comparative taxon name, sediment sample number), type number, date and scanner's initials should be written on the photo. This makes it much easier to use the desk-top aid and to compare one type directly to others.

Publication of results of sediment or comparative analysis should include detailed descriptions and clear photographs of any new diagnostic phytolith types, as well as discussion of their precision of identification, *ie*, is the new type diagnostic at the species, genus or family level?

USING THE CLASSIFICATION SYSTEM

We present an example of how the hierarchical classification system functions using a portion of Category 10, epidermal quadrilaterals.

The overall structure of category 10 can be described as follows:

10 epidermal quadrilateral
- I rectilinear long (length 2-4 times width)
- II rectilinear very long (length more than 4 times width)
- III striated bodies, many striations parallel to short edge, very long.

For purposes of illustration, we examine category 10I further:

- 10IA four smooth edges,
- 10IB sinuous edges,
- 10IC two serrated edges and two smooth to sinuous edges,
- 10ID four serrated edges,
- 10IE one long edge serrated, three edges sinuous.

Figures 3.2, 3.4 and 3.6 illustrate in flow-chart format the relationships between types included in categories 10IA, 10IB, and 10IC. Individual types are illustrated in Figures 3.3, 3.5 and 3.7.

The Missouri classification system is arranged in a photo identification aid so that a phytolith scanner has immediate access to every correct number (and its significance) while scanning. Diagnostic types are flagged for easy recognition. Using the system, phytoliths in each sediment sample and plant comparative specimen scanned can be documented numerically and entered into a system where they are immediately comparable to like-phytoliths.

Scanning at the Missouri laboratory frequently takes the form of a Quick Scan, a term coined by Rovner. A sample is "Quick Scanned" by beginning at one corner of the slide and listing each distinctive phytolith type encountered. The abundance of each type on the slide is estimated using abundance codes developed at the lab: VR (very rare, one or two specimens on the slide); R (rare, less than one specimen every 30 fields, or 1.5 rows, 3-15 per slide); M (moderate, about one specimen every 10-30 fields, or 0.5-1.5 rows, 15-40 per slide); C (common, several specimens per row, 40-220 per slide), A (abundant, one specimen in almost every field); VA (very abundant, more than one specimen in every field). Quick Scans are fast because a great deal of estimation occurs. The slide is scanned only until phytolith occurrence reaches redundancy; this frequently happens in less than three rows. The number of phytoliths of a particular type is considered representative of the entire slide and multiplied by the percentage of slide scanned to estimate abundance. The scanner then examines the remainder of the slide quickly in order to determine whether the portion scanned was indeed representative.

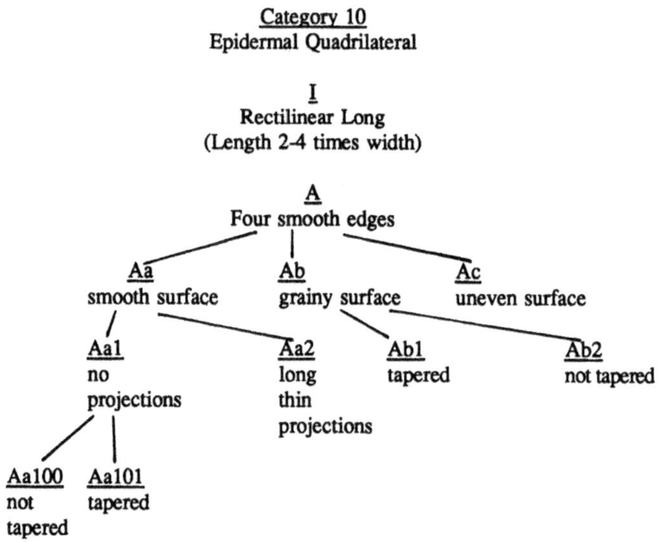

Fig 3.2. Organization of Category 10IA.

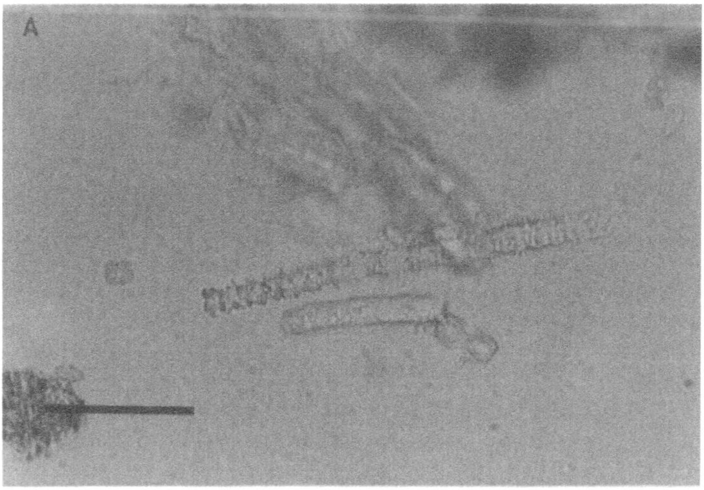

A. 10IAa2. Edges with long, thin projections, smooth surface.

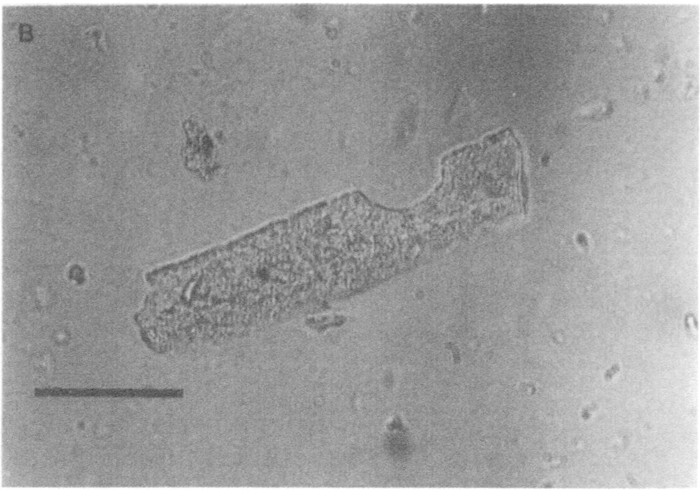

B. 10IAb2. Quadrilateral with a grainy surface, outline not tapered.

Fig 3.3A–D. Examples of Category 10IA: long rectilinear epidermal quadrilaterals with four smooth sides. Bar is 50 microns.

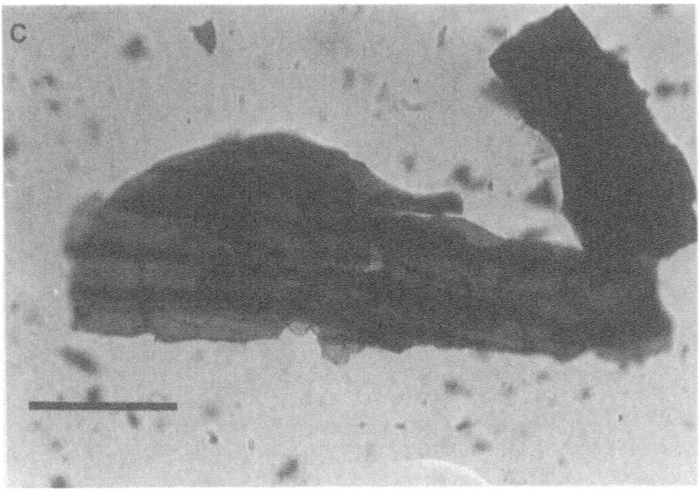

C. 10IAa100. Quadrilateral with a smooth surface, no projections, outline not tapered.

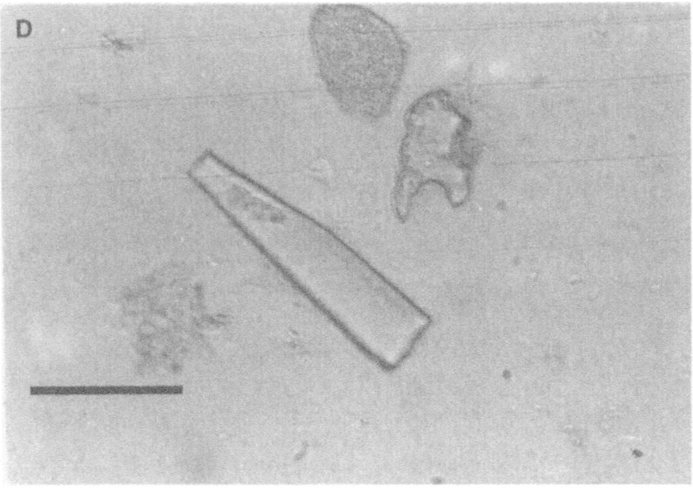

D. 10IAa101. Tapered quadrilateral with smooth surface and no projections.

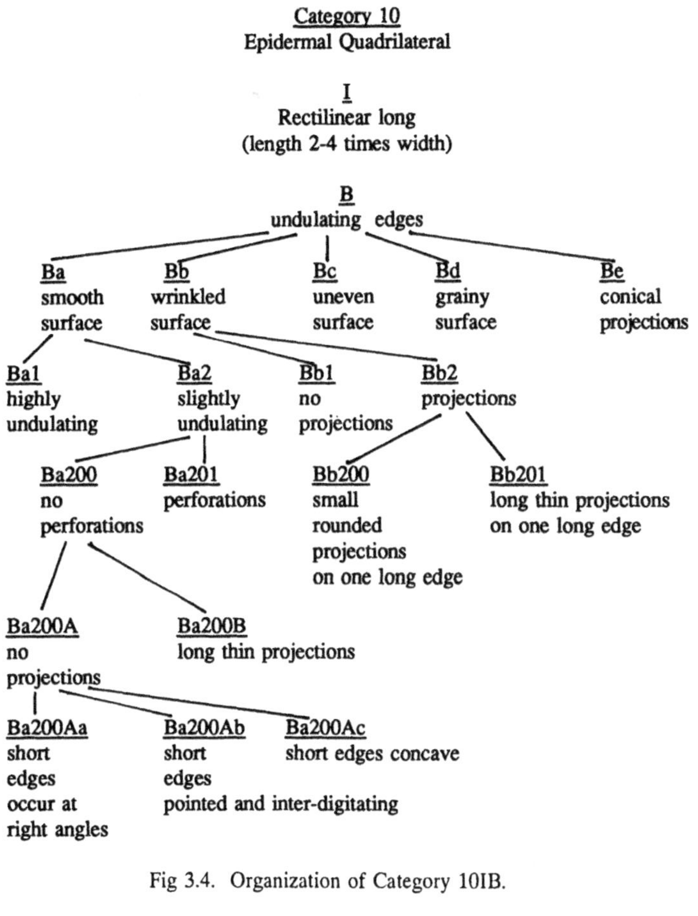

Fig 3.4. Organization of Category 10IB.

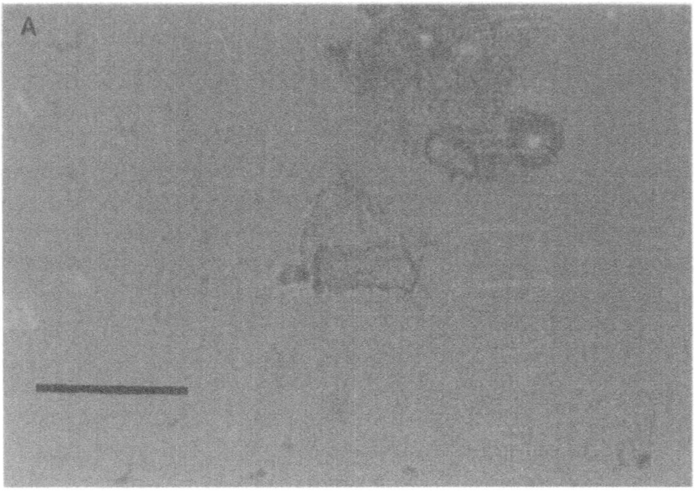

A. 10IBa1. Highly undulating quadrilateral with smooth surface.

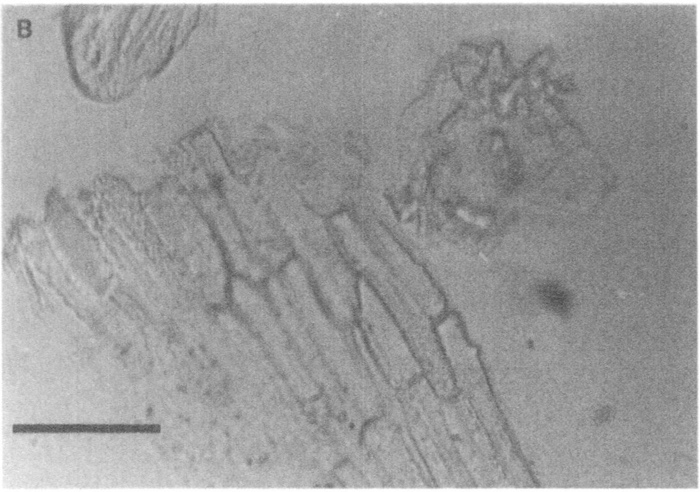

B. 10IBa200Aa. Slightly undulating quadrilateral with smooth surface, no perforations, no projections, and right angles on the short edges.

Fig 3.5A–I. Examples of Category 10IB:long rectilinear epidermal quadrilaterals with undulating edges. Bar is 50 microns.

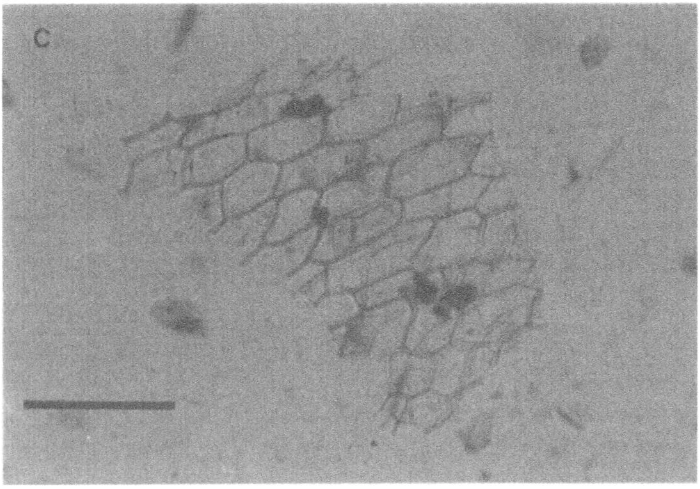

C. 10IBa200Ab. Slightly undulating quadrilateral with smooth surface, no perforations, no projections, and pointed, inter-digitating short edges.

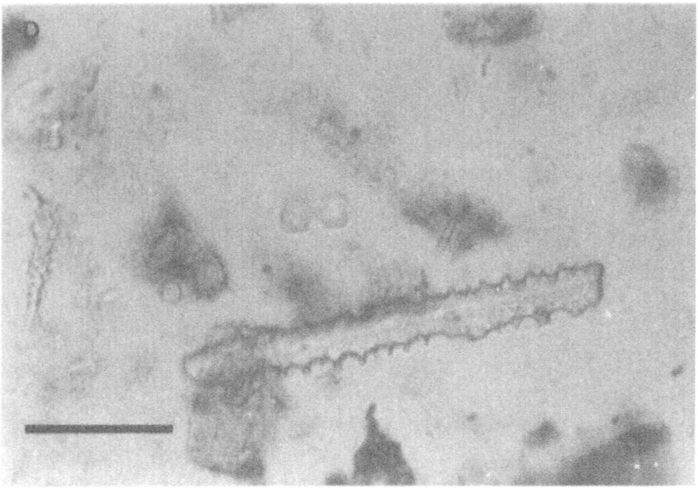

D. 10IBa200B. Slightly undulating quadrilateral with smooth surface, no perforations, and long, thin projections.

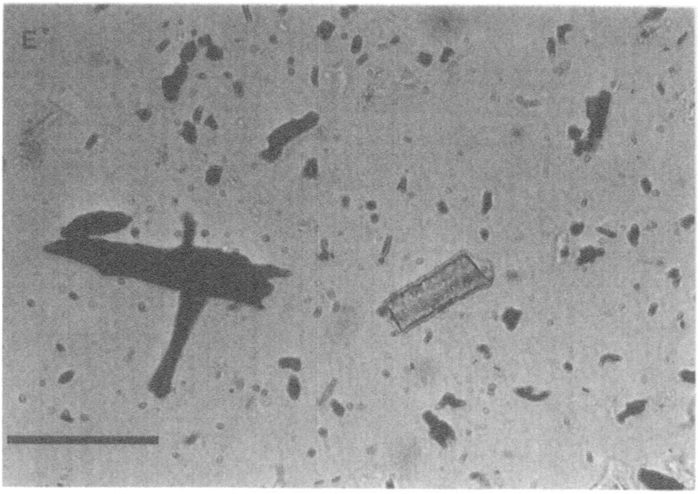

E. 10IBa201. Slightly undulating quadrilateral with smooth surface and perforations.

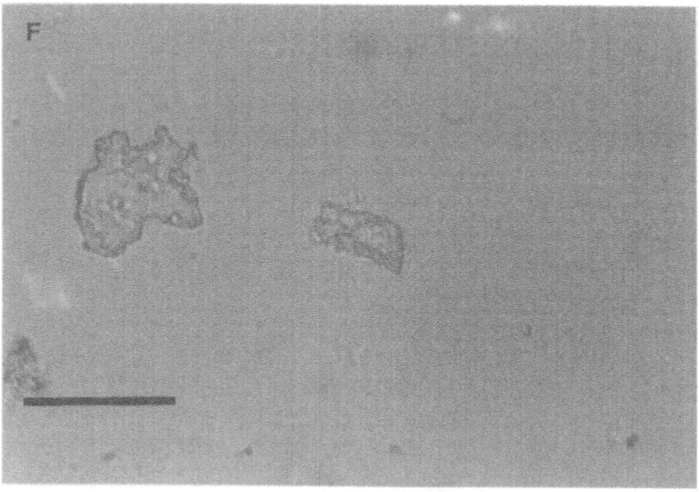

F. 10IBb. Wrinkled surface.

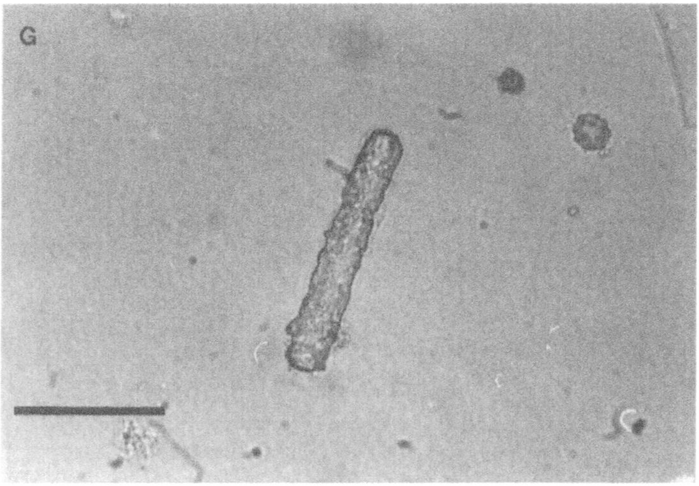

G. 10IBc. Uneven surface.

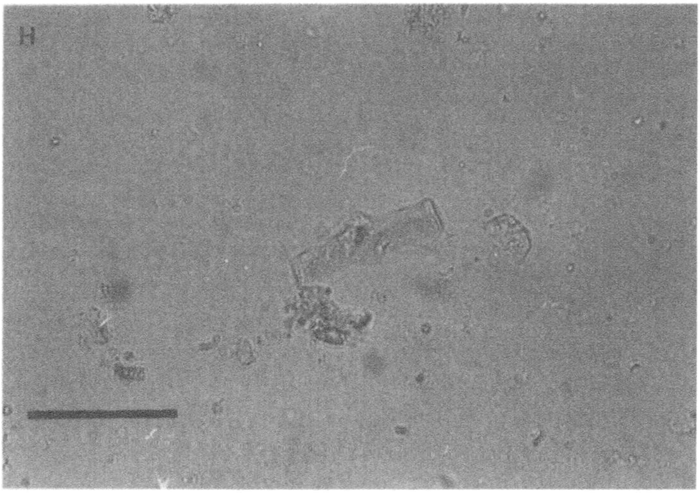

H. 10IBd. Grainy surface.

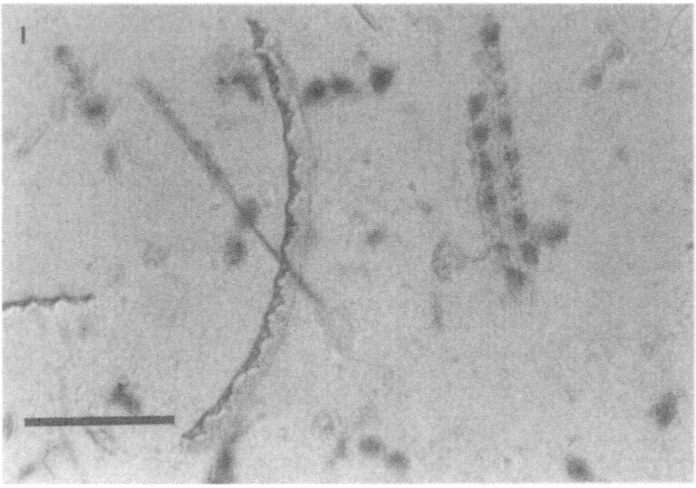

I. 10IBe. Conical projections.

Category 10
Epidermal Quadrilateral

I
Rectilinear long
(length 2-4 times width)

C
Two serrated edges and
Two smooth to undulating edges

- **Ca** long edges serrated, short edges smooth to undulating
  - **Ca1** highly serrated
    - **Ca100** rounded serrations
    - **Ca101** pointed serrations
      - **Ca101A** smooth short edges
      - **Ca101B** inwardly curving short edges
  - **Ca2** slightly serrated
  - **Ca3** very highly serrated
- **Cb** conical surface projections
- **Cc** perforations

Fig 3.6. Organization of Category 10IC.

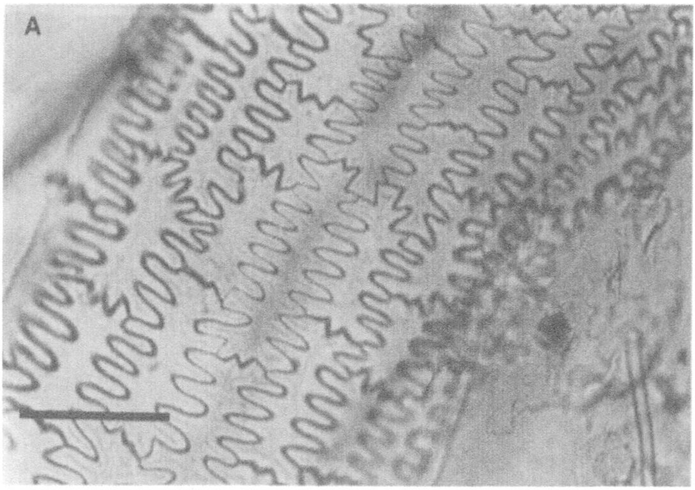

A. 10ICa100. Quadrilateral with undulating short edges, highly serrated long edges, serrations rounded.

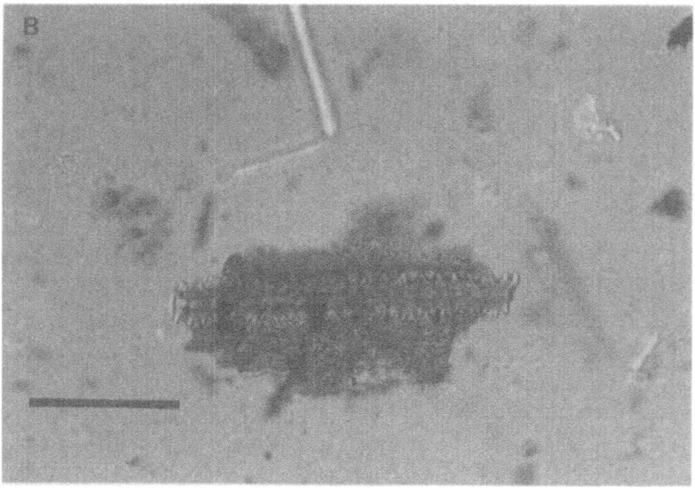

B. 10ICa101A. Quadrilateral with smooth short edges, highly serrated long edges, serrations pointed.

Fig 3.7A–G. Examples of Category 10IC: long rectilinear quadrilaterals with two serrated edges and two smooth to sinuous edges. Bar is 50 microns.

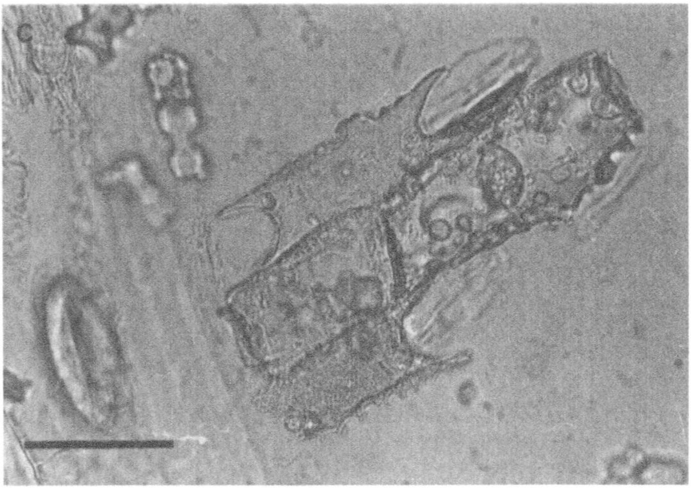

C. 10ICa101B. Quadrilateral with smooth short edges, inwardly curving, highly serrated long edges, and pointed serrations.

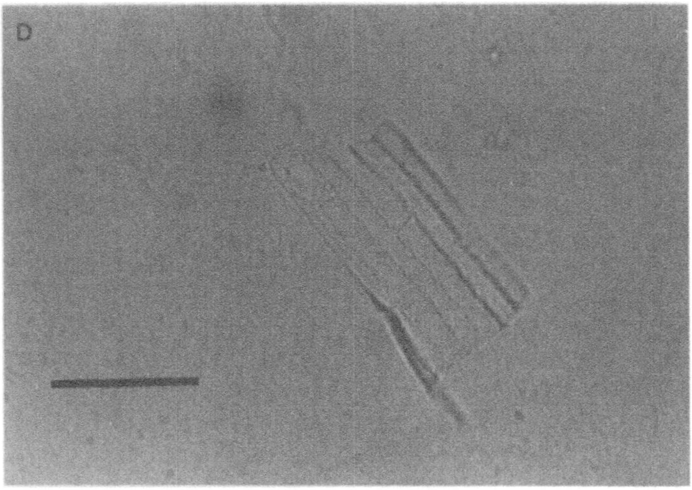

D. 10ICa2. Quadrilateral with smooth short edges and slightly serrated long edges.

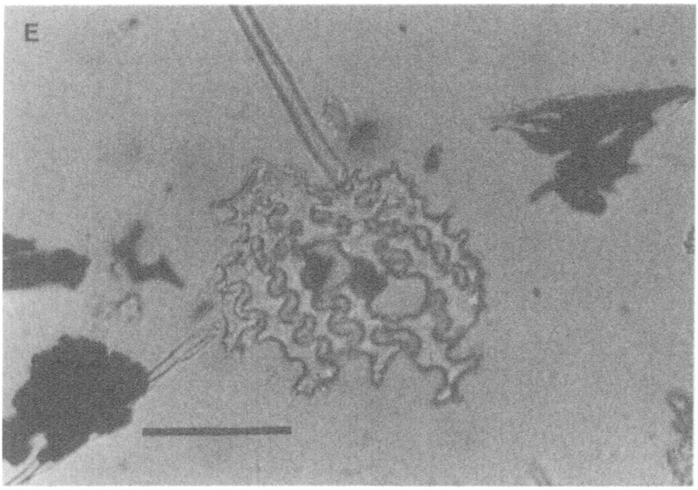

E. 10ICa3. Quadrilateral with smooth short edges and very highly serrated long edges.

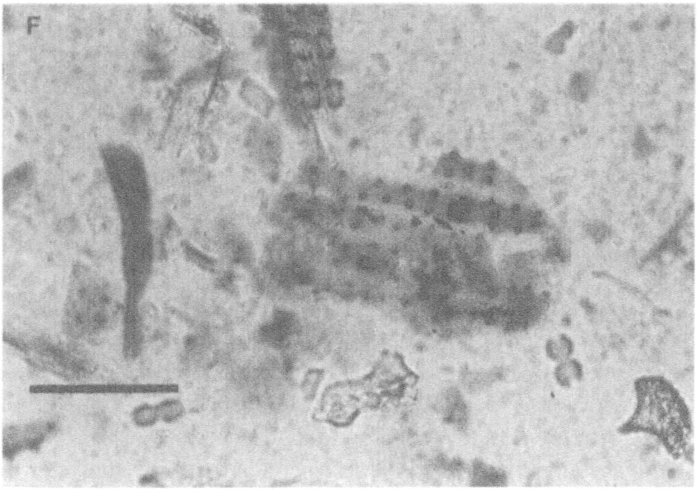

F. 10ICb. Serrated appearance produced by conical surface projections.

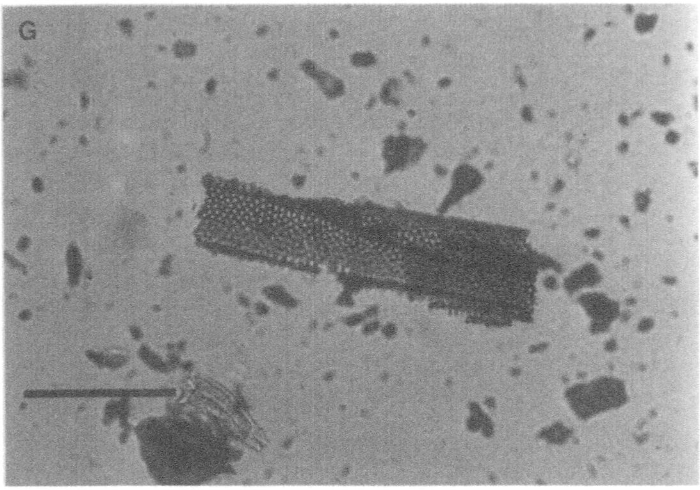

G. 10ICc. Quadrilateral with serrated long edges, perforated surface.

We also use phytolith counts to characterize relative abundance of various types of phytoliths. We have found three types of phytolith sums useful: short cells, diagnostics and crosses (Pearsall 1989). For a short cell count, we usually tally a total of 200 grass short cells. In a diagnostic count, a procedure patterned after the approach used by Piperno (1988), we include a variety of discrete phytolith forms (*ie*, non-fractured, "countable" types) that have been identified to plant taxon. Grasses are often included in this sum, which usually totals 500. The cross count is used for identifying maize; crosses are tallied by size and variant after Piperno (1984).

ISSUES IN PHYTOLITH CLASSIFICATION

The focus of this paper has been on the practical, rather than the theoretical, level. Our goal was to share some of our experiences in developing a working phytolith classification system for day-to-day laboratory use, in the hope that other researchers might find our approach useful. However, there are two broader issues that arise from this discussion of our efforts at phytolith classification. These are, 1) the utility of classifying and tallying unknowns and redundant forms, and 2) the way phytoliths should be examined and measured.

Our approach to phytolith classification has been to develop a system whereby we can classify all phytoliths seen in a sediment or comparative plant sample. New types are photographed, described, and given a number, whether or not we can identify them to plant taxon. Thus our system includes both known and unknown phytoliths, both diagnostic and highly redundant forms. When a plant specimen or sediment sample is quick-scanned, a tally is made of all phytoliths present, including unknown and redundant types.

We recognize, however, that counting known, diagnostic phytoliths is usually the most useful approach for reconstructing subsistence or vegetation patterning. It is for this reason that we routinely employ three sums of "known" phytoliths (short cells, diagnostics, crosses), and flag all diagnostic phytolith types in the system.

By including all phytoliths in quick scanning, we believe that we can accurately "track" the occurrence of new, unknown types. If an unknown is identified, we know in what sediment samples or plant specimens it has been observed. Including redundant phytolith forms in quick scans of sediment samples may be useful in making broad characterizations of the source of phytoliths in the sample. For example, certain redundant forms may help separate grassland and forest formations. Having a mechanism to keep track of all phytoliths is especially useful for a new researcher or an experienced analyst beginning a new regional or taxonomic focus. As diagnostic forms are discovered, unknowns and redundant forms decline in importance, and may eventually be dropped from the classification system.

The other issue we would like to address briefly is how one should examine and measure phytoliths. At the Missouri lab we use light microscopy, usually at magnifications of 312× or 400×, for examining and measuring phytoliths (the lab has two different microscopes; measurements are made in eyepiece micrometer units and converted to microns). Magnification of 500× is used for making some fine distinctions, such as among cross variants. All photographs are taken at 400×. Phytoliths are rotated in the mount to examine their three-dimensional structures. This basic approach is used by many researchers, some of whom also use scanning electron microscopy for documenting phytolith form (refer to Pearsall 1989 for a review of phytolith laboratory procedures).

Russ and Rovner (1989; Rovner & Russ, this volume) advocate the use of computer-assisted image analysis in phytolith research. They have found this technology useful for separating phytoliths produced by maize and teosinte comparative specimens, for example. In computer-assisted image analysis, the phytolith images, captured in a photograph or video, are analyzed as two-dimensional images. A variety of measurements of size and shape are made automatically, and can then be used to distinguish among the phytoliths represented by the images.

Some may question whether phytolith descriptions and measurements made using the technology of the light microscope and eyepiece micrometer are as accurate as those made by computer-assisted image analysis. Since the "traditional" technology is the backbone of most phytolith classification, this issue should be addressed. In a recent side-by-side comparison of measurements made by computer-assisted image analysis and by the "traditional" technology, it has been shown that equally accurate results can be obtained using either method (Piperno & Pearsall, nd).

This test was made using five grass leaf specimens, two maize races and three wild grasses. Using light microscopy and eyepiece micrometer measurements, cross width was measured and crosses were classified by three-dimensional variant using the methods of Pearsall (1979, 1989) and Piperno (1984, 1988). Crosses from the same slides were then videotaped and the images analyzed by computer-assisted image analysis. It was found that two of the grasses overlapped in cross size with one maize race at the 95% confidence interval. *Identical* results were obtained using each method, "traditional" and image-analysis. Size and two-dimensional shape analysis of the crosses did not permit separation of the wild grasses and maize. Only when three-dimensional characteristics (*ie*, cross variant classification) were used could the separation be made; no overlap between the three wild grasses and the two maize races occurred when both size and three-dimensional structure of the crosses were assessed. Although computer-assisted image analysis is a useful technology in phytolith research, it should not be considered a substitute for careful study of the three-dimensional structure of phytoliths.

CONCLUSIONS

We have discussed our phytolith classification system to illustrate how phytoliths can be accurately described and tallied on a day-by-day basis in a working laboratory. Our classification system is not complete and is subject to ongoing revisions. One of its strengths is that it does not have to be finished to achieve its goal: to document phytolith occurrences in a controlled, replicable manner.

We have also addressed briefly two general issues of phytolith classification, the usefulness of describing and tallying redundant and unknown phytolith types, and the way phytoliths should be described and measured. "Keeping track" of all phytolith forms is advisable for any new researcher, and probably useful for anyone beginning a new geographic or taxonomic focus. Careful description of the three-dimensional morphology of phytoliths, coupled with size measurements, seems the most useful approach to distinguishing among closely related phytolith forms.

We have several avenues of future research planned to improve our classification system. As work on comparative material progresses, we are learning which branches in the system represent taxonomic divisions. Within category 10, for example, there are groups of types that can be characterized as grass long cells, while other types occur more widely in the monocotyledons. Some divisions within category 20 are exclusively dicotyledon epidermal cells, whereas other non-quadrilateral types occur more widely in the higher plants. Where necessary we plan to restructure the system to reflect real taxonomic divisions more clearly, allowing easier use of the system for interpretation of sediment assemblages.

Another goal is to refine the criteria used to distinguish among similar phytolith types. We are in the process of analyzing how we classify phytoliths by modelling classification using the expert systems approach. We ultimately hope to produce a series of easy-to-use regional identification aids useful to other researchers.

ACKNOWLEDGMENTS

Early work on the Missouri classification system was carried out by Michael K. Trimble. Marcelle Umlauf refined various aspects of the system. Laboratory facilities and administrative support for this research are provided by the American Archaeology Division, Michael J. O'Brien, Director. John Amann and Howard Wilson of the College of Veterinary Medicine of the University of Missouri provided facilities and assistance for the image analysis study. Funding has come, in part, from the National Science Foundation (BNS 8709649). An earlier version of this paper was presented at the Third Phytolith Workshop, Columbia, Missouri, January 1988.

REFERENCES

Bozarth, SR 1986 Morphologically distinctive *Phaseolus, Cucurbita*, and *Helianthus annuus* phytoliths. *In* Rovner, I, ed, Plant opal phytolith analysis in archaeology and paleoecology: Proceedings of the 1984 Phytolith Research Workshop, North Carolina State University, Raleigh, North Carolina, *Occasional Papers of The Phytolitharien* No 1:56–66.

____1987 Diagnostic opal phytoliths from rinds of selected *Cucurbita* species. *American Antiquity* 52: 607–615.

Brown, DA 1984 Prospects and limits of a phytolith key for grasses in the central United States. *Journal of Archaeological Science* 11: 345–368.

Pearsall, DM 1978 Phytolith analysis of archeological soils: Evidence for maize cultivation in Formative Ecuador. *Science* 199: 177–178.

____(ms) 1979 The application of ethnobotanical techniques to the problem of subsistence in the Ecuadorian Formative. PhD dissertation, University of Illinois-Urbana: 267 p.

____(ms) 1981 Phytolith analysis at Natural Trap Cave. Ms on file, American Archaeology Division, University of Missouri-Columbia.

____1982 Phytolith analysis: Applications of a new paleoethnobotanical technique in archeology. *American Anthropologist* 84: 862–871.

____1989 *Paleoethnobotany: A Handbook of Procedures*. San Diego, Academic Press: 470 p.

Pearsall, DM and Trimble MK 1983 Phytolith analysis of soil samples: Report 18. *In* Clark, JT and Kirch, PV, eds, Archaeological investigations of the Mudlane-Waimea-Kawaihae Road Corridor, Island of Hawaii: An interdisciplinary study of an environmental transect, *Bernice P. Bishop Museum Departmental Report Series* 83–1: 472–497.

____1984 Identifying past agricultural activity through soil phytolith analysis: A case study from the Hawaiian Islands. *Journal of Archaeological Science* 11: 119–133.

Piperno, DR 1984 A comparison and differentiation of phytoliths from maize and wild grasses: Use of morphological criteria. *American Antiquity* 49: 361–383.

____1985 Phytolith analysis and tropical paleo-ecology: Production and taxonomic significance of siliceous forms in New World plant domesticates and wild species. *Review of Paleobotany and Palynology* 45: 185–228.

____1988 *Phytolith Analysis: An Archaeological and Geological Perspective*. San Diego, Academic Press: 280 p.

Piperno, DR and Pearsall, DM (ms) The nature and status of phytolith analysis. *In* Pearsall, DM and Piperno, DR, eds, *Current Research in Phytolith Analysis: Applications in Archaeology and Paleoecology*. In preparation.

Russ, JC and Rovner, I 1989 Stereological identification of opal phytolith populations from wild and cultivated *Zea*. *American Antiquity* 54: 784–792.

Twiss, PC, Suess, E and Smith, RM 1969 Morphological classification of grass phytoliths. *Soil Science Society of America, Proceedings* 33: 109–115.

4

A MORPHOLOGICAL CLASSIFICATION OF GRASS SILICA-BODIES

SUSAN C. MULHOLLAND and GEORGE RAPP, JR.

ABSTRACT

Classifications of grass silica-bodies are constructed for various purposes ranging from botany to archaeology. Specific organizational details vary with the experience of the phytolith analyst and the condition of the material to be classified. The first step in classification of disaggregated phytoliths, however, must be consistent identification of types based on characteristics that survive burial, *ie*, morphology. The defined types then need to be correlated with plant taxa at as low a level as possible.

This paper presents a standard terminology for classification of silica phytoliths, focusing on grass silica-bodies. Morphology forms the basis of the organization, although it is correlated to anatomical origin where possible. Grass silica-bodies exhibit three general geometric shapes; eight types are defined based on outline of the base. In addition, basic three-dimensional structures are defined. The classification presents standardized definitions of the "classic" grass phytolith shapes reported in the literature.

INTRODUCTION

Various types of grass phytolith classifications have been proposed over the last several decades. Each reflected the then current state of knowledge about phytoliths, size of reference collection and intended application of the data. The influence of this last factor, type of application, is quite significant. The specific objective of a study often determines the structure used in a classification. For example, regional floras include only general information on seed morphology because botanists examine entire plants (not just seeds) when making an identification. In contrast, much greater morphological detail is required for identification of seeds recovered from paleoenvironmental contexts; the characteristics of plant parts that survive burial in sediments must be emphasized (Renfrew 1973: 9).

Phytolith Systematics, Edited by G. Rapp, Jr. and S.C. Mulholland, Plenum Press, New York, 1992

Phytolith classification for paleoenvironmental objectives (paleoecology, archaeology, pedology, etc.) should be based largely or entirely on morphology. Some classifications discussed below incorporate anatomical characteristics, such as orientation of phytoliths in plant tissue. However, most phytoliths recovered from sediments are discrete grains rather than part of articulated tissue fragments (but see Palmer 1976 for an important exception). Disaggregated phytoliths can be classified only by characteristics that survive burial. Particle morphology is often the only type of information available (Bombin 1984: 22).

In many cases, the cell type from which the phytolith derived can be identified by morphology. Some classifications are based on the original cell type, whereas others are based solely on particle shape (Piperno 1988: 53). Elements of both approaches have been combined in a few classifications (Bombin 1984: 58). Whether the nomenclature is anatomical or morphological, classifications suitable for disaggregated material must ultimately be constructed on the basis of particle morphology.

Two general types of classifications have been distinguished: taxonomic and nontaxonomic (Piperno 1988: 52). Both are based on particle morphology, but taxonomic schemes also emphasize the correlation of phytoliths to plant taxa. Nontaxonomic schemes have been characterized as emphasizing particle morphology without any consideration of plant origin. However, a classification that stresses morphology does not have to be nontaxonomic in nature. Morphological characteristics (size, outline, surface texture, etc.) that display limited taxonomic distribution can be emphasized rather than those that are redundant.

Correlation of phytoliths to plant taxa at as low a level as possible is essential for paleoenvironmental objectives. The more precise the identification, the more accurate the reconstruction. However, taxonomic information can be useful at many levels, not just species. Characteristics indicative of a genus or family may yield significant information. For example, Piperno (1988: 155) identifies arboreal, nonherbaceous taxa on the basis of silicified sclereids.

This paper presents a morphological classification of grass phytoliths. The major units are based on original cell types indicated by particle morphology. Seven phytolith types have been identified to date, six of which have been observed in other families. Grass silica-bodies, however, are considered characteristic of the Poaceae (Gramineae) (Esau 1977: 85). Major morphological characteristics are incorporated here in a hierarchical classification.

PREVIOUS SYSTEMATIC STUDIES

Grass phytolith classifications have followed two general approaches: botanical (for intact tissue fragments) and morphological (for discrete particles). The botanical approach incorporates characteristics of epidermal structure as well as morphology. Metcalfe's (1960) work, based on examination of more than 200 grass species, contains at least one phytolith type defined in part by orientation within leaf tissue. The *Oryza* type is a dumbbell shape that forms with the longest dimension perpendicular to the cell rows, rather than parallel as is the case for most other dumbbell shapes. Other types are also defined at least partly by orientation, although named by shape. The type Tall and Narrow, for example, again refers to phytoliths from silica-cells with the longest dimension perpendicular to cell rows. Conversely, the two Horizontally Elongated types are longest parallel to cell rows. The organization of this classification reflects Metcalfe's purpose: to describe grass anatomy in a consistent manner and to record diagnostic characteristics for use in taxonomic studies of grasses. Metcalfe wanted to provide agrostologists with information about grass micromorphology for use in taxonomic and identification studies. Consequently, he examined sections of leaf epidermis and recorded data for in situ phytoliths.

Later studies expanded on the botanical approach. Ellis (1976, 1979) proposed a standardized, comprehensive procedure and set of keys for description of grass leaf epidermal tissue. The numerous anatomical characteristics used in the classification included shape, size and distribution of long cells, short cells, hairs, stomata and other specialized cells. Silica-cells are one of the types of specialized cells described. Ellis's key expands on Metcalfe's classification, using orientation within the epidermis as a primary distinguishing characteristic. Clifford and Watson (1977) used a modified version of Metcalfe's system when describing grasses of Australia. Palmer and Tucker (1981, 1983); Palmer, Gerbeth-Jones and Hutchinson (1985); and Palmer and Gerbeth-Jones (1986, 1988) also follow the botanical type of classification for identification of charred grass fragments from East African lakes.

The botanical approach to phytolith classification is designed for identification of intact plant material, but is largely unsuitable for identification of disarticulated material. Although tissue fragments are sometimes recovered, most sediments contain discrete phytoliths only. Orientation in plant tissue cannot be identified once the structure has broken apart (unless indicated by morphological characteristics). In addition, mixing of phytoliths from many taxa occurs in disarticulated material. Distinctive groupings of phytoliths and cells are not available for use in identification. Classification of disarticulated material needs to be based on a morphological approach.

Twiss, Suess and Smith (1969) proposed a grass phytolith classification based solely on morphology. The 4 classes and 26 types in their hierarchical scheme were differentiated based on examination of 17 species. The phytolith units were based on Metcalfe's (1960) groups, but because the objective was to identify discrete bodies, emphasis was placed on morphology rather than orientation in growth position. Three of the four classes were named after the grass subfamilies in which they appeared, because the subfamilies were thought to be characterized by the respective phytolith classes. The fourth class, Elongates, was found in all species.

Brown (1984) proposed an expanded version of the classification of Twiss, Suess and Smith. Based on data from 112 grass species, it included 8 major categories and well over 100 individual units. The major categories correspond in part to Twiss's four classes, with some classes divided into separate categories. Additional categories representing silicifications of other epidermal cell types were also defined. The cells are anatomical elements described by Ellis (1979) and others; Brown's categories refer to the silicifications of these cells. Division of the major categories into lower units is based on morphological characteristics, following a hierarchical organization.

Pearsall (1978, 1979) and Piperno (1983, 1984) also dealt with disarticulated material, using a mixture of Twiss's and Metcalfe's units for the grasses. One major objective of these studies was identification of corn (maize) by phytolith content. Pearsall (1978: 178) focused on the cross shape, as more abundant and larger in *Zea mays* than other Ecuadorian grasses. Piperno (1984: 370) observed distinctive three-dimensional morphology patterns in maize. All three characteristics—cross abundance, cross size and cross morphology—were identified as distinctive of corn.

Several phytolith classifications have been constructed and tested at the Archaeometry Laboratory during the last ten years (Mulholland, Rapp and Gifford 1982; Mulholland *et al* 1983; Collins *et al* 1986). All are based primarily on data from grass species, although specimens from other families have been examined. A morphological approach was followed; epidermal orientation and cell origin were not included.

A slightly different organization is followed in more recent classifications (Mulholland & Rapp 1985, 1989; Mulholland 1986, 1989). Major categories are based on the type of plant cell silicified (Table 4.1): trichomes, including hairs and papillae (Fig 4.1), stomatal complexes (Fig 4.2), bulliform cells (Fig 4.3), epidermal groundmass cells (Fig 4.4), rods (Fig 4.5), rectangles/squares (Fig 4.6) and grass silica-bodies (Figs 4.9–4.15). "Silica-body" is here defined following Ellis (1979: 658) as the phytolith that forms inside a grass silica-cell. Bombin (1984: 49) clearly distinguishes between phytoliths from grass silica-cells (template-phytoliths) and those from other cell types (cast-phytoliths). The two types of phytoliths display different optical appearances, possibly caused by different processes of silicification. Categories 1–6 of Table 4.1 are here correlated to Bombin's cast-phytoliths; category 7 contains template-phytoliths.

Table 4.1 can be expanded easily to accommodate additional cell types that are silicified in other families. The seven types listed in Table 4.1 have been observed to date only during examination of grasses and a limited number of other species. However, only silica-bodies have been observed in all grass specimens. The other six cell types are not universally silicified, although the cell types may be present in the plant. Silica cells, on the other hand, consistently accumulate silica and the resulting silica-bodies are characteristic of grasses (Esau 1977: 85).

Specimens from nongraminaceous families were examined to determine whether phytoliths similar to grass silica-bodies are produced in other taxa. In certain families, "silica bodies" have been reported in the literature. However, the reports from the Cyperaceae (Metcalfe 1971: 13) are silicifications of papillae and are classified here as trichomes. "Silica bodies" reported for leaves and culms of the Restionaceae and Thurniaceae (Cutler 1969: 6) are described as spheroidal and nodular, varying to conform with the cell lumen. They lack the morphological diversity and do not exhibit the shapes characteristic of Poaceae (Gramineae) silica-bodies.

TABLE 4.1. Phytolith Categories Based on Cell Type

1. Trichomes—Hairs and papillae. Spherical to ovoid base with a conical top.
2. Stomatal complex—Guard and/or subsidiary cells of the stomatal complex. If entirely silicified, the complex is ovoid in shape. Guard cells are shaped like a telephone receiver. Subsidiary cells are ovoid to trianguloid.
3. Bulliform cells—Enlarged thin-walled epidermal cells. Keystone shapes.
4. Epidermal groundmass cells—Unspecialized epidermal cells. Various rectangular box shapes with interlocking edges. Thin.
5. Rods—Fibers, sclereids, xylem cells and other cylindrical cells.
6. Rectangles/Squares—Large blocky cells. Cube to rectangular box. Thicker than groundmass cells or silica-bodies.
7. Silica-bodies—Phytoliths from specialized grass silica accumulating cells. Truncated to beveled pyramids, cones, rectangular boxes, and cylinders. At least one broad face (base) is present. Note that although silica-bodies are equated with short cells in botanical texts, some very long bodies are included here with the shorter ones. The long bodies are consistently silicified and resemble the other silica-bodies in surface texture (unlike groundmass cells that become silicified).

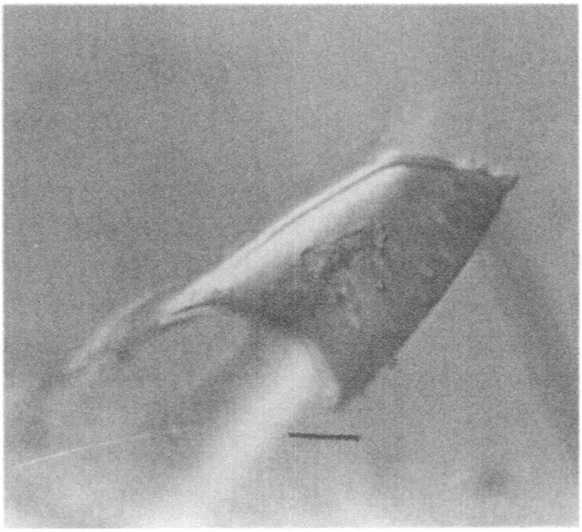

Fig 4.1. Trichome, *Arundo donax* L., leaf. Bar is 10 microns.

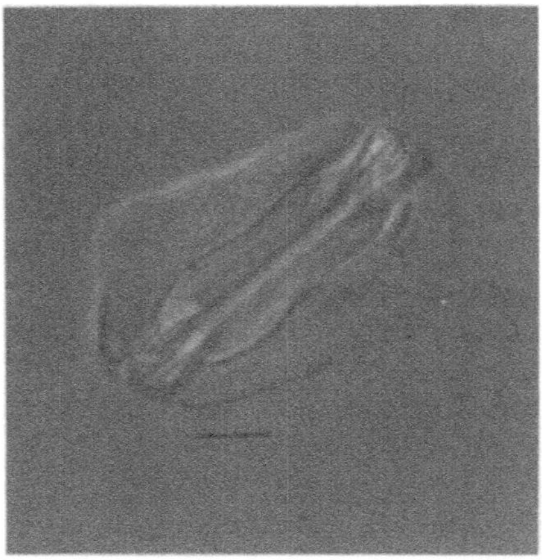

Fig 4.2. Stomatal complex, *Zea mays* L., leaf. Bar is 10 microns.

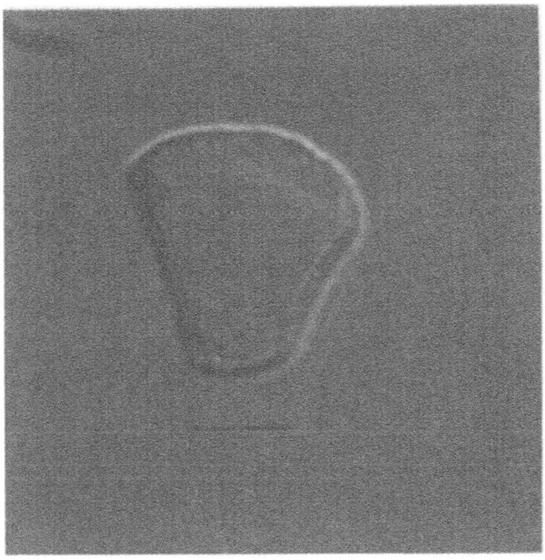

Fig 4.3. Bulliform cell, *Zea mays* L., leaf. Bar is 10 microns.

Fig 4.4. Epidermal groundmass cells, *Phalaris arundinacea* L., leaf. Bar is 10 microns.

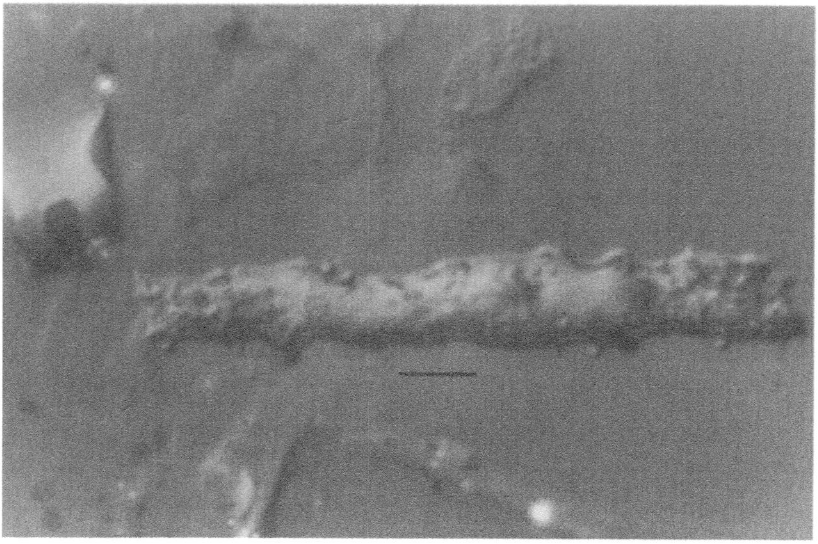

Fig 4.5. Rod, *Phalaris arundinacea*, leaf. Bar is 10 microns.

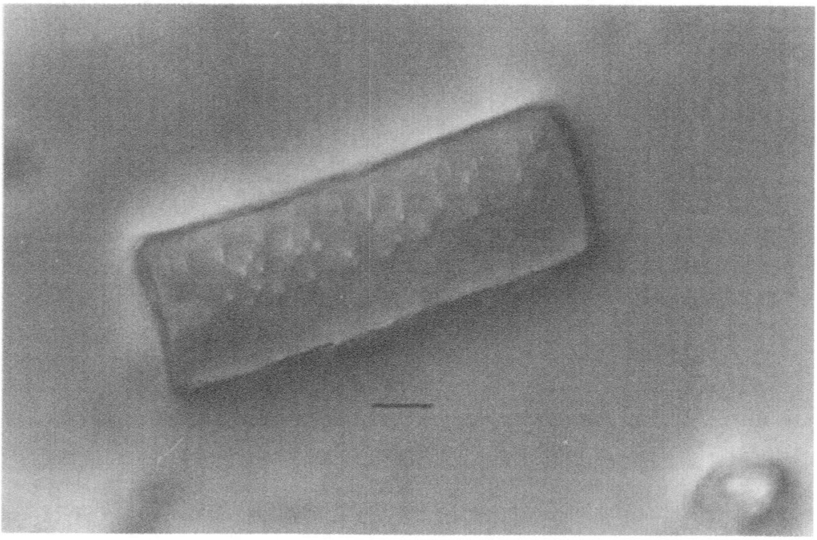

Fig 4.6. Rectangle/square, *Triticum* sp, root. Bar is 10 microns.

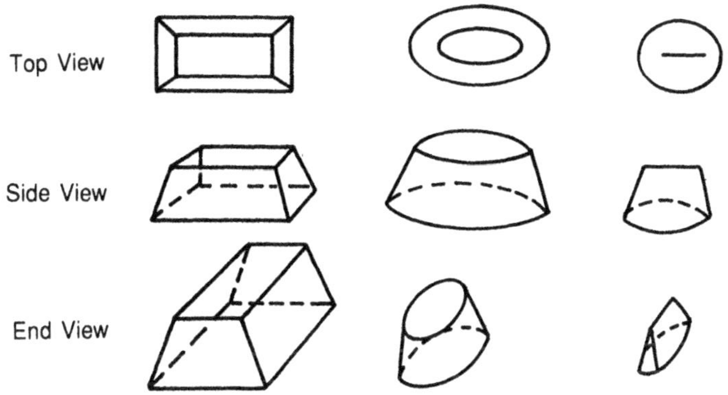

Fig 4.7. Definitions and orientations of grass silica-bodies.

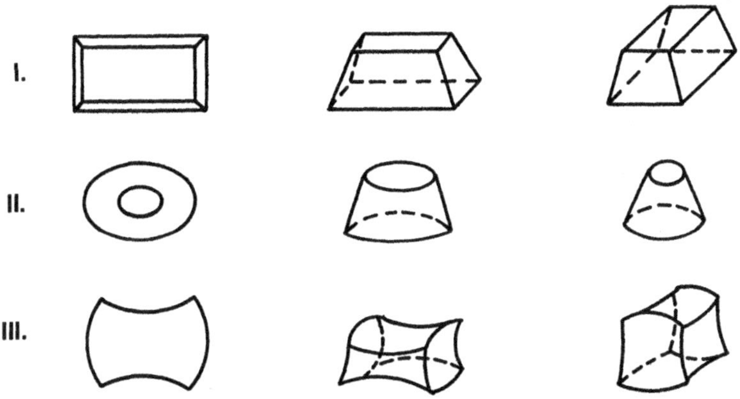

Fig 4.8. Geometric silica-body classes. I. Body is a rectangular box to truncated or beveled pyramid; outline of base approximately rectangular to square or other polygon (base may have lobes but general outline is a polygon); top is a flat to slightly concave or convex face or elevated ridge(s). II. Body is a short cylinder to truncated or beveled cone; outline of base approximately oval to circular or other curved shape (base may have concave or flat segments but general outline is curved shape); top is a flat to slightly concave or convex face or elevated ridge(s). III. Body is saddle-like; outline of top or both top and base has two opposite convex edges that flare outward from the face surface and two opposite lower edges that are usually concave; top is concave.

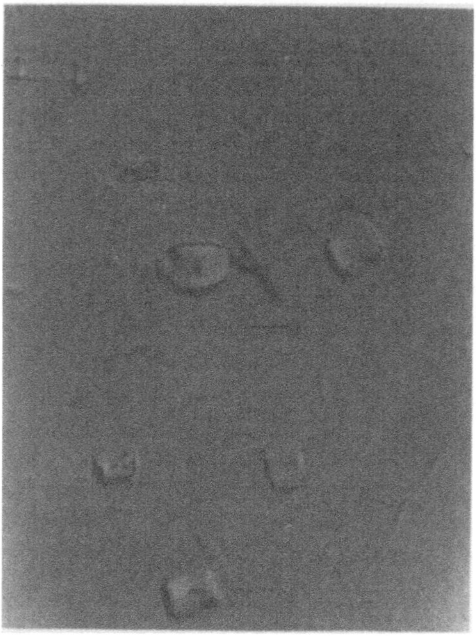

Fig 4.9. Saddle, *Bouteloua curtipendula* Torr., leaf. Bar is 10 microns.

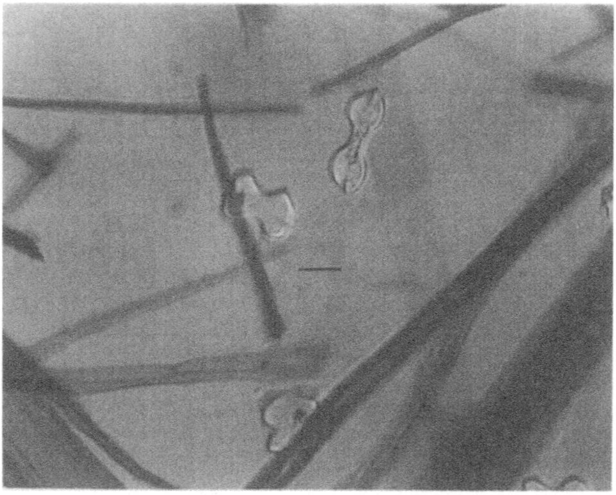

Fig 4.10. Dumbbell, *Panicum capillare* L., leaf. Bar is 10 microns.

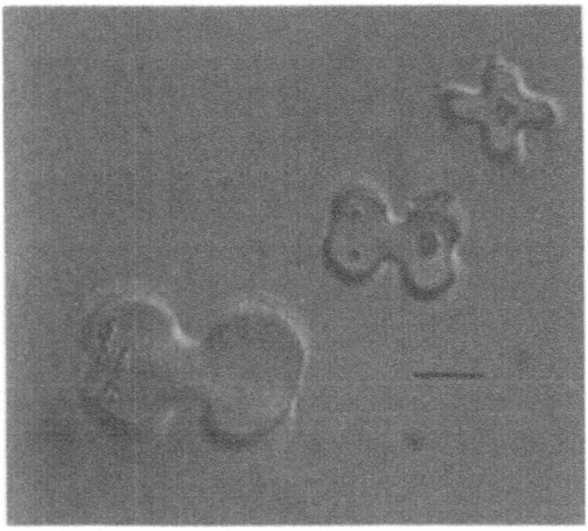

Fig 4.11. Dumbbell-cross sequence, *Arundo donax* L., leaf. Bar is 10 microns.

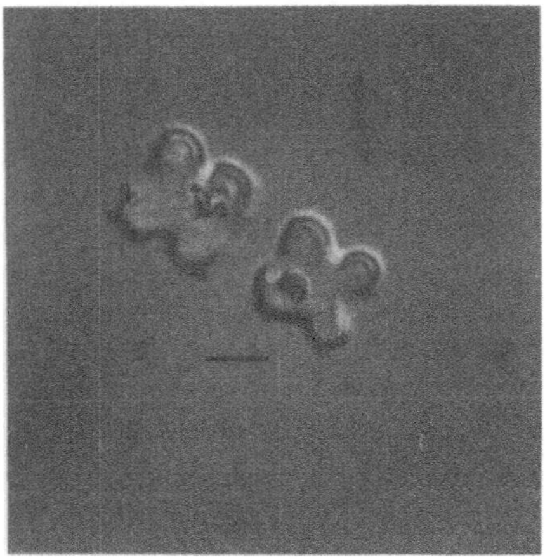

Fig 4.12. Cross, *Panicum capillare* L., leaf. Bar is 10 microns.

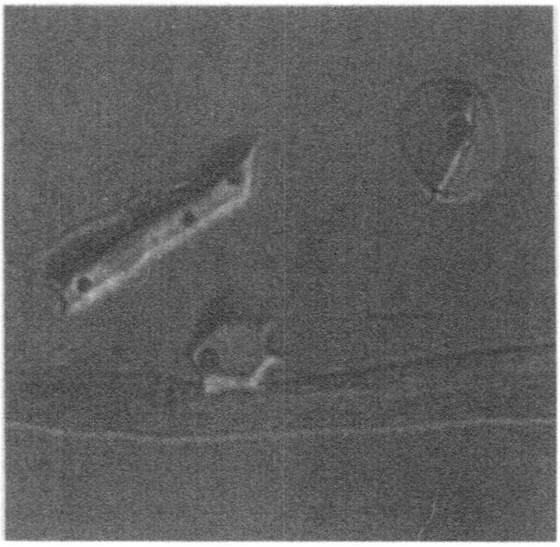

Fig 4.13. Rondel, *Dactylis glomerata* L., leaf. Bar is 10 microns.

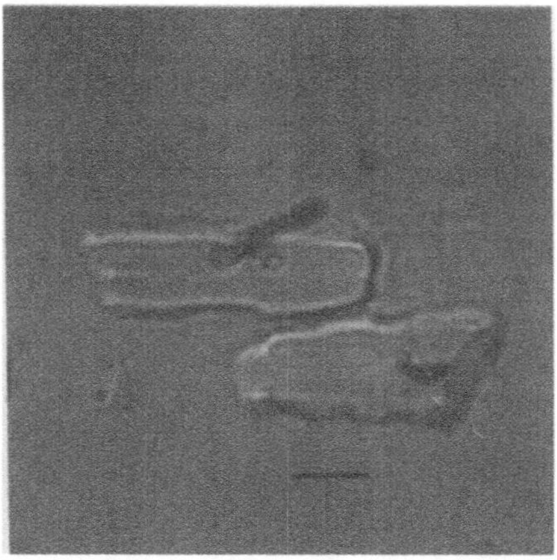

Fig 4.14. Rectangle, *Phalaris arundinacea* L., leaf. Bar is 10 microns.

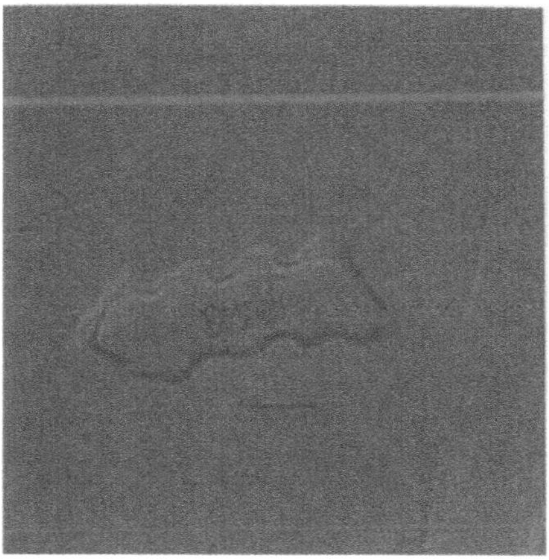

Fig 4.15. Sinuate, *Phleum pratense* L., leaf. Bar is 10 microns.

SILICA-BODY CLASSIFICATION

The following silica-body classification incorporates a standard orientation and terms. Silica-bodies are three-dimensional; that is, relief or height is present as well as an outline of the body in top view (the most commonly described aspect). Relief is generally fairly low with respect to length and width. All silica-bodies observed to date possess at least one face that is broad and relatively flat, defined here as the base. Some forms have two opposite broad faces, in which case the larger one is defined as the base. If the faces are equal in size, the one without ornamentation is arbitrarily designated as base. The side opposite the base, whether a thin ridge or a broad face, is defined as the top. The outline of the body at the edge of the base is used as a major shape characteristic (previously defined as cross section of the base, Mulholland 1989). Long silica-bodies typically orient on a microscope slide with the base (and top) parallel to the slide, showing the outline of the base. This orientation is defined as top view or planar view (Pearsall 1982, "viewed from the planar surface"). Short silica-bodies may also orient in top view but are more likely than long silica-bodies to present a different orientation (Fig 4.7).

The four lateral faces connecting the top and base may be equal or unequal in length. If unequal, the longest pair (a pair comprises two faces opposite each other) is defined as sides; the other pair is defined as ends. If the four faces are roughly equal in length, the two parallel to major ornamentation are arbitrarily designated as sides. If no ornamentation is present, any pair may be designated as sides. Silica-bodies, especially short ones, may be oriented on either edge in side view or end view. Note that measurements are defined in relation to these faces (length is parallel to the sides and width to the ends).

The present classification is based on a detailed comparison of 100 silica-bodies from leaf blades of 19 grass species (Table 4.2). Only those bodies in top or planar view are considered, since tilted bodies obscure the outline of the base. The major units (saddle, dumbbell, etc.) were proposed by Mulholland and Rapp (1985); in turn, these units represent groupings distinguished in most major taxonomic studies (Mulholland 1989: 495). Subdivisions of these units are based on distinctive morphological characteristics.

Silica-bodies with a trapezoidal side view ("Trapezoids," Brown 1984; "Festucoids," Twiss, Suess & Smith 1969) have posed a special problem in classification: a qualitative difference between short bodies with rounded outlines and longer bodies with sinuous sides had not been satisfactorily defined. A previously proposed length:width ratio of 3:1 was arbitrary and untested (Mulholland *et al* 1983). A preliminary study of these phytoliths was therefore conducted on phytoliths from eight species (Table 4.2). Maximum and minimum lengths and widths were measured. Qualitative characteristics including condition of the sides (sinuous, indented, entire), number of lobes or indentations and three-dimensional shape were recorded. Frequency diagrams of the measurements reveal a normal distribution. Absence of polymodal distribution indicates there is no clear evidence for a division based on length or width. Consequently, the classification was based largely on qualitative morphological characteristics. Definitions of morphological units include only a few measurement ratios.

Silica-bodies may be divided into three geometric classes based on general shape (Fig 4.8). Class I includes variations of cubes, rectangular boxes, truncated pyramids and parallelopipeds (see Rich 1963: 204–206 for definitions). Class II consists of cylinders and beveled cones. Class III represents silica-bodies with mostly concave faces and a saddle-shaped base outline. The three classes are actually idealized groupings. Silica-bodies vary considerably and most faces are not completely flat; shapes are not strictly rectangular boxes, pyramids, cones or other geometric figures. In addition, there is much variation in outline of the base, degree of face concavity and general configuration. However, the three arbitrary morphological classes describe the majority of silica-bodies observed to date.

The classification schemes discussed above all recognize some common morphological units. Twiss, Suess and Smith (1969) proposed three major divisions: dumbbell/cross, saddle and a group of geometrically simple forms that included rectangular, oblong and circular forms. Most of Metcalfe's (1960) 20 shapes correlate to these divisions. He specifically recognized dumbbells, crosses and saddles; other shapes such as Tall and Narrow are similar to Twiss's geometrical-

TABLE 4.2. Grass Species Used in This Study

Taxon*	Accession Number
POOIDEAE: POEAE	
Bromus inermis Leyss.**	10496
POOIDEAE: AVENEAE	
Avena sativa L.**	10544
POOIDEAE: PHALARIDEAE	
Phalaris arundinacea L.**	10463
POOIDEAE: TRITICEAE	
Agropyron cristatum (L.) Gaertn.**	10453
A. riparium Scribn. & Sm.**	10574
A. trachycaulum (Link) Malte.**	10452
Hordeum jubatum L.**	10516
Triticum aestivum L.**	10543
POOIDEAE: STIPEAE	
Stipa viridula Trin.	10468
ARUNDINOIDEAE: ARISTIDEAE	
Aristida longiseta Steud.	10576
CHLORIDOIDEAE: ERAGROSTEAE	
Eragrostis cilianensis (All.) Lutati	10503
CHLORIDOIDEAE: CHLORIDEAE	
Bouteloua curtipendula (Michx.) Torr.	10585
B. gracilis (HBK) Lag.	10550
Spartina pectinata	10085, 10210
PANICOIDEAE: PANICEAE	
Setaria lutescens (Weigel) F.T. Hubb	10402
PANICOIDEAE: ANDROPOGONEAE	
Andropogon gerardi Vitman	10525
Zea mays L. cultivar Mandan Yellow Flour[†]	10672, 10853 (chaff)
Zea mays cultivar Mandan Clay Red[†]	10655
ORYZOIDEAE: ZIZANIEAE	
Zizania aquatica L.	10403

*Taxonomy following Pohl 1978.
**Species used in Trapezoid study.
[†]From USDA Plant Introduction Station: PI 213794 and 213807, respectively.

ly simple forms. Brown (1984) included phytoliths from other cell types (Table 4.1, categories 1–6) but also distinguished crosses, dumbbells (his bilobates and polylobates), saddles and trapezoids (similar to Twiss's group of simple forms). The second step in the classification presented here is to define the major silica-body shape types, using previous classifications as a guide (Table 4.3). The three geometric classes may be divided into eight common shapes by outline of the base (Table 4.3, Fig 4.16).

TABLE 4.3. Key to Major Shape Types of Grass Silica-Bodies

I. Body is a rectangular box to truncated or beveled pyramid; outline of base in planar view approximately rectangular to square or other polygon (base may have lobes but general outline is a polygon); top is a flat to slightly concave or convex face or elevated ridge(s).	
A. Nonlobate: sides of base lack definite lobes	
1. Base has 3 sides	TRIANGLE
2. Base has 4 sides	RECTANGLE
3. Base has 5 sides	PENTAGON
B. Lobate: sides of base have definite lobes	
1. Minimal diameters of base approx. equal	CROSS
2. Minimal diameters of base unequal	
a. Bilobate: Maximum of 2 lobes per side	
(1) Shaft/lobe ratio >2/3	SINUATE
(2) Shaft/lobe ratio <2/3	DUMBBELL
b. Polylobate: More than 2 lobes per side	
(1) Shaft/lobe ratio >2/3	SINUATE
(2) Shaft/lobe ratio <2/3	DUMBBELL
II. Body is a short cylinder to truncated or beveled cone; outline of base approximately oval to circular or other curved shape (base may have concave or flat segments but general outline is curved shape); top is a flat to slightly concave or convex face or elevated ridge(s).	
A. Entire: edges of base all convex	RONDEL
B. Flattened: some edges of base straight	RONDEL
C. Indented: some edges of base concave	RONDEL
III. Body is saddle-like: outline of top or both top and base has two opposite convex edges that flare outward from the face surface and two opposite lower edges that are usually concave; top is concave.	
A. Two concave edges (base and/or top)	SADDLE
B. One concave edge (base and/or top)	SADDLE
C. No concave edges (base)	SADDLE

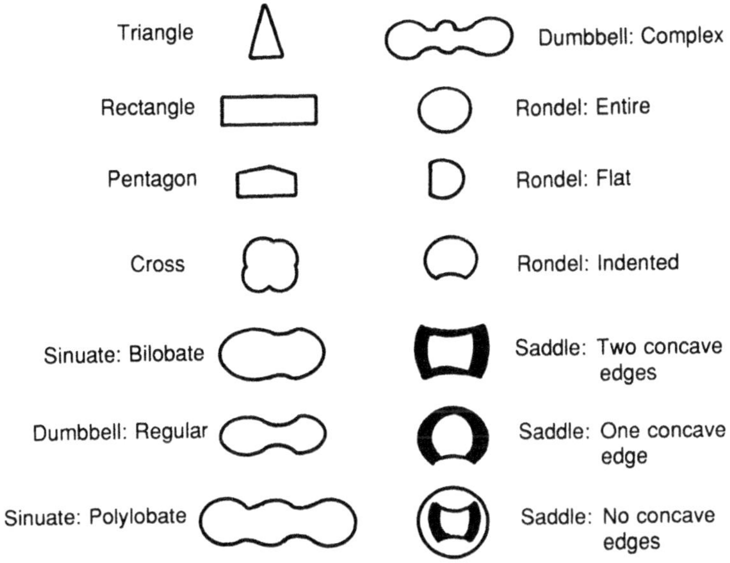

Fig 4.16. Eight silica-body shape types (outline of base in planar view).

Saddles have been identified consistently as distinctive silica-bodies. Twiss, Suess and Smith (1969) distinguished saddles as the Chloridoid type. The description, "battle-axes with double edges," refers to the outline of the base when the body is oriented in top view. Metcalfe (1960) identified two shapes that probably correlate to saddles: saddle-shaped and crescent-shaped. Suess (1966: 20) also correlated Metcalfe's cuboid shape with the Chloridoid (saddle) type. Brown (1984) distinguished saddles as a separate category, with 11 types. Most illustrations from the studies show the typical "battle-ax" outline of the base of two opposite convex edges and two opposite concave edges. However, some specimens have only one concave side or lack them entirely (see Brown 1984: 363). In our classification, geometric class III is composed entirely of saddles. The distinguishing characteristic chosen here is the upward and outward flare of the two convex edges from the main body (Fig 4.9). Most also have the traditional saddle or battle-ax outline of the base, although variations lacking this characteristic have been observed.

Dumbbells and crosses also traditionally have been differentiated from other silica-bodies. The Panicoid division in Twiss's classification is composed of 11 types of dumbbells and crosses. One description of dumbbells is given as "halter-shaped" (Twiss, Suess & Smith 1969: 112). Dumbbells are illustrated as having an outline of the base with two lobes connected by a shaft; some (complex types) have one or two additional lobes on the shaft. Crosses are shorter, nearly equidimensional bodies with four distinct lobes (or four indentations). Metcalfe (1960) also recognized variation within this type. He identified three types of dumbbells, two "dumbbell-shaped" units, one cross and one intermediate between cross and dumbbell. In addition, "nodular" is a form with many small lobes rather than two large ones. Brown (1984) recognized dumbbells (his bilobates), complex dumbbells (his polylobates) and crosses. In developing our classification, dumbbells and crosses were observed to be fairly tabular in overall shape. Although the outline of the base is distinctly lobed, the body is more similar to a rectangular box than a cylinder or cone. Also, considerable variation exists in the outline of the base, from slightly sinuous sides to deep lobes. Therefore, all lobed bodies are placed into geometric class I. Dumbbells are distinguished from other lobed class I bodies by a shaft-to-lobe width of less than 2/3 (Fig 4.10). This ratio was selected by comparing lobed bodies from Panicoid and Pooid species; most from the Panicoids were less than 2/3, whereas most from the Pooids were greater than 2/3. Also, most lobed Panicoids were bilobate, whereas most lobed Pooids were polylobate.

Crosses have been the subject of several detailed studies. Pearsall (1978) defined a cross as "having a dimension of N by N to N by (N+2), with indentations on at least three sides." The quantitative portion of this definition means that a cross may be up to 9.16 microns (two of Pearsall's microscope units) longer than it is wide (Piperno 1983: 53). The length and width measurements refer to maximum dimensions (Pearsall 1979: 146), although it should be noted that Pearsall's size categories are based on the shorter of these two measurements (*ie*, width rather than length, Pearsall 1982).

The definition of cross used in this study differs from that described above. A considerable range of variation (in shapes, measurements and ratios of measurements) was observed in the data. Some specimens fit Pearsall's definition of a cross but were clearly dumbbell-like in that only two lobes rather than four were present (Fig 4.11, central phytolith). Most had concave ends, so the stipulation of "indentations on at least three sides" was met. Conversely, some specimens with four distinct lobes had one that was greatly elongated, technically a dumbbell in maximum dimensions. A definition of roughly equal minimum measurements ensures that the four lobes are approximately equally distinct (Fig 4.13). Ultimately, the range of variation should probably be described by a series of morphological units, rather than just two (dumbbells and crosses).

Twiss, Suess and Smith (1969) placed circular, elliptical, oblong and rectangular forms in the Festucoid class (terms refer to outline of the base). The forms were described as "geometrically simple," although one oblong type has sinuous (shallowly lobed) sides. Metcalfe (1960) defined several types that correspond to these forms: Tall and Narrow, Round, Elliptical, Oblong, Horizontally

Elongated (with smooth or sinuous outlines), Acutely Angled and "Fitting into a concavity in a cork cell." Brown (1984) proposed the Trapezoid class for bodies with a trapezoidal side view, including elongates with sinuous and smooth base outlines and rounded bodies. Although used extensively in our previous classifications (Mulholland *et al* 1983), the "Trapezoid" class is heterogeneous in terms of the outline of the base.

The rather heterogenous grouping of simple shapes is redefined here. Geometric class II consists of beveled and truncated cones and cylinders with curved basal outlines (Fig 4.12). The oval to circular small bodies of Twiss, Metcalfe and Brown correspond to class II, here called rondels (Rondel: from Middle English, from Old French, "small circle," from ronde, rounde, round. *American Heritage Dictionary* 1970: 1127.) Rondels correspond to Short Trapezoids in some of our previous classifications (Mulholland *et al* 1983; Mulholland 1986). Tabular forms (truncated pyramid, rectangular box, parallelopiped) are placed in geometric class I. Basal outline in planar view is a polygon (Fig 4.14), or approximates a polygon in the case of sinuous and lobed forms (Fig 4.15). The distinguishing characteristic is basal outline; class I is generally rectangular to square, whereas class II is generally circular to oval.

Eight shape types are defined by this approach (Table 4.3, Fig 4.16). Six are in geometric class I (two of which are rare) and one each in classes II and III. The somewhat arbitrary divisions are based on examination of grass species (particularly Table 4.2 but also incorporating past studies) and on previous classifications by the various researchers discussed in the previous section. All morphological characteristics were found to exhibit variation to a greater or lesser extent; our previous classifications that based type definitions on various single characteristics have not been satisfactory (Collins 1979; Collins *et al* 1986). Study of additional modern grass specimens will undoubtedly lead to revision of this classification in light of ever more detailed descriptions of more specimens.

The third step in this classification is subdivision of the eight shape types. Details of base outline, ornamentation and three-dimensional structure are the major characteristics incorporated in this classification. Basal outline details include straight vs undulating sides, number and regularity of lobes (if present) and shape of ends. Ornamentation refers mainly to secondary ridges found on top of some types. Three-dimensional structure refers to the shape of the top as compared to the base (Piperno 1984: 368). Three major types are illustrated in Mulholland 1985. The top and base of a tabular type have roughly the same shape and size. This structure is found on rectangular boxes, parallelopipeds, cylinders and tabular saddles. The top of a plateau type is smaller than or a different shape than the base. These bodies include truncated pyramids, cones and plateau type saddles. A ridge type has the top reduced to one or more ridges that slope to the base (not to be confused with a secondary ridge on a tabular top). Beveled pyramids and cones possess this structure (no ridge saddles have yet been observed). Several forms of each structure are defined, based on morphological details. Table 4.4 lists the complete classification, organized following the key to major shapes (Table 4.3).

As an example, silica-bodies illustrated in this paper can be classified following Tables 4.3 and 4.4. Figure 4.10 contains two phytoliths that, while obviously lobed, approximate a rectangular box in convex perimeter of the base. They are obviously not oval to circular nor do they have a saddle-like outline. Therefore Section 1 of Table 4.3 is the proper classification. The lobate nature of the outline leads to section B—Lobate; the fact that one minimum diameter is definitely longer than the other indicates that cross is not the proper choice. There are only two lobes on each side (bilobate) and the shaft/lobe ratio is <2/3. Therefore, these phytoliths are classified as dumbbells.

TABLE 4.4. Major Shape Types of Grass Silica-Bodies

I. Body is a rectangular box to truncated or beveled pyramid; cross section of base approximately rectangular to square or other polygon (base may have lobes but general outline is a polygon); top is a flat to slightly concave or convex face or elevated ridge(s).	
A. Nonlobate: sides of base lack definite lobes	
1. Base has 3 sides	TRIANGLE
a. Sides of base are straight	
(1) Top is tabular (same size and shape as base)	
(a) Secondary ridge present on top	
(b) Secondary ridge absent	
(2) Top is plateau (smaller than or different shape from base)	
(a) Top is same shape as base	
(b) Top is different shape from base	
(3) Top is ridge(s)	
(a) Single ridge	
(b) Multiple ridges	
b. Sides of base are undulating	
Subcategories b1, b2, b3 same as a1, a2, a3	
2. Base has 5 sides	PENTAGON
Categories a, b same as Triangle a, b	
3. Base has 4 sides	RECTANGLE
a. Sides of base are straight	
(1) Top is tabular (same size and shape as base)	
(a) Secondary ridge present on top	
(b) Secondary ridge absent	
(2) Top is plateau (smaller than or different shape from base)	
(a) Top same shape as base	
(b) Top different shape from base	
(3) Top is ridge(s)	
(a) Single ridge	
(b) Multiple ridges	

TABLE 4.4. (Continued)

b. Sides of base are undulating	
Subcategories b1, b2, b3 same as a1, a2, a3	
B. Lobate: sides of base have definite lobes	
1. Minimal diameters of base approx. equal	CROSS
a. Four lobes present on base	
(1) Lobes approximately equal size	
(a) Top is tabular (same size and shape as base)	
i) Secondary ridge present on top	
ii) Secondary ridge absent	
(b) Top is plateau (smaller than or different shape from base)	
i) Top is same shape as base	
ii) Top is different shape from base	
(c) Top is ridge(s)	
i) Single ridge	
ii) Multiple ridges	
(2) Lobes unequal in size	
Subcategories 2a, 2b, 2c same as 1a, 1b, 1c	
b. Three or five lobes present on base	
Subcategories b1, b2 same as a1, a2	
2. Minimal diameters of base unequal	
a. Bilobate: Maximum of 2 lobes per side	
(1) Shaft/lobe ratio >2/3	SINUATE
(a) Bilobate (maximum of 2 lobes per side)	
Subcategories ai, aii, aiii same as Rectangle a1, a2, a3	
(2) Shaft/lobe ratio <2/3	DUMBBELL
(a) Regular (maximum 2 lobes per side of base)	
Subcategories ai, aii, aiii same as Rectangle a1, a2, a3	
b. Polylobate: More than 2 lobes per side	
(1) Shaft/lobe ratio >2/3	SINUATE
(a) Polylobate (more than 2 lobes per side)	
Subcategories a1, a2, a3 same as Rectangle b1, b2, b3	
(2) Shaft/lobe ratio <2/3	DUMBBELL
(a) Complex (more than 2 lobes per side of base)	
Subcategories ai, aii, aiii same as Rectangle b1, b2, b3	
II. Body is a short cylinder to truncated or beveled cone; cross section of base approximately oval to circular or other curved shape (base may have concave or flat segments but general outline is curved shape); top is a flat to slightly concave or convex face or elevated ridge(s).	RONDEL
A. Entire (edges of base all convex)	
1. Top is tabular (same size and shape as base)	
a. Secondary ridge present on top	
b. Secondary ridge absent	

(Continued)

TABLE 4.4. (Continued)

2. Top is plateau (smaller than or different shape from base)
    a. Top is same shape as base
    b. Top is different shape from base
3. Top is ridge(s)
    a. Single ridge
    b. Multiple ridges

B. Flattened (some edges of base straight)
1. 1 edge straight
    a. Top is tabular (same size and shape as base)
        (1) Secondary ridge present on top
        (2) Secondary ridge absent
    b. Top is plateau (smaller than or different shape from base)
        (1) Top is same shape as base
        (2) Top is different shape from base
    c. Top is ridge(s)
        (1) Single ridge
        (2) Multiple ridges
2. 2 edges straight—adjacent sides
    Subcategories B2a, B2b, B2c same as B1a, B1b, B1c
3. 2 edges straight—opposite sides
    Subcategories B3a, B3b, B3c same as B1a, B1b, B1c
4. 3 edges straight
    Subcategories B4a, B4b, B4c same as B1a, B1b, B1c
5. 4 edges straight
    Subcategories B5a, B5b, B5c same as B1a, B1b, B1c

C. Indented (some edges of base concave)
1. 1 edge concave
    Subcategories C1a, C1b, C1c same as B1a, B1b, B1c
2. 2 edges concave—adjacent sides
    Subcategories C2a, C2b, C2c same as B1a, B1b, B1c
3. 2 edges concave—opposite sides
    Subcategories C3a, C3b, C3c same as B1a, B1b, B1c
4. 3 edges concave
    Subcategories C4a, C4b, C4c same as B1a, B1b, B1c
5. 4 edges concave
    Subcategories C5a, C5b, C5c same as B1a, B1b, B1c

III. Body is saddle-like: cross section of top or both top and base has two opposite convex edges that flare outward from the face surface and two opposite lower edges that are usually concave; top is concave. SADDLE

A. Two concave edges (base and/or top)
1. Tabular (top same as base)
2. Plateau (top smaller than or different shape from base)
3. Ridge

B. One concave edge (base and/or top)
    1, 2, 3 Same as above

C. No concave edges (base)
    1, 2, 3 Same as above

CONCLUSIONS

The phytolith classification described in this paper constitutes a hypothesis as to which morphological characteristics will differentiate between grass taxa. It is by no means a finished identification guide nor have all the characteristics been tested. One test on grass species from North Dakota concentrated on the basic outline of the base in planar view (Mulholland 1989). The major groups identified by previous classifications (Twiss, Suess & Smith 1969) were found *generally* to correlate to grass subfamilies. Major deviations include significant production of dumbbells by some Pooideae (Festucoideae) tribes and presence of rondels throughout the Poaceae (Gramineae). In addition, saddles are found in minor amounts in the Pooideae. Although the correlation of dumbbells to Panicoideae, saddles to Chloridoideae and sinuates to Pooideae is valid on a statistical average, it may not be accurate with small numbers of phytoliths. "Abnormal" phytolith production (production of shapes in other subfamilies) is more widespread than previously reported in the literature.

Further tests of the classification will include three-dimensional characteristics as well as additional characteristics of the base outline. Mulholland (1989) suggests that dumbbells from three subfamilies may be differentiated by three-dimensional shape. Changes and additions to the classification are expected as more material is examined; characteristics that do not differentiate between taxa will be discarded and replaced by those that do. Variation within taxa may indicate that some classificatory divisions are not meaningful.

This classification is based for the most part on examination of grasses from the Upper Midwest. Phytoliths produced by grasses from other geographic areas may not be well represented by the characteristics stressed here; however, it is hoped that the general morphological units (Tables 4.1 and 4.3) will prove fairly useful. Non-graminaceous specimens have not been examined in great detail. Categories 1–6 (Table 4.1) therefore are not subdivided to any extent. More information from all families can be incorporated once the appropriate material is examined.

ACKNOWLEDGMENTS

We would like to thank Dr. Robert Bright for much helpful comment and criticism of this paper in its many drafts. The plant material was identified by Dr. John LaDuke, University of North Dakota, and Dr. Gary Clambey, North Dakota State University; the staff of the Knife River Indian Villages National Historic Site and Michael Holm, area manager, provided assistance in the collection. Native varieties of corn were provided by the USDA North Central Plant Introduction Station at Ames, Iowa. Laboratory processing was done by Jeanne

Daniels, Richard Sundell, Mark Endrizzi and Lance Campbell. Judy Holz edited the manuscript; drawings were drafted by Sharon Hongell. Stephen Mulholland assisted with all aspects of the project.

REFERENCES

American Heritage Dictionary of the English Language 1970 New York, American Heritage Pub Co, Inc.

Bombin, M (ms) 1984 On phytoliths, Late Quaternary evolution of Beringia, and information evolutionary theory. PhD dissertation, University of Alberta: 164 p.

Brown, DA 1984 Prospects and limits of a phytolith key for grasses in the central United States. *Journal of Archaeological Science* 11: 345–368.

Clifford, HT and Watson, L 1977 *Identifying Grasses: Data, Methods, and Illustrations.* St. Lucia, University of Queensland Press: 146 p.

Collins, SM (ms) 1979 Phytoliths as indicators of plant use at ancient Troy. MS thesis, University of Minnesota: 181 p.

Collins, S, Rapp, G, Jr, Gifford, JA, Rondina, D and Thomson, M 1986 Phytolith analyses of samples from Voyageurs National Park, Minnesota. *In* Lynott, MJ, Richner, JJ, and Thompson, M, eds, Archeological Investigations at Voyageurs National Park: 1979 and 1980, *Midwest Archeological Center, Occasional Studies in Anthropology* No 16: 363–398.

Cutler, DF 1969 *Anatomy of the Monocotyledons. IV. Juncales.* Oxford, Clarendon Press: 357 p.

Ellis, RP 1976 A procedure for standardizing comparative leaf anatomy in the Poaceae. I. The leaf blade as viewed in transverse section. *Bothalia* 12: 65–109.

____1979 A procedure for standardizing comparative leaf anatomy in the Poaceae. II. The epidermis as seen in surface view. *Bothalia* 12: 641–671.

Esau, K 1977 *Anatomy of Seed Plants.* Second edition, New York, John Wiley and Sons: 550 p.

Metcalfe, CR 1960 *Anatomy of the Monocotyledons. I. Gramineae.* Oxford, Clarendon Press: 731 p.

____1971 *Anatomy of the Monocotyledons. V. Cyperaceae.* Oxford, Clarendon Press: 597p.

Mulholland, SC (ms) 1985 Trapezoid grass silica cells: Distribution and morphology. Paper presented at the Second Phytolith Conference, Duluth, Minnesota. Abstract, *Phytolitharien Newsletter* 3(3): 7.

____1986 Classification of grass silica phytoliths. *In* Rovner, I, ed, Plant opal phytolith analysis in archaeology and paleoecology: Proceedings of the 1984 Phytolith Research Workshop, Raleigh, North Carolina, *Occasional Papers of The Phytolitharien* No 1: 41–52.

____1989 Phytolith shape frequencies in North Dakota grasses: A comparison to general patterns. *Journal of Archaeological Science* 16: 489–511.

Mulholland, SC and Rapp, G, Jr 1985 Grass silica phytoliths. *Society for Archaeological Sciences Newsletter* 8(2): 5–6 (Research Report 5).

____1989 Characterization of grass phytoliths for archaeological analysis. *Materials Research Society Bulletin* 14(3): 36–39. Errata, *MRS Bulletin* 14(5): 50.

Mulholland, SC, Rapp, G, Jr and Gifford, JA 1982 Phytoliths. *In* Rapp, G, Jr and Gifford, JA, eds, *Troy: The Archaeological Geology*, Supplementary Monograph 4. Princeton, New Jersey, Princeton University Press: 117–137.

Mulholland, SC, Rapp, G, Jr, Huber, JK, and Selness, D (ms) 1983 Results of a preliminary phytolith analysis of sediment samples from 22TS954. Unpublished report, Archaeometry Laboratory, University of Minnesota, Duluth.

Palmer, PG 1976 Grass cuticles: A new paleoecological tool for East African lake sediments. *Canadian Journal of Botany* 54: 1725–1734.

Palmer, PG and Gerbeth-Jones, S 1986 A scanning electron microscope survey of the epidermis of East African grasses: IV. *Smithsonian Contributions to Botany*, No 62, Washington, DC, Smithsonian Institution Press: 120 p.

Palmer, PG and Gerbeth-Jones, S 1988 A scanning electron microscope survey of the epidermis of East African grasses: V. and West African supplement. *Smithsonian Contributions to Botany* No 67, Washington, DC, Smithsonian Institution Press: 157 p.

Palmer, PG, Gerbeth-Jones, S, and Hutchinson, S 1985 A scanning electron microscope survey of the epidermis of East African grasses: III. *Smithsonian Contributions to Botany*, No 55, Washington, DC, Smithsonian Institution Press: 136 p.

Palmer, PG and Tucker, AE 1981 A scanning electron microscope survey of the epidermis of East African Grasses: I. *Smithsonian Contributions to Botany*, No 49, Washington DC, Smithsonian Institution Press: 84 p.

_____1983 A scanning electron microscope survey of the epidermis of East African grasses: II. *Smithsonian Contributions to Botany*, No 53, Washington DC, Smithsonian Institution Press: 72 p.

Pearsall, DM 1978 Phytolith analysis of archaeological soils: Evidence for maize cultivation in Formative Ecuador. *Science* 199: 177–178.

_____(ms) 1979 The application of ethnobotanical techniques to the problem of subsistence in the Ecuadorian Formative. PhD dissertation, University of Illinois: 267 p.

_____1982 Maize phytoliths: A clarification. *Phytolitharien Newsletter* 1(2): 3.

Piperno, DR (ms) 1983 The application of phytolith analysis to the reconstruction of plant subsistence and environments in prehistoric Panama. PhD dissertation, Temple University: 459 p.

_____1984 A comparison and differentiation of phytoliths from maize and wild grasses: Use of morphological criteria. *American Antiquity* 49:361–383.

_____1985 Phytolith analysis and tropical paleoecology: Production and taxonomic significance of siliceous forms in New World plant domesticates and wild species. *Review of Palaeobotany and Palynology* 45: 185–228.

_____1988 *Phytolith Analysis: An Archaeological and Geological Perspective*. San Diego, Academic Press: 280 p.

Pohl, RW 1978 *How to Know the Grasses*. Third edition, Dubuque, Iowa, Wm C Brown Pub Co: 200 p.

Renfrew, JM 1973 *Palaeoethnobotany: The Prehistoric Food Plants of the Near East and Europe*. New York, Columbia University Press: 245 p.

Rich, B 1963 *Principles and Problems of Plane Geometry*. Schaum's Outline Series in Mathematics, New York, McGraw-Hill Book Co: 231 p.

Rovner, I 1971 Potential of opal phytoliths for use in paleoecological reconstruction. *Quaternary Research* 1: 343–359.

Suess, E (ms) 1966 Opal phytoliths. MS thesis, Kansas State University: 77 p.

Twiss, PC, Suess, E and Smith, RM 1969 Morphological classification of grass phytoliths. *Soil Science Society of America, Proceedings* 33: 109–115.

5

TOWARD A CLASSIFICATION SCHEME OF SEDGE (CYPERACEAE) PHYTOLITHS

AMY L. OLLENDORF

ABSTRACT

Sedge cone phytoliths appear to be morphologically distinct from those of other families. As a result, these phytoliths from Cyperaceae may be of taxonomic importance and may be useful to archaeologists and paleoecologists. This study is a compilation of information regarding sedge phytoliths from widely separated geographic areas. Included here are previously published descriptions along with newly interpreted material. It is hoped that all the data presented will be of use in the future development of a classification scheme for phytoliths of Cyperaceae.

INTRODUCTION

The study of sedge phytoliths can be traced back to an 1872 report by Duval-Jouve on epidermal silica cells in Cyperaceae (Mehra & Sharma 1965). Since that time, sedge phytolith data obtained from widely separated geographic areas have been reported in the scientific literature. Although not an exhaustive inventory, Appendix 5.1 is a compilation of more than 176 species of sedges for which phytolith information has been published. Despite its lack of high-resolution photographs, Metcalfe's (1971) monograph in his series on the monocotyledons remains the most comprehensive description of sedge phytoliths.

The purpose of this brief report is to synthesize previously published descriptions and to present suggestions for future research. It is hoped that all of the data presented will be of use in the future development of a classification scheme for phytoliths of Cyperaceae.

SEDGE PHYTOLITHS: TERMINOLOGY AND MORPHOLOGY

"The terminology and classification systems applied to phytoliths have not as yet been standardized" (Piperno 1988: 52). This lack of uniformity is noticeable

Phytolith Systematics, Edited by G. Rapp, Jr. and
S.C. Mulholland, Plenum Press, New York, 1992

within publications on sedge phytoliths. For instance, many investigators reported cones (Metcalfe 1971), "conical-shaped" (Piperno 1985), "hat-shaped" (Piperno 1988) and "cyperaceous type" (Mehra & Sharma 1965) to be the phytolith characteristic of sedge species (Raeside 1970; Dahlgren & Clifford 1982; Thomasson 1983; Andrejko & Cohen 1984; Ollendorf, Mulholland & Rapp 1987). Although the distinction between conical- and hat-shaped is not quantified or even explicitly described, it appears that Piperno (1985, 1988) used "hat-shaped" to describe the cone-like phytoliths with gently sloping central apices in Palmae and Marantaceae. "Conical-shaped" appears to be a designation reserved for phytoliths with more distinct central apices in Cyperaceae and Orchidaceae; the phytoliths in these families may have more pronounced surficial relief (*ie*, from the apex to the base of the cone) than hat-shaped phytoliths.

In addition to variation in the names of these apparently distinctive sedge phytoliths, several types of phytoliths have been found in sedges. Rod-shaped phytoliths similar to those from grasses were noted by several authors (Witty & Knox 1964; Raeside 1970; Metcalfe 1971; Norgren 1973; and Piperno 1983). "Hookbases" (Norgren 1973) and "hair bases" (Piperno 1988), apparently different names for identical features that occur in grasses with trichomes or "hair cells" (Piperno 1983), also have been noted for sedges. The most potentially troublesome sighting, however, consists of obscure views of the common dumbbell-shaped grass phytolith in the sedge *Epischoenus villosus* reported by Metcalfe (1971: 16). It should be noted that only very small (less than 10 microns), very pointed, individual cones (Table 5.1) were seen by the present author in a recent examination of *Epischoenus villosus* leaf samples removed from Missouri Botanical Garden herbarium vouchers. Because cones seem to be the most consistent and characteristic phytoliths of sedges, this paper focuses solely on the details of cone morphology.

There is also variation in descriptive terminology regarding phytolith surface ornamentation, decoration or sculpturing. Piperno (1985) observed and defined several kinds of surface ornamentation. Where Piperno uses adjectives such as smooth and verrucose, I have opted to use "psilate" and "verrucate," terms common to standard pollen analysis (Faegri & Iversen 1975). The following definitions of "psilate" and "verrucate" are used in this paper: "psilate" refers to a smooth or virtually smooth surface, whereas "verrucate" sculpturing yields projections with non-constricted bases and non-pointed tops (approximate to definitions given by McAndrews, Berti & Norris 1973). These slight modifications constitute an effort aimed at standardization of phytolith terminology whenever possible; such an approach has been similarly employed by Bozarth (this volume).

Cones from sedges are readily distinguishable from those in other families (eg, Palmae, Marantaceae, Orchidaceae and Poaceae or Gramineae). This distinction may be the result of the locus of silica deposition. "For every species in the

TABLE 5.1. Characteristics of Sedge Cone Phytoliths among Selected Species

Species	Phytolith Description	Author
Asterochaete glomerata	RPXSI, RPXSM & RZVWI	Metcalfe 1971
Bulbostylis barbata Kunth.	RPXWM	Mehra & Sharma 1965
B. capillaris Kunth.	RPXWI	Mehra & Sharma 1965
Carex walteriana Bailey	RPXWI, RPXSM & RPXSI	Andrejko & Cohen 1984
C. kobomugi	RDXSM	Kondo & Sase 1986
C. canescens	NKXWI	Toivonen & Timonen 1976
C. diandra	APXWI	"
C. echinata	RPXSM & RZXSM	"
C. elongata	AKXSI	"
C. exilis	APXWI	"
C. mackenziei	NKXWI	"
C. maritima	AKXWI	"
C. paniculata	APXSM & APXWM	"
C. remota	APXWI	"
C. ursina	APXWI & AKXWI	"
C. comosa	ADXWI	Walter 1975
C. hystericina	ADXWI	"
C. lurida	APXWI	"
C. oligosperma	AKXWI	"
C. pseudo-cyperus	AKXWI	"
C. retrorsa	APXWI	"
C. rostrata	AKXWI	"
C. schweinitzii	ADXWI	"
C. tuckermanii	APXWI	"
C. vesicaria	APXWI	"
Cephalocarpus sp.	RPXWM, RZXWI, AZXWI & AZXWM	Metcalfe 1971
Cyperus conglomeratus	AKXWI, AKVWI, RPXWI & NPXWM	Ollendorf, Mulholland & Rapp 1987
C. longus	RPXSI	"
C. mucronatus	APXSI & RDXWI	"
Cladium jamaicensis Crantz.	RDXWI & RDXWM	Andrejko & Cohen 1984
C. mariscus	RPXWM & NPXWM	Metcalfe 1971
Desmoschoenus frondosus	RPXWM	Raeside 1970
Dulichium arundinaceum (L) Britton	RPXSM	Andrejko & Cohen 1984
Eleocharis plantaginea R. Br.	RPXWM	Mehra & Sharma 1965
Epischoenus villosus	APXWI & RPXWI	Ollendorf (quick scan 1991)
Eriophorum comosum Wall. ex Nees	RPXWM	Mehra & Sharma 1965
E. comosum	RPXWI, RPXWM & RPXSM	Ollendorf (quick scan 1991)
E. vaginatum	RDXSI	Kondo & Sase 1986
E. comosum Wall.	AKXWI	Schuyler 1971
E. virginicum L.	APXSI	Schuyler 1971
Everardia montana	RPXWI, NPXWM & NPXWI	Metcalfe 1971
Ficinia indica	RPXWM, NPXWM & NPXWI	Metcalfe 1971
Fimbristylis junciformis Kunth.	RPXWI	Mehra & Sharma 1965
F. miliacea Vahl.	RPXSM	Mehra & Sharma 1965

(Continued)

TABLE 5.1. (Continued)

Species	Phytolith Description	Author
Mariscus ustulatus	AKVWI & RPXWM	Andrejko & Cohen 1984
Pycreus sanguinolentus Nees.	RPXSI	Ollendorf, Mulholland & Rapp 1987
Rhynchospora inundata (Oakes) Fernald	RPXSI & RPXSM	Andrejko & Cohen 1984
Scirpus holoschoenus	NXWI	Ollendorf, Mulholland & Rapp 1987
S. wichurae	AKXSI	Kondo & Sase 1986
S. wichurae	APXSM, ADXSM, APXSI & ADXSI	Ollendorf (quick scan 1991)
S. cernuus Vahl.	AKVWI	Schuyler 1971
S. divaricatus Ell.	NKXWI	"
S. koilolepis (Steud.) Gleas.	APXWI	"
S. lacustris L.	NKVWI & AKVWI	"
S. tabernaemontanii Gmel.	NKXWI & NKXWM	"
S. wichurae Bocklr.	NKXWI	"

Key to Table 5.1:
Shape: A = angular shape; R = rounded shape; N = sinuous shape
Apex: Z = multiple apices; K = knobby apex; P = pointed apex; D = rounded apex
Sculpturing: V = verrucate sculpturing; X = psilate sculpturing
Satellites: S = satellites present; W = satellites absent
Platelets: M = occurs in platelets; I = occurs as individual

Example: NKXWI is a cone with a sinuous shape, a knobby apex, psilate sculpturing, no satellite cones present, and occurring as an individual cone.

Palmae, Orchidaceae, and Marantaceae, subepidermal (mesophyll) tissue is virtually the sole locus of silica deposition, while in the Cyperaceae and Gramineae, silicification is predominant in the epidermis" (Piperno 1988: 39). Although there are occasional exceptions, "the epidermis of the Cyperaceae, unlike that of the Gramineae, is hardly ever differentiated into a mixture of 'long-cells' and 'short-cells'" (Metcalfe 1971: 41).

Piperno (1988) cited the pointed apices and smooth or nearly smooth surfaces of the cone-shaped phytoliths of Cyperaceae as their characteristic features. At least partly as a result of the different epidermal structure, sedges yield cones that are distinct from those in grasses. Sedge cones differ from the cone-shaped phytoliths prevalent in grasses, which fall into two categories (Mulholland, personal communication 1989): those that are long, hollow and sharply pointed (Fig 5.1A, B) and those that are short, solid and resemble "Hershey's kisses," with arched but sharply pointed apices (Fig 5.2A, B). Each of these varieties is very different in appearance from the cones of Cyperaceae. Although "rondels" are described by Mulholland (this volume) as "beveled and truncated cones and cylinders with curved basal outlines," they also are easily distinguishable from sedge cones because they lack a central apex.

Andrejko and Cohen's (1984) sighting of rounded conical shapes was regarded by Piperno (1988) as genus-level sculpturing rather than important fea-

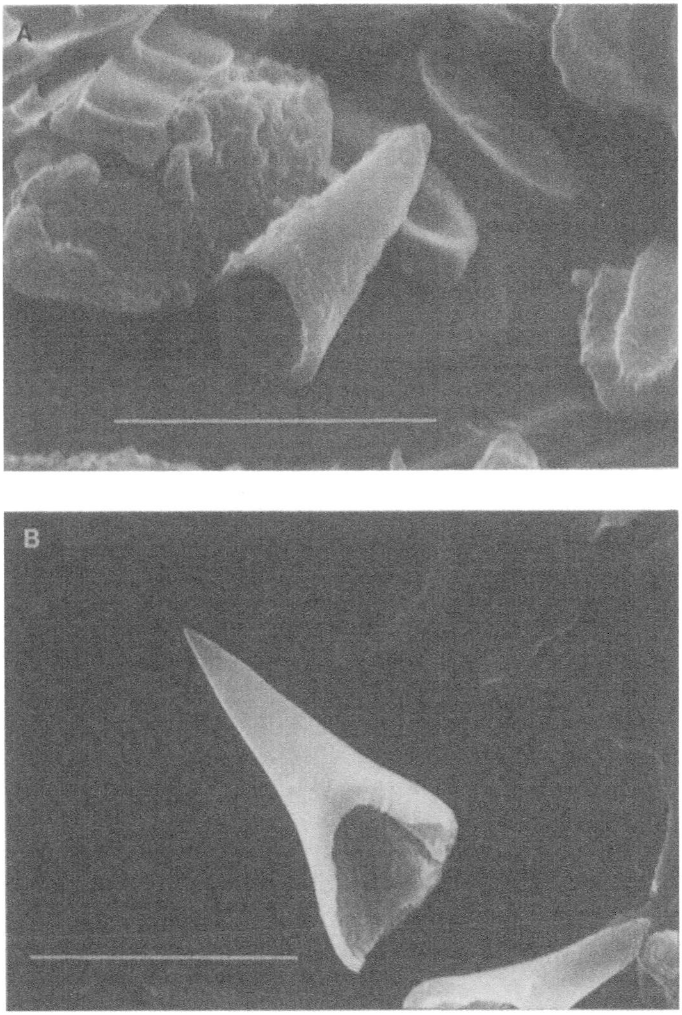

Fig 5.1. Scanning electron micrographs of long, hollow and sharply pointed cones from Poaceae (Gramineae) (Archaeometry Laboratory photos). Bar is 10 microns. A. From inflorescence of *Agrostis scabra* (10391). B. From inflorescence of *Triticum dicoccoides* (10410).

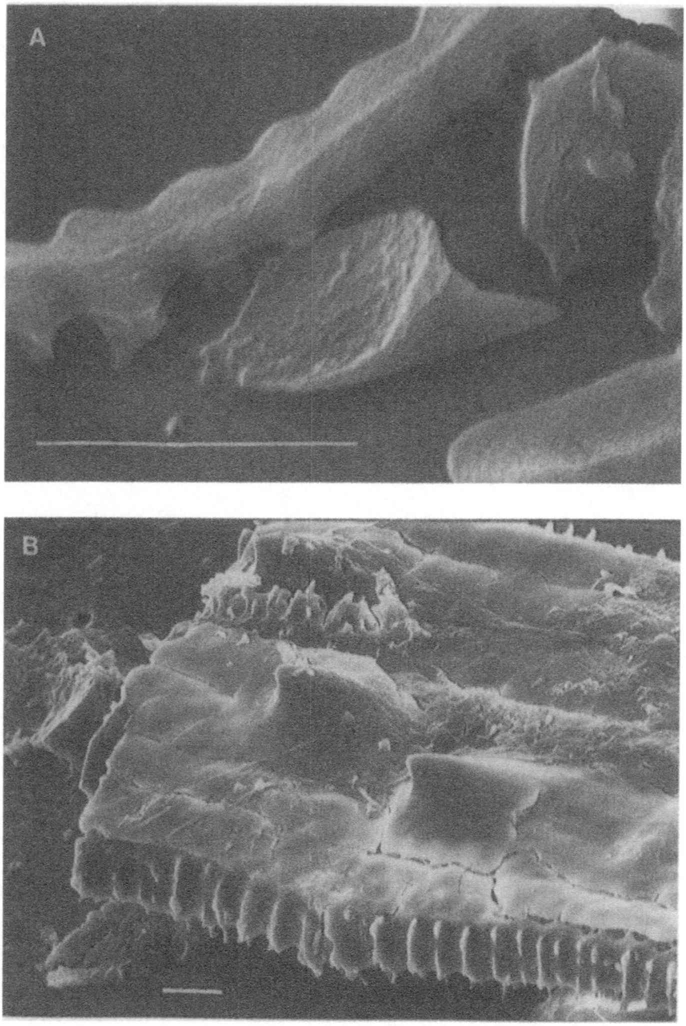

Fig 5.2. Scanning electron micrographs of short, solid cones resembling "Hershey's kisses," from Poaceae (Gramineae) (Archaeometry Laboratory photos). Bar is 10 microns. A. From inflorescence of *Agrostis scabra* (10391). B. From inflorescence of *Triticum dicoccoides* (10410).

tures for familial identification. Examination of sedge phytoliths reported in the literature, however, has shown a great deal of variation in apical shape of cones. For instance, cones from the same genus (Fig 5.3A, B) vary noticeably as do those from the same species (Fig 5.4A, B).

In general, as illustrated in Figures 5.3 and 5.4, apices seem to range from knobby to pointed. The surfaces of cones also range from verrucate to psilate, with or without satellites (*ie*, secondary apices surrounding the central apex). Cones are found as solitary phytoliths or in multiples on platelets. The reader is advised to consult Metcalfe's (1971) monograph for sketches, and articles by Andrejko and Cohen (1984) along with Ollendorf, Mulholland and Rapp (1987) for high-resolution photographs depicting a range of sedge phytolith morphology. Figure 5.5 is a schematic representation of phytolith morphologies to which the text and tables refer.

SEDGE CONE PHYTOLITHS: RESULTS

Table 5.1 is a compilation of the characteristics of cone phytoliths from selected sedge species as noted from the published scientific literature. Multiple characterizations could be the result of differential silicification within species, or identification errors resulting from insufficient photographs (eg, planar views only or low magnification) and generalized descriptions within publications.

The category present in the greatest number of species in Table 5.1 is the rounded, psilate cone with a pointed apex without satellites (*ie*, RPXW-). Species yielding such cones include the following: *Bulbostylis barbata*, *B. capillaris*, *Carex walteriana*, *Cephalocarpus* sp, *Cladium mariscus*, *Cyperus conglomeratus*, *Desmoschoenus frondosus*, *Eleocharis plantaginea*, *Eriophorum comosum*, *Everardia montana*, *Ficinia indica*, *Fimbristylis junciformis* and *Mariscus ustulatus*. Some of these species yield individual cones (*Carex walteriana*, *Cyperus conglomeratus*, *Everardia montana* and *Fimbristylis junciformis*), whereas the others produce cones on platelets.

Phytoliths with very different morphologies are reported from different plants of two species, *Eriophorum comosum* and *Scirpus wichurae*. For *Eriophorum comosum*, rounded cones with pointed apices (RPXWM) are reported in Mehra and Sharma (1965), whereas angular cones with knobby apices (AKXWI) are shown in Schuyler (1971). My recent quick scans of leaf samples from *E. comosum* from Missouri Botanical Garden herbarium vouchers (Table 5.1) yielded smooth, rounded cones with pointed apices as individuals (RPXWI) or on platelets with or without satellites (RPXSM & RPXWM). Angular cones with satellites (AKXSI) are illustrated in Kondo and Sase (1986) for *Scirpus wichurae*, and sinuous cones without satellites (NKXWI) appear in Schuyler's (1971) report. My recent quick scans of a leaf sample from *S. wichurae* from a Missouri Botan-

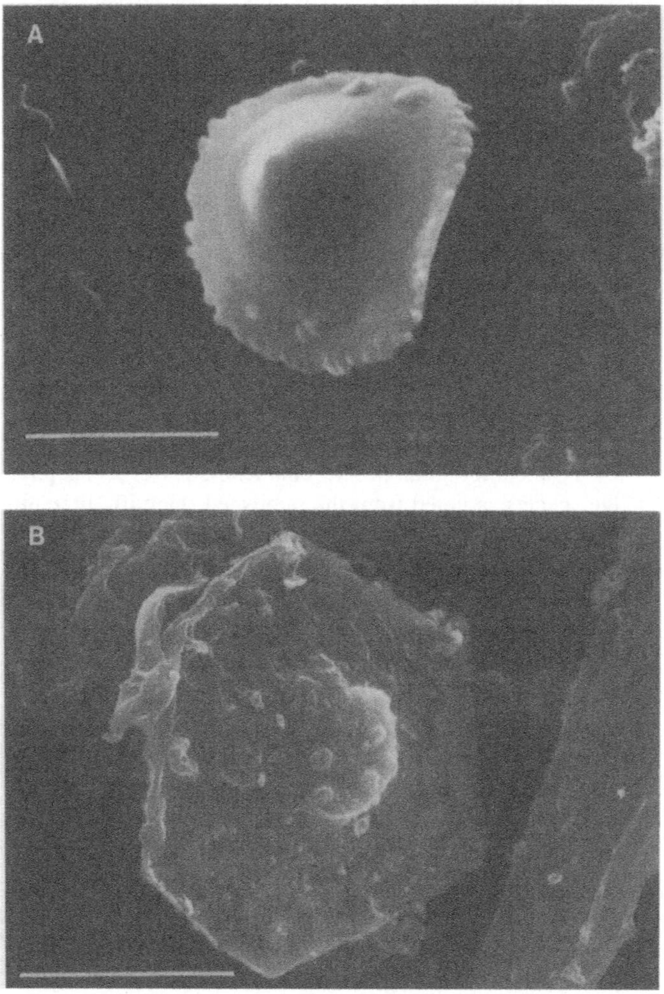

Fig 5.3. Range of cones from the same genus. Bar is 10 microns. A. Light micrograph from *Cyperus mucronatus* (Archaeometry Laboratory photo). B. Scanning electron micrograph from inflorescence of *Cyperus conglomeratus* (10120) (Appeared in *Israel Journal of Botany* 36: 129. Reprinted by permission).

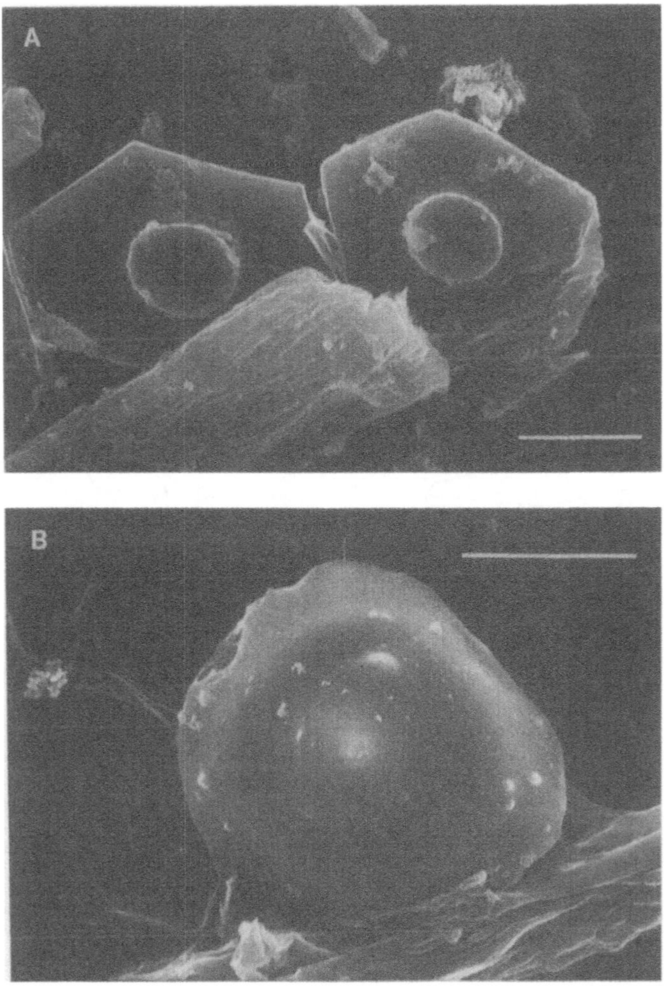

Fig 5.4. Range of cones from the same species (Appeared in *Israel Journal of Botany* 36: 129. Reprinted by permission). Bar is 10 microns. A. Scanning electron micrograph from inflorescence of *Cyperus conglomeratus* (10120). B. Scanning electron micrograph from stem of *Cyperus conglomeratus* (10120).

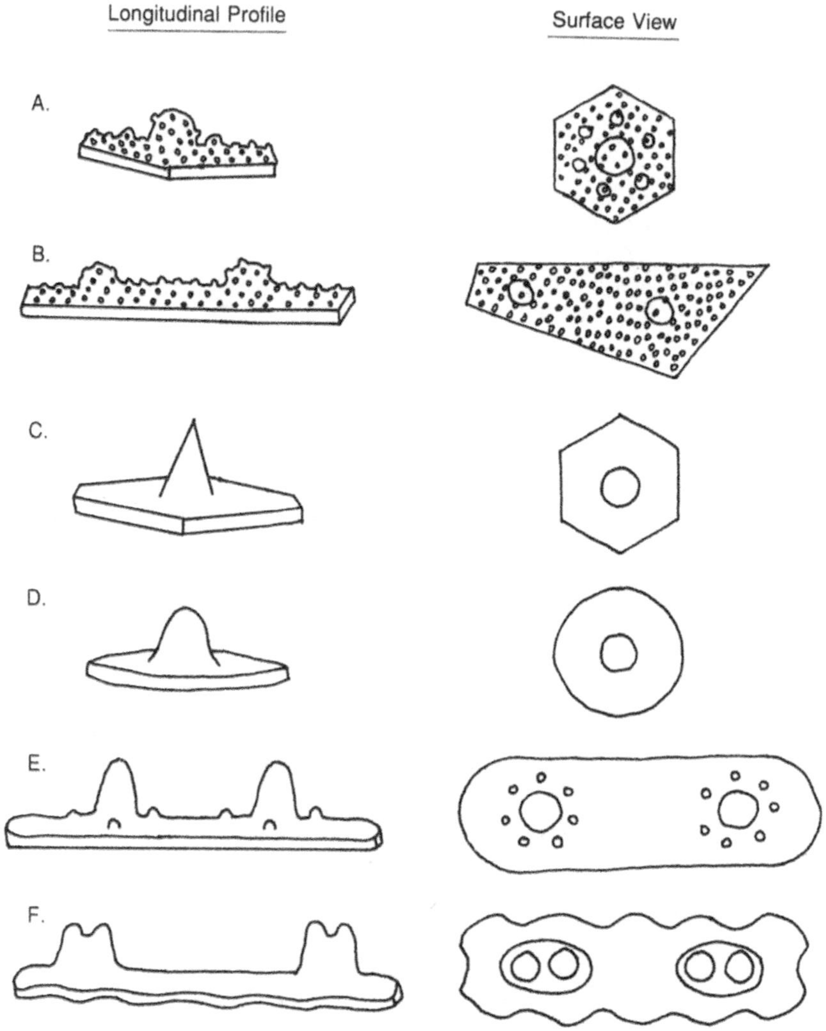

Fig 5.5. Schematic illustrations of phytolith morphologies mentioned in text and tables. A. Individual, angular, verrucate, knobby apex with satellites; B. Angular, verrucate, knobby apex without satellites on platelet; C. Individual, angular, psilate, pointed apex without satellites; D. Individual, rounded, psilate, rounded apex without satellites; E. Rounded, psilate, rounded apex with satellites on platelet; F. Sinuous, psilate, multiple apices with satellites on platelet.

ical Garden herbarium voucher yielded angular cones with psilate surfaces and satellites on platelets or as individuals (Table 5.1). The apices of the cones I observed in *S. wichurae* ranged from rounded to very pointed. The variation in phytolith morphology may be the result of geographic location.

Another type of intraspecific variation, the occurrence of two cone varieties in one specimen, is apparent in sixteen samples (Table 5.1). All but three publications (Mehra & Sharma 1965; Walter 1975; Kondo & Sase 1986) have examples of species that contain several categories of cones. For instance, Metcalfe's (1971) sketches depict individual rounded, psilate cones with pointed apices and satellites in *Asterochaete glomerata* (RPXSI), as well as individual rounded, verrucate cones with multiple apices without satellites (RZVWI). Similarly, *Carex walteriana* in Andrejko and Cohen's (1984) paper yields individual rounded, psilate cones with pointed apices without satellites (RPXWI) along with rounded, psilate cones with pointed apices with satellites on platelets (RPXSM). The species showing the most variation in its cone phytoliths seems to be *Cyperus conglomeratus* (Ollendorf, Mulholland & Rapp 1987). This species not only has individual angular, psilate or verrucate cones with knobby apices without satellites (AKXWI or AKVWI), but also yields individual rounded, psilate cones with pointed apices without satellites (RPXWI) and sinuous, psilate cones with pointed apices without satellites on platelets (NPXWM). Lanning and Eleuterius (1981: 50) noted silica occurring "in a sheetlike pattern and in rows of oval particles" in leaves of *Scirpus validus*. Whether such intraspecific variation is the result of examination of different plant parts (eg, *Cyperus conglomeratus* inflorescence and stem) or different individuals (eg, perhaps *Carex walteriana* or *Scirpus validus* leaves) is difficult to discern. Many of the sources do not explicitly state what type of material was examined.

Table 5.2 illustrates the range of variation in cones grouped by sedge genus. Almost all sedge genera display a wide range of variation for cone phytoliths. The commonly studied genera *Carex*, *Cyperus* and *Scirpus* show particularly wide variation. Most frequently, angular psilate cones with pointed apices and without satellites (APXW-) are produced in *Carex*. These cones are followed in frequency by the angular, psilate, knobby apex without satellites variety (AKXW-). The most frequent variety of cone in *Scirpus* is that which is sinuous, psilate, with knobby apex without satellites (NKXW-). *Cyperus* displays a very broad range of cones particularly, as noted above, as a result of the variation in cones from *Cyperus conglomeratus*. At least six lesser studied genera also seem to be characterized by a wide range in cone variation. These genera are: *Asterochaete*, *Cephalocarpus*, *Cladium*, *Eriophorum*, *Everardia* and *Ficinia*. Although many more species need to be examined and documented to determine the validity of this preliminary conclusion, the genera yielding the least variable types of cones seem to be *Bulbostylis* and *Rhynchospora*.

TABLE 5.2. Variation in Cone Phytoliths from Sedge Genera*

Genus	Phytolith Description (number of species in parentheses)
Asterochaete	RPXSI(1), RPXSM(1) & RZVWI(1)
Bulbostylis	RPXWM(1) & RPXWI(1)
Carex	APXWI(8), APXWM(1), AKXWI(5), ADXWI(3), RPXSM(2), RPXSI(1), RPXWI(1), NKXWI(2), RZXSM(1), AKXSI(1), APXSM(1) & RDXSM(1)
Cephalocarpus	RPXWM(1), RZXWI(1), AZXWI(1) & AZXWM(1)
Cladium	RDXWI(1), RDXWM(1), RPXWM(1) & NPXWM(1)
Cyperus	AKXWI(1), AKVWI(1), RPXWI(1), NPXWM(1), RDXWI(1), RPXSI(1) & APXSI(1)
Desmoschoenus	RPXWM(1)
Dulichium	RPXSM(1)
Eleocharis	RPXWM(1)
Epischoenus	APXWI(1) & RPXWI(1)
Eriophorum	RPXWM(1), RPXWI(1), RPXSM(1), RDXSI(1), AKXWI(1) & APXSI(1)
Everardia	RPXWI(1), NPXWM(1) & NPXWI(1)
Ficinia	RPXWM(1), NPXWM(1) & NPXWI(1)
Fimbristylis	RPXWI(1) & RPXSM(1)
Mariscus	AKVWI(1) & RPXWM(1)
Pycreus	RPXSI(1)
Rhynchospora	RPXSI(1) & RPXSM(1)
Scirpus	NKXWI(4), NKXWM(1), AKVWI(2), AKXSI(1), APXWI(1), NKVWI(1), APXSM(1), APXSI(1), ADXSM(1) & ADXSI(1)

*Based on data from Table 5.1.

Key to Table 5.2:

Key to Table 5.2:	Shape:	A = angular shape	Sculpturing:	V = verrucate sculpturing
		R = rounded shape		X = psilate sculpturing
		N = sinuous shape	Satellites:	S = satellites present
	Apex:	Z = multiple apices		W = satellites absent
		K = knobby apex	Platelets:	M = occurs in platelets
		P = pointed apex		I = occurs as individual
		D = rounded apex		

As discussed in the previous paragraphs, variation in phytolith morphology may be related to taxonomic group, geographic location or plant part. An example of possible variation by plant part is seen in the phytoliths from inflorescences of some sedge species. In particular, sedge phytoliths that are angular and verrucate with knobby apices may be restricted to inflorescences (Piperno, personal communication 1988); the present author has not observed this category of cone in other plant parts. Species yielding such cone phytoliths from inflorescences include: *Cyperus conglomeratus* (Ollendorf, Mulholland & Rapp 1987), *Scirpus cernuus* and *Scirpus lacustris* (Schuyler 1971). *Mariscus ustulatus* also produces angular and verrucate cones with knobby apices, but it is unclear in Raeside's (1970) paper from which plant part his samples were derived.

It should be noted that phytoliths from sedge inflorescences are not restricted to the angular and verrucate with knobby apex variety; they may be angular and psilate with pointed apices, such as *Carex diandra, C. exilis, C. paniculata, C. remota, C. ursina* (Toivonen & Timonen 1976); *Carex lurida, C. retrorsa, C. tuckermanii, C. vesicaria* (Walter 1975); *Eriophorum virginicum* and *Scirpus koilolepis* (Schuyler 1971); or angular and psilate with knobby apices, such as *Carex elongata, C. maritima, C. ursina* (Toivonen & Timonen 1976); *Carex oligosperma, C. pseudo-cyperus, C. rostrata* (Walter 1975); and *Eriophorum comosum* (Schuyler 1971), for instance.

SUGGESTIONS FOR FUTURE RESEARCH

Following Metcalfe's (1971) lead, studies of sedge phytoliths involving a rigorous, systematic approach with especial attention to detail must be conducted to devise a classification scheme. Since cones seem to be the phytolith type diagnostic of Cyperaceae, taxonomists interested in developing a classification scheme should devote their attention to these silica bodies.

Future studies should include a number of sedge species, documented with high-resolution photographs taken from a variety of angles (*ie*, planar and cross-sectional views), and should take into account the possible effects of individual variation, plant part and geographic location. Such cautionary factors were discovered through phytolith analysis of grass species, particularly maize varieties. For instance, Mulholland, Rapp and Ollendorf (1988) reported that even different leaf parts significantly affect phytolith assemblages! Although Piperno (1988) reported constancy of phytolith size and morphology of replicate wild grass and maize samples collected across Central America, the range in sedge phytolith morphology compiled in the present study should motivate future researchers to be aware that variability may be introduced by different geographic proveniences.

Since the "units produced by means of systematics require explicit definitions" (Dunnell 1971: 25), microscopic examination and photomicroscopy should involve quantification of sedge phytolith attributes. To differentiate securely between the phytoliths of sedges and certain grasses, in addition to distinguishing sedges from one another, quantification is prerequisite.

Quantification of size and three-dimensional morphology has been useful in grass phytolith taxonomy (Pearsall 1978; Piperno 1984; Mulholland, Ollendorf & Rapp 1989, 1990). Descriptions of the apex, periphery, interapical area and satellites (Fig 5.6A, B) should be quantified. For instance, quantitative data should be gathered for phytolith size and shape, apex angle, number of apices per platelet and number of satellites. Because several investigators have observed that a platelet of cones can be fractured (Metcalfe 1971; Andrejko & Cohen 1984; Ollendorf, Mulholland & Rapp 1987), differentiation between taxa by the number of apices should *not* be considered a sole criterion for taxon identification.

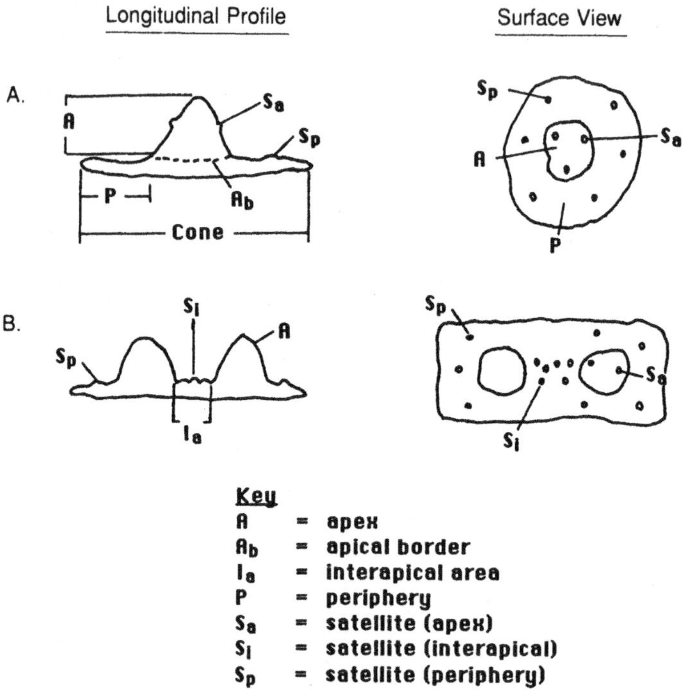

Fig 5.6. Parts of sedge phytoliths (after Andrejko & Cohen 1984: 489). A. Single-apex; B. Multiple-apex.

Because relatively few plant taxa have been isolated as yielding diagnostic phytoliths, it is generally believed that "descriptions of phytoliths and constructions of phytolith keys will be more complicated than in pollen analysis" (Piperno 1988: 108). The Cyperaceae, however, seem to be one family with distinctive phytoliths. The development of a key to sedge genera no doubt will be a complicated task (as hinted by the variation of cones listed in Table 5.2) and a key to sedge species will be even more complex (as illustrated by the diversity of cones listed in Table 5.1). However, a classification scheme potentially could aid in paleoenvironmental reconstructions and in the understanding of past lifeways as illustrated by the following examples.

Andrejko and Cohen (1984) suggested using sedge phytoliths to differentiate degraded peats derived from closely related sedge genera, such as *Carex* and *Cladium*. They proposed that phytolith analysis may be useful even when all pollen and identifiable plant tissues have been destroyed (eg, from natural peat fires and drought). Piperno (1988) mentioned that sedges are frequent components of secondary vegetation worldwide and their presence, along with *Heliconia* and grasses, in phytolith assemblages could provide evidence for anthropogenically disturbed vegetation and swidden agriculture. A classification scheme for sedge phytoliths might also be applied in the study of early literate communities associated with the Nile region. Since papyrus was produced from the pith of *Cyperus papyrus*, phytolith analysis may aid archaeologists in isolating the occurrence of papyrus phytoliths in cultural horizons. Whether or not finds of papyrus phytoliths correlate to the use of writing would need to be verified by independent means; however, this is only one example of how the potential of phytolith analysis is constrained by the lack of a classification scheme. It remains to be seen what creative applications will result from the development of a classification scheme for sedge phytoliths.

ACKNOWLEDGMENTS

I thank the editors of this volume, Susan Mulholland and George Rapp, Jr., for introducing me to phytolith analysis; their comments and patience over the years have been educational and much appreciated. Helpful suggestions for improvements of the first and subsequent drafts were made by the editors and two anonymous reviewers. I have accommodated these suggestions whenever possible, but remain responsible for the content of the paper. Toshiko Yamada-Nevells volunteered her time and skills to draft Figure 5.5. I am also indebted to the Missouri Botanical Garden for providing herbarium vouchers of *Epischoenus villosus* (MO#2657359 and MO#2261414), *Eriophorum comosum* (MO#883656 and MO#90022), and *Scirpus wichurae* (MO#107504). The provenience of these samples is as follows: samples of *E. villosus* were originally collected in the Cape Province (South Africa); MO#883656 was collected in the Himalayas (India); MO#90022 was collected in Yunnan (China); and *S. wichurae* was collected in Yunnan (China). I thank Dr. Anita Cholewa (University of Minnesota Herbarium) for obtaining the herbarium vouchers and securing their safe return to Missouri.

REFERENCES

Andrejko, MJ and Cohen, AD 1984 Scanning electron microscopy of silicophytoliths from the Okefenokee Swamp-Marsh Complex. *In* Cohen, AD, Casagrande, DJ, Andrejko, MJ and Best, GR, eds, *The Okefenokee Swamp: Its Natural History, Geology, and Geochemistry*. Los Alamos, Wetland Surveys: 468-491.

Dahlgren, RMT and Clifford, HT 1982 *The Monocotyledons: A Comparative Study*. London, Academic Press: 378 p.

Dunnell, RC 1971 *Systematics in Prehistory.* New York, Free Press: 214 p.

Faegri, K and Iversen, J 1975 *Textbook of Pollen Analysis.* Third edition, New York, Hafner Press: 295 p.

Kondo, R and Sase, T 1986 Opal phytoliths, their nature and application. *Daiyonki Kenkyu* (Quaternary Research) 25: 31-63.

Lanning, FC and Eleuterius, LN 1981 Silica and ash in several marsh plants. *Gulf Research Reports* 7: 47-52.

McAndrews, JH, Berti, AA and Norris, G 1973 *Key to the Quaternary Pollen and Spores of the Great Lakes Region.* Toronto, Royal Ontario Museum: 61 p.

Mehra, PN and Sharma, OP 1965 Epidermal silica cells in the Cyperaceae. *Botanical Gazette* 126: 53-58.

Metcalfe, CR 1971 *Anatomy of the Monocotyledons. V. Cyperaceae.* Oxford, Clarendon Press: 597 p.

Mulholland, SC, Ollendorf, AL and Rapp, G, Jr (ms) 1989 Maize phytolith assemblages: Data from the northern Plains, USA. In preparation.

Mulholland, SC, Rapp, G, Jr and Ollendorf, AL 1988 Variation in phytoliths from corn leaves. *Canadian Journal of Botany* 66: 2001-2008.

_____1990 Variation in phytoliths within a population of corn (cv. Mandan Yellow Flour). *Canadian Journal of Botany* 68: 1638-1645.

Norgren, JA (ms) 1973 Distribution, form and significance of plant opal in Oregon soils. PhD dissertation, Oregon State University: 165 p.

Ollendorf, AL, Mulholland, SC and Rapp, G, Jr 1987 Phytoliths from some Israeli sedges. *Israel Journal of Botany* 36: 125-132.

Pearsall, DM 1978 Phytolith analysis of archaeological soils: Evidence for maize cultivation in Formative Ecuador. *Science* 199: 177-178.

Piperno, DR (ms) 1983 The application of phytolith analysis to the reconstruction of plant subsistence and environments in prehistoric Panama. PhD dissertation, Temple University: 459 p.

_____1984 A comparison and differentiation of phytoliths from maize and wild grasses: Use of morphological criteria. *American Antiquity* 49: 361-383.

_____1985 Phytolith analysis and tropical paleo-ecology: Production and taxonomic significance of siliceous forms in New World plant domesticates and wild species. *Review of Palaeobotany and Palynology* 45: 185-228.

_____1988 *Phytolith Analysis: An Archaeological and Geological Perspective.* San Diego, Academic Press: 280 p.

Raeside, JD 1970 Some New Zealand plant opals. *New Zealand Journal of Science* 13: 122-132.

Schuyler, AE 1971 Scanning electron microscopy of achene epidermis in species of *Scirpus* (Cyperaceae) and related genera. *Proceedings of the Academy of Natural Sciences, Philadelphia* 123: 29-52.

Thomasson, JR 1983 *Carex graceii* sp. N., *Cyperocarpus eliasii* sp. N., *Cyperocarpus terrestris* sp. N., and *Cyperocarpus pulcherrima* sp. N. (Cyperaceae) from the Miocene of Nebraska. *American Journal of Botany* 70: 435-449.

Toivonen, H and Timonen, T 1976 Perigynium and achene epidermis in some species of *Carex*, subg. *Vignea* (Cyperaceae), studied by scanning electron microscopy. *Annales Botanici Fennici* 13: 49-59.

Walter, KS 1975 A preliminary study of the achene epidermis of certain *Carex* (Cyperaceae) using scanning electron microscopy. *Michigan Botanist* 14: 67-72.

Witty, JE and Knox, EG 1964 Grass opal in some chestnut and forested soils in north central Oregon. *Soil Science Society of America, Proceedings* 28: 685-688.

APPENDIX 5.1. Sedges Reported in the Scientific Literature

Species	Author
Asterochaete glomerata	Metcalfe 1971
Bulbostylis barbata Kunth.	Mehra & Sharma 1965
B. capillaris Kunth.	"
Carex secta	Raeside 1970
C. walteriana Bailey	Andrejko & Cohen 1984
C. kobomugi	Kondo & Sase 1986
C. spp.	"
C. condensata Nees.	Mehra & Sharma 1965
C. filicina Nees.	"
C. longicruris Nees.	"
C. remota L.	"
C. wallichiana Presc. ex Nees.	"
C. appropinquata	Toivonen & Timonen 1976
C. brunnescens	"
C. canescens	"
C. chordorrhiza	"
C. diandra	"
C. dioica	"
C. disperma	"
C. echinata	"
C. elongata	"
C. exilis	"
C. glareosa	"
C. heleonastes	"
C. interior	"
C. lachenalii	"
C. lapponica	"
C. localiacea	"
C. mackenziei	"
C. maritima	"
C. paniculata	"
C. parallela	"
C. remota	"
C. tenuiflora	"
C. ursina	"
C. comosa	Walter 1975
C. hystericina	"
C. lurida	"
C. oligosperma	"
C. pseudo-cyperus	"
C. retrorsa	"
C. rostrata	"
C. schweinitzii	"
C. tuckermanii	"
C. vesicaria	"

(Continued)

APPENDIX 5.1. (Continued)

Species	Author
Cephalocarpus sp.	Metcalfe 1971
Cyperus conglomeratus	Ollendorf, Mulholland & Rapp 1987
C. lanceus?	"
C. longus	"
C. mucronatus	"
C. papyrus	"
C. chorizanthus	Piperno 1985
C. aristatus Rottb.	Mehra & Sharma 1965
C. bulbosus Vahl.	"
C. compressus L.	"
C. cuspidatus HBK	"
C. difformis L.	"
C. digitatus Roxb.	"
C. eleusinoides Kunth.	"
C. flavidus Retz.	"
C. iria L.	"
C. niveus Retz.	"
C. pilosus Vahl.	"
C. rotundus L.	"
C. tegetum Roxb.	"
Cladium jamaicensis Crantz.	Andrejko & Cohen 1984
C. mariscus	Metcalfe 1971
Desmoschoenus frondosus	Raeside 1970
Dulichium arundinaceum (L) Britton	Andrejko & Cohen 1984
Eleocharis atropurpurea Kunth.	Mehra & Sharma 1965
E. palustris R. Br.	"
E. plantaginea R. Br.	"
E. nodulosa	Piperno 1985
Epischoenus villosus	Metcalfe 1971
Eriophorum comosum Wall. ex Nees.	Mehra & Sharma 1965
E. vaginatum	Kondo & Sase 1986
E. alpinum L.= *Scirpus hudsonianus* (Michx.) Fern.	Schuyler 1971
E. comosum Wall.	"
E. crinigerum (Gray) Beet.= *Scirpus criniger* Gray	"
E. gracile Koch.	"
E. japonicum Maximow. = *Scirpus maximowiczii*	"
E. latifolium Hoppe.	"
E. microstachyum Bocklr.	"
E. spissum Fern.	"
E. virginicum L.	"
Everardia montana	Metcalfe 1971

APPENDIX 5.1. (Continued)

Species	Author
Ficinia indica	Metcalfe 1971
Fimbristylis annua	Piperno 1985
F. complanata Link.	Mehra & Sharma 1965
F. dichotoma (L.) Vahl.	"
F. diphylla Vahl.	"
F. ferruginea Vahl.	"
F. junciformis Kunth.	"
F. miliacea Vahl.	"
F. monstachya Hassk.	"
F. quinquangularis Kunth.	"
F. schoenoides Vahl.	"
F. tenera Roem. & Schult.	"
Juncellus alopecuroides (Rottb.) Clarke	"
J. laevigatus Clarke.	"
J. serotinus Clarke.	"
Kobresia capillifolia Clarke.	"
K. laxa Boeck.	"
K. nitens Clarke.	"
K. royleana Nees. ex Boeck.	"
Kyllinga brevifolia Rottb.	"
K. monocephala Rottb.	"
K. squamulata Vahl.	"
K. triceps Rottb.	"
Mariscus microcephalus Presl.	"
M. paniceus Vahl.	"
M. ustulatus	Raeside 1970
Pycreus globosus Reich.	Mehra & Sharma 1965
P. pumilus (L.) Turrill	"
P. sanguinolentus Nees.	"
Rhynchospora barbaya	Piperno 1985
R. inundata (Oakes) Fernald	Andrejko & Cohen 1984
Scirpus holoschoenus L.= *Holoschoenus vulgaris* L. Frisch	Ollendorf, Mulholland & Rapp 1987
S. maritimus	"
S. tabernaemontani	"
S. wichurae	Kondo & Sase 1986
S. caricis Retz.	Mehra & Sharma 1965
S. juncoides Roxb.	"
S. lacustris L.	"
S. mucronatus L.	"
S. roylei (Nees.) Parker	"
S. setaceus L.	"
S. squarrosus L.	"
S. triqueter L.	"

(Continued)

APPENDIX 5.1. (Continued)

Species	Author
S. acutus Muhl. ex Bigel.	Schuyler 1971
S. americanus Pers.	"
S. ancistrochaetus x *atrovirens* or *S. ancistrochaetus* x *hattorianus*	"
S. ancistrochaetus Schuyl.	"
S. asper Presl.	"
S. atrocinctus Fern.	"
S. atrovirens x *hattorianus*	"
S. badius Presl.	"
S. californicus (Mey.) Steud.	"
S. cernuus Vahl.	"
S. cespitosus L.	"
S. clementis Jones.	"
S. clintonii Gray.	"
S. congdonii Britt.	"
S. conglomeratus HBK	"
S. cyperinus (L.) Kunth.	"
S. deltarum Schuyl.	"
S. diffusus Schuyl.	"
S. divaricatus Ell.	"
S. erismanae Schuyl.	"
S. expansus Fern.	"
S. flaccidifolius (Fern.) Schuyl.	"
S. georgianus Harp.	"
S. grossus L.	"
S. hattorianus Mak.	"
S. heterochaetus Chase.	"
S. karuizawensis Mak.	"
S. koilolepis (Steud.) Gleas.	"
S. lacustris L.	"
S. lineatus Michx.	"
S. littoralis Schrad.	"
S. microcarpus Presl.	"
S. mitsukurianus Mak.	"
S. pallidus (Britt.) Fern.	"
S. pendulus Muhl.	"
S. polyphyllus Vahl.	"
S. purshianus Fern.	"
S. radicans Schkuhr.	"
S. rollandii Fern.	"
S. setaceus L.	"
S. smithii Gray.	"
S. subcapitatus Thw.	"
S. supinus L.	"
S. sylvaticus L.	"

APPENDIX 5.1. (Continued)

Species	Author
S. tabernaemontanii Gmel.	Schuyler 1971
S. ternatanus Reinw. ex Miq.	"
S. validus Vahl.	"
S. verecundus Fern.	"
S. wichurae Bocklr.	"
S. wilkensii Schuyl.	"
Scleria tessellata Willd.	Mehra & Sharma 1965

Notes for Appendix 5.1:

Ollendorf, Mulholland & Rapp (1987): all plants are from Israel; some plants are from Tel Aviv University Herbarium, some from the vicinity of Tel Michal.

Kondo & Sase (1986): all plants are Japanese; exact provenience is unknown since article is printed in Japanese.

Piperno (1985): leaves of plants from herbaria of Harvard University, Cambridge; Academy of Natural Sciences, Philadelphia; Missouri Botanical Garden, St. Louis; Smithsonian Tropical Research Institute, Panama City; field collections of some species (Panama).

Andrejko & Cohen (1984): field collections of plants (Okefenokee Swamp-Marsh Complex); University of South Carolina Herbarium.

Toivonen & Timonen (1976): all collected in Finland(?); preserved in the Botanical Museum of the University of Helsinki.

Walter (1975): collected plants in Kalamazoo, Chippewa, Van Buren, Lake, Mackinac, Schoolcraft, Cheboygan and Keweenaw counties in Michigan.

Metcalfe (1971): plants from various locations and probably all curated at Kew Herbarium.

Schuyler (1971): species of *Scirpus* collected in United States (Virginia, Georgia, Pennsylvania, New York, Vermont, Oregon, California, Alabama, Texas, Florida, Ohio, Mississippi, Louisiana), Argentina, Rumania, Poland, India, Korea, Japan, Taiwan, Quebec, North Wales, Spain, Borneo, France, Colombia, Egypt, Ceylon and Peru. Species of *Eriophorum* collected in California, Delaware, Maine, France, India, China and Japan.

Raeside (1970): sedges from grassland or its coastal margin in New Zealand.

Mehra & Sharma (1965): Indian Himalayas from plains to 3700 m altitude; stored in Panjab University Herbarium.

6

PREDICTED WORLD DISTRIBUTION OF C_3 AND C_4 GRASS PHYTOLITHS

PAGE C. TWISS

ABSTRACT

The shapes of individual opal phytoliths from grasses, silica bodies formed in the epidermal cells of the leaf, are indicators of the C_3 and C_4 photosynthetic pathways. Pooid phytoliths are circular, rectangular, elliptical, crescent or oblong forms and occur primarily in C_3 grasses, which are concentrated in high latitudes or high elevations. Chloridoid phytoliths are saddle-shaped, and occur dominantly in C_4 grasses that grow in warm, arid to semiarid regions in which the available soil moisture is very low. Panicoid phytoliths include many variations of crosses and dumbbells, and occur largely in C_4 grasses that flourish in warm, tropical to subtropical regions with a moderate amount of available soil moisture.

Identifiable grass phytoliths, concentrated from soils, paleosols, sediment, archaeological sites and sedimentary rock, can be used to reconstruct the Cenozoic environment of a site or region. Taxa of the subfamily Pooideae grow in the higher latitudes and elevations. Taxa of subfamilies Chloridoideae and Panicoideae are restricted to the tropics and subtropics. The majority of extensive grasslands occur in continental interiors and are the dominant sources of identifiable phytoliths. High ratios of pooid to total pooid, chloridoid and panicoid phytoliths imply cool temperatures in the region. High ratios of chloridoid to total chloridoid and panicoid phytoliths imply a warm arid or semiarid region. Because the sources of phytoliths may be local or remote, the locations and directions of major wind belts and ocean currents must be considered in any interpretation.

INTRODUCTION

Twenty years have elapsed since we proposed a morphological classification of grass phytoliths (Twiss, Suess & Smith 1969), which was developed to determine the sources of atmospheric dust deposited east of the Rocky Mountains

(Smith & Twiss 1965). Leaves from 17 species of grass from the short and tall grass prairies were examined along with descriptions from Metcalfe (1960) to develop the classification. Although opal-A bodies occur to some degree in most plants, they are especially abundant, diverse and distinctive in the Poaceae (Gramineae). The discrete silica bodies are relatively insoluble, except at high pH, and have been recovered from soils, paleosols, sediment and rocks of the Cenozoic Era. They also strongly indicate whether the parent plant was C_3 or C_4 (Twiss 1980).

Grass systematics has also developed during this period. Understanding of the C_3 and C_4 pathways has increased (Smith & Brown 1973; Hattersley & Watson 1975; Brown 1977). Watson (1987: 343–351) has developed a data base of the grasses that uses 341 characters derived from general morphology, leaf anatomy including silica bodies, physiology, fruit and embryo structure, seedling form, cytology, geography and ecology. More than 735 genera have been included. Additional information on the geographic distribution of grasses has accumulated.

This paper summarizes the general data on the distribution of C_3 and C_4 grasses so that the global distribution of discrete opal phytoliths preserved in soil, sediment and rock of the Cenozoic Era can be predicted.

CLASSIFICATION OF GRASS PHYTOLITHS

Existing classifications of grass phytoliths are based on the micromorphology of discrete silica bodies, which is independent of the orientation of the bodies in silica cells in the various vegetative parts of the individual grass plants. A great range of shape of silica bodies can occur within a single grass plant, a single species, and among species, genera, tribes and subfamilies, a characteristic termed "multiplicity" by Rovner (1971). Furthermore, several shapes of individual phytoliths occur in many grass taxa, a phenomenon called "redundancy" by Rovner (1971).

One characteristic of an active scientific discipline is the development and refinement of different classification systems. Several classifications of discrete grass opal phytoliths have been proposed and expanded. Ehrenberg (1854) and Deflandre (1963) used a binomial taxonomic system with parataxa. Locker and Martini (1986) adopted this system in describing the phytoliths recovered from deep sea cores east of Australia that extend from the Middle Miocene to Early Pleistocene. Other classifications are largely morphological and have been used by most workers (Twiss, Suess & Smith 1969; Bertoldi de Pomar 1971; Sase & Kondo 1974; Brown 1984; Piperno 1988; Mulholland & Rapp 1989).

Currently, I am using the following basic forms for the grasses while recognizing that taxonomic exceptions do exist for each (Fig 6.1). The first three are termed "regular"; other forms can be added as needed:

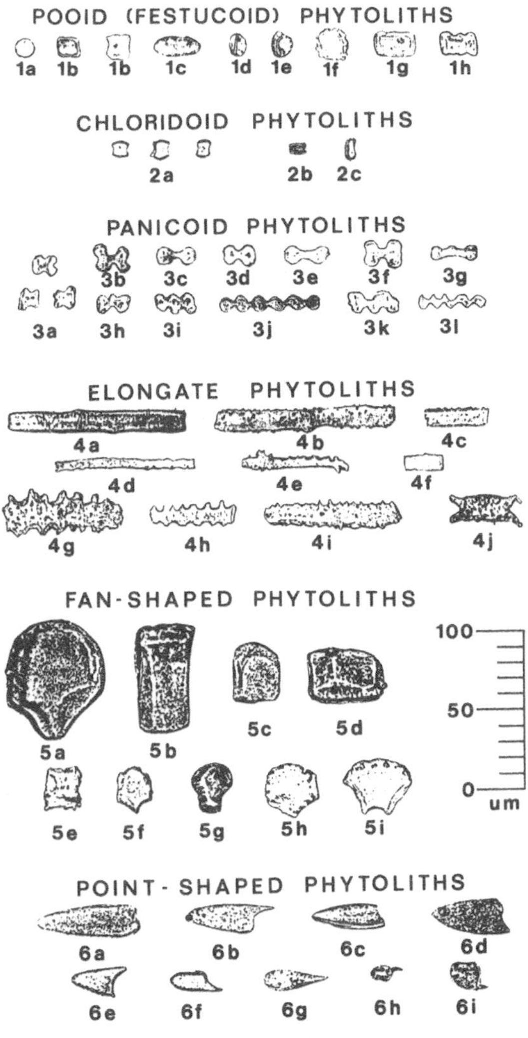

Fig 6.1. Classification of grass phytoliths. Pooid (festucoid) phytoliths, Class 1; Chloridoid phytoliths, Class 2; Panicoid phytoliths, Class 3; Elongate phytoliths, Class 4; Fan-shaped phytoliths, Class 5; Point-shaped phytoliths, Class 6. Classes 5 and 6 are from Sase and Kondo (1974).

1. Pooid (festucoid of Twiss, Suess & Smith 1969): circular, rectangular, elliptical, crenate and oblong; some double outlines and some trapezoids of Brown (1984); rondels of Mulholland and Rapp (1989);
2. Chloridoid: normal and thin chloridoid; saddles of Brown (1984) and Mulholland and Rapp (1989);
3. Panicoid: several types of crosses and dumbbells; bilobates, polylobates and some crosses of Brown (1984);
4. Elongate: smooth, sinuous and spiny walls; elongate plates of Brown (1984);
5. Fan-shaped: Smithson (1958), Sase and Kondo (1974);
6. Point-shaped: Sase and Kondo (1974), some trichomes of Brown (1984);
7. Unidentified opal fragments.

Relating phytolith types to contemporary grass taxa and their photosynthetic pathways will aid in paleoenvironmental interpretations throughout the globe.

GEOGRAPHIC DISTRIBUTION OF SUBFAMILIES OF POACEAE (GRAMINEAE)

World Distribution of Subfamily Pooideae

The subfamily Pooideae (Festucoideae) contains about 160 genera, distributed between two supertribes and among seven tribes (Macfarlane 1987). Supertribe Poodae contains tribes Poeae (Festuceae), Seslerieae, Aveneae (including Agrostideae) and Meliceae (including Glycerieae). Supertribe Triticodae contains tribes Triticeae (Hordeae), Brachypodieae and Bromeae.

Hartley (1973) showed that the Pooideae (Festucoideae) are best represented, relative to the total grass flora, at high latitudes and at other places at high elevations (Fig 6.2). The Pooideae dominate the grass flora in the temperate zones of the northern and southern hemispheres and, if present in the tropics, occur only in high altitudes. Figure 6.2 shows a general absence between the Tropics of Cancer and Capricorn, except in the Andes of the tropics and subtropics. The Pooideae seem to favor the temperate and cold parts of the globe. Although Hartley (1973) included some genera and species that are now excluded from the Pooideae, they do not significantly affect his conclusions (Macfarlane 1987).

Macfarlane (1987: 275) reported that members of the supertribe Poodae occur everywhere that do pooids, even Antarctica. Supertribe Triticodae, representing about 20 percent of the genera, is "less common in the Southern Hemisphere and absent from Antarctica and the subantarctic islands."

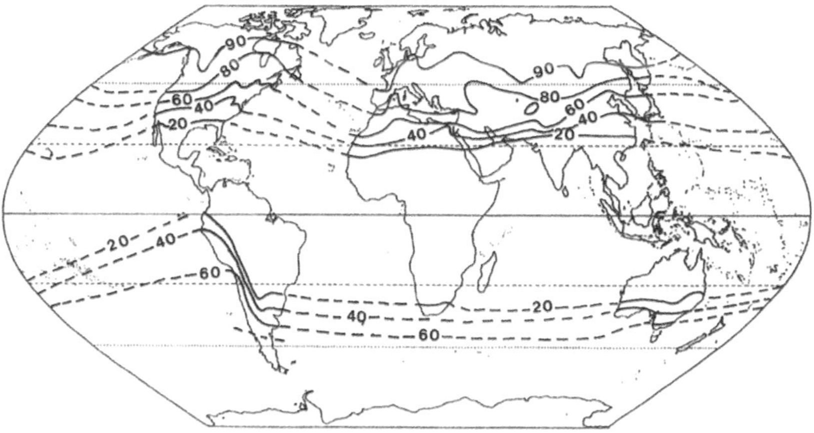

Fig 6.2. World distribution of subfamily Pooideae (Festucoideae) expressed in percent of species in total grass flora (after Hartley 1973).

All members of the Pooideae are non-Kranz, C_3 plants (Ellis 1987). Their silica bodies are horizontally elongated, oblong and nodular; the intercostal long cells are fusiform, or rectangular straight-walled or sinuous. Pooid phytoliths, which are characteristic of the subfamily, are circular, rectangular, elliptical, crescent or oblong; no chloridoid saddles or panicoid dumbbells or crosses occur. The Pooideae include the cereal grains; barley, rye, oats and wheat, which are especially useful to archaeologists.

World Distribution of the Subfamily Chloridoideae

The subfamily Chloridoideae contains tribes Chlorideae, Eragrostideae, Sporoboleae, Zoysieae, Pappophoreae, Orcuttieae, Triodieae and Pommereulleae (Ellis 1987), and is nearly a homogeneous subfamily (Jacobs 1987). Hartley and Slater (1960) and Hartley (1964), who also included the tribe Aristideae in the subfamily, plotted the relative specific frequency of the subfamily Chloridoideae (Eragrostoideae). All high concentrations of the subfamily occur in arid regions of Africa, Australia and India where precipitation occurs in the summer, or where no distinct rainy season exists (Fig 6.3). Hartley and Slater (1960) reported that chloridoid species are most common where the mean temperature of the coldest month is above 10°C.

Jacobs (1987: 284) reported that "most chloridoid species, and their greatest contribution to grass floras, occur south of the equator in areas of highly seasonal rainfall or areas where rainfall is highly irregular and rather low." They are abun-

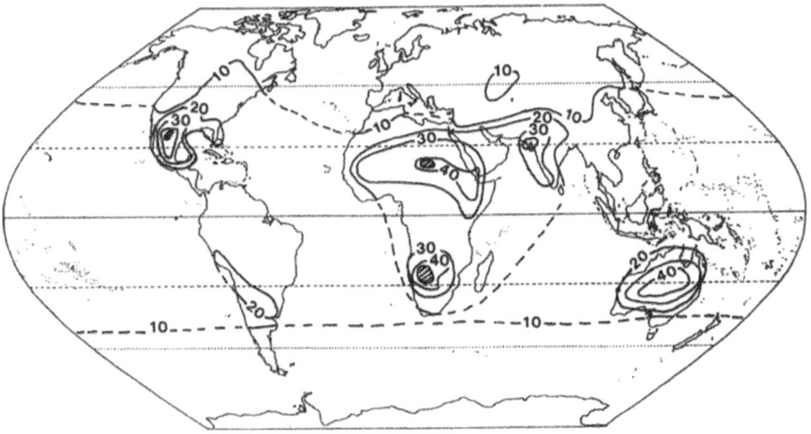

Fig 6.3. World distribution of subfamily Chloridoideae as expressed in percent of species in total grass flora (after Hartley & Slater 1960).

dant in savannas or open grasslands and seem to tolerate the extremes of high temperatures and aridity better than the other grasses.

Subfamily Chloridoideae is entirely C_4 with Kranz anatomy (Ellis 1987; Hattersley 1987; Jacobs 1987). Costal and intercostal zones of the epidermis are distinct. The intercostal zones contain elongated sinuous cells that are frequently interspersed with silica-cork cell pairs. The costal silica cells range from saddles to crosses and dumbbells, but saddles are most typical (Jacobs 1987). Although these kinds of phytoliths are most characteristic of the Chloridoideae, they also occur in subfamily Bambusoideae and rarely in subfamily Arundinoideae. Chloridoid phytoliths are abundant in the short grasses (*Buchloe dactyloides* and *Bouteloua gracilis*) of the western prairies of the United States. No major cereal grains occur in this subfamily.

World Distribution of the Subfamily Panicoideae

Subfamily Panicoideae consists of two supertribes: Andropogonodae and Panicodae (Ellis 1987). Taxa of the former are all C_4, Kranz plants, whereas the latter contains both C_3 and C_4 species. Three genera of the Panicodae, *Panicum*, *Alloteropsis* and *Neurachne*, contain both photosynthetic pathways (Ellis 1987); *Panicum* includes species of all three C_4 subtypes (NADP-malic enzyme, NAD-malic enzyme and PEP-carboxykinase (Brown 1977), C_3, and C_3-C_4 intermediates (Zuloaga 1987). *Panicum* is one of the largest genera, containing more than 600 species distributed throughout the world (Zuloaga 1987), and one emphasized by Ellis (1987) as needing much more study.

Hartley (1958a, 1958b) prepared world distribution maps of the relative species distribution for the tribes Andropogoneae and Paniceae. Species of the Andropogoneae are concentrated in the tropics and subtropics. Their maximum abundance is in India and Malaysia, the eastern hemisphere having a higher concentration than the western. The tribe seems to favor warm temperatures and high summer moisture.

The Tribe Paniceae has a wider geographical distribution (Hartley 1958b) and more varied photosynthetic pathways than the Andropogoneae. Its species are common in tropical and subtropical regions. Some have adapted to temperate climate, but none are common above 40° latitudes.

The distribution of subfamily Panicoideae was constructed by combining Hartley's (1958a, 1958b) maps of the Andropogoneae and Paniceae. The tropical and subtropical concentration of species of the subfamily is emphasized (Fig 6.4). In North America, Gould and Shaw (1983: 110, Fig 4–6) show that species of Panicoideae are especially successful in moist, humid, tropical and subtropical habitats, but that very few species have become adapted to arid regions.

The epidermal costal zones of the Panicoideae contain horizontally elongate, dumbbell, cross and nodular silica bodies; the intercostal zones contain rectangular, sinuous-walled long cells and some crenate cells. Dumbbells and crosses are characteristic of the panicoid grasses, but they also occur in the bambusoid, arundinoid and a few chloridoid grasses. Most of the dominant species of the tall grass prairie in the Great Plains of the US are members of the Andropogoneae (*Schizachyrium scoparium, Andropogon gerardi, Sorghastrum nutans*), are C_4, and

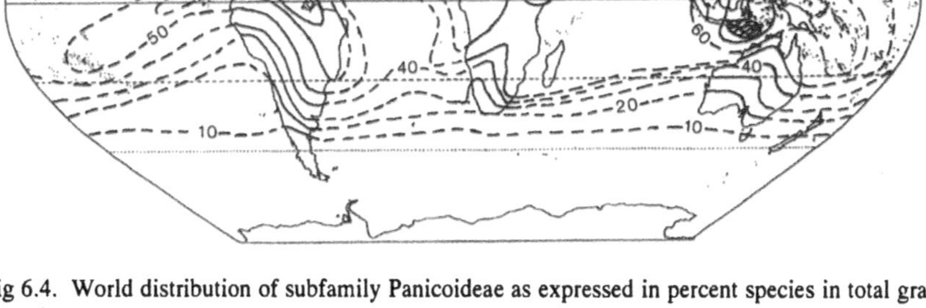

Fig 6.4. World distribution of subfamily Panicoideae as expressed in percent species in total grass flora (compiled from maps prepared by Hartley 1958a, 1958b).

contain dumbbell and cross-shaped phytoliths. Subfamily Panicoideae includes maize, sorghum and sugarcane.

World Distribution of Subfamily Bambusoideae

Watson, Clifford and Dallwitz (1985) proposed subfamily Bambusoideae and two supertribes: Oryzanae and Bambusanae. They emphasized the shapes of the silica bodies among the taxa. The Oryzanae contain mostly cross- to dumbbell-shaped phytoliths, some oryzoid (Metcalfe 1960) and very few saddles. The Bambusanae contain saddles and sometimes oryzoid silica bodies. Crosses, dumbbells and nodular forms are uncommon.

Soderstrom and Ellis (1987) showed the distribution of the sub-tribes of the Bambuseae; Figure 6.5 is generalized from their map. The Bambusineae are concentrated in the tropics and subtropics in central and southern Africa and Madagascar, India and the Far East, and in Central and South America; they are absent from the Arctic, but a few genera occur in Western Eurasia and the USSR, the Mediterranean and Australia.

Subfamily Bambusoideae contain vertically elongated dumbbell-, saddle-, cross- and olyroid-shaped silica bodies in the costal regions. Rectangular and sinuous-walled long cells occur in the intercostal zones. Discrete bambusoid phytoliths may be saddles, dumbbells and/or crosses; they cannot be identified unless other parts of the plant are preserved. All are non-Kranz, C_3 plants.

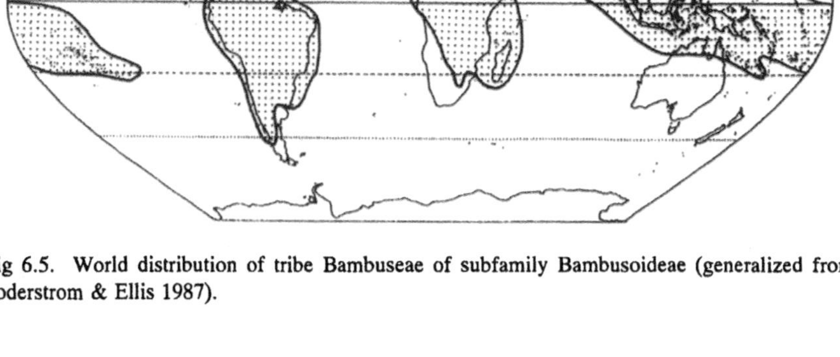

Fig 6.5. World distribution of tribe Bambuseae of subfamily Bambusoideae (generalized from Soderstrom & Ellis 1987).

World Distribution of the Subfamily Arundinoideae

Subfamily Arundinoideae is the most heterogenous group because it lacks specialized and reliable diagnostic characteristics (Ellis 1987). As Conert (1987) emphasized, the genera and tribes are widely distributed over several continents, and no monograph has been completed. He included only three tribes: Arundineae, Danthonieae and Cortaderieae. Most are C_3, but a few genera are C_4. Silica bodies are mostly square, round and oblong, but some are saddles, crosses and dumbbells.

Barkworth and Everett (1987) assigned the tribe Stipeae to the Arundinoideae as did Watson, Clifford and Dallwitz (1985). Most silica bodies are round, oval or crenate; all are non-Kranz, C_3 plants.

Tribe Aristideae is included in the Arundinoideae by Watson, Clifford and Dallwitz (1985) and Hattersley (1987), the latter listing two of three genera as C_4. Other tribes included in the subfamily by different workers are Centotheceae, Ehrharteae, Lygeae, Thysanolaeneae and Steyermarkochloeae (Ellis 1987).

Subfamily Arundinoideae contains horizontally elongated, square, oblong, cross-, saddle- and dumbbell-shaped silica bodies in the costal zones and sinuous-walled long cells in the intercostal zones. The subfamily is distributed predominantly in the southern hemisphere (Conert 1987); genera are common in the temperate regions of the world, tropical regions of Africa and Asia, Central and South America, Australia, New Zealand, New Guinea, western Europe and North America. Most of the taxa are C_3, but a few C_4 plants occur.

Metcalfe (1960: 542–551) described 12 species of *Arundinaria* and recorded saddle-, barrel-, cross- and dumbbell-shaped silica bodies over the veins, and in the intercostal zones, tall and narrow and crenate silica bodies. Ollendorf, Mulholland and Rapp (1988) reported that *Arundo donax* contains dumbbells and crosses, whereas *Phragmites communis* contains saddle- and short trapezoid-shaped phytoliths; both species are C_3 plants. Thomasson (1984: 208) described several leaf fragments from the Miocene of Nebraska; all contained "finger-like, bicellular microhairs, saddle-shaped silica bodies, and typical C_3 (non-Kranz) leaf anatomy with arm cells present and fusoid cells absent." He assigned them to the arundinoid group. Identification of arundinoid phytoliths is possible only with an abundance of associated epidermal segments.

GEOGRAPHIC DISTRIBUTION OF C_3 AND C_4 GRASS PHYTOLITHS

Even though the silica bodies produced in the silica cells of the leaf epidermis are only a small fraction of the total siliceous residue of grasses, they are the most diagnostic paleoenvironmental indicators recovered from ancient sediment. Although some exceptions exist, pooid phytoliths tend to be produced in the leaf epidermis of C_3 grasses, whereas chloridoid and panicoid phytoliths are produced dominantly in the leaf epidermis of C_4 grasses.

Pooid phytoliths are derived largely from the subfamily Pooideae, which flourishes in the temperate zones or in high elevations elsewhere. Chloridoid phytoliths form in the C_4 subfamily Chloridoideae whose taxa tolerate warm, arid and semiarid regions where the rainfall is low, or where precipitation occurs in the summer. Panicoid phytoliths are common in the subfamily Panicoideae, which is most successful in warm and humid tropical and subtropical regions. Taxa of supertribe Andropogonodae are C_4, but nearly half the species in supertribe Panicodae are C_3.

The deficiencies of a morphological classification of individual silica bodies (phytoliths) derived from the grasses that were emphasized in 1969 (Twiss, Suess & Smith) are still pertinent. Multiplicity (Mulholland, Rapp & Ollendorf 1988) of pooid and panicoid phytoliths in *Zea* and *Aristida* are only two of many examples. Redundancy (Mulholland, Rapp & Ollendorf 1988) is a major problem with several taxa. Panicoid phytoliths, although overwhelmingly abundant in the subfamily Panicoideae, occur in taxa within subfamilies Bambusoideae, Arundinoideae and Chloridoideae, and as vertically elongated silica bodies in the supertribe Oryzanae (Watson, Clifford & Dallwitz 1985; Ellis 1987). Chloridoid phytoliths also occur in some taxa of the subfamilies Bambusoideae and Arundinoideae (Ellis 1987). These problems emphasize that opal phytoliths are only one indicator of the kind of grass; preserved segments of the plants are necessary for identifying taxa.

Both C_3 and C_4 grass plants occur together in many localities. The C_3 plants can flourish early and late in the growing season when the weather is cool and moist, only to become dormant when the temperature increases during the summer months. The C_4 plants then become active.

Teeri and Stowe (1976) tabulated the percentage of the ratio of C_4 grasses in the total grass flora at 32 locations in North America. They showed the highest concentration of C_4 grasses in the lower latitudes, more than 80 percent, and that C_4 grasses were absent from the extreme northern latitudes. They reported that the percent of C_4 grasses in the continental United States is most accurately predicted by a linear combination of the normal July minimum temperature, mean annual degree-days and the log of the length of the freeze-free period. Winter temperatures of about 10°C, or lower, tend to limit growth of C_4 grasses, but do not affect C_3 grasses. In the Great Plains of the US, the percent of C_4 grasses decreases from about 70 percent in Texas to less than 20 percent in the Prairie Provinces of Canada. Teeri and Stowe warned that, although the percentage of C_4 grasses may be high in harsh environmental areas, such as Death Valley, California, and other desert regions, the total number of C_4 species may be very low.

In an altitudinal transect in Kenya, Tieszen *et al* (1979) showed that C_4 grasses are dominant at low and intermediate elevations and at relatively low available soil moistures, whereas the C_3 grasses are dominant at higher elevations and at lower temperatures where the available soil moistures are high (Fig 6.6). Moreover, the C_4 tribes Chlorideae, Eragrosteae and Sporoboleae (which contain dominantly chloridoid phytoliths), and Aristideae (which produces mostly panicoid types), are most abundant at low elevations and very low available soil moistures. The Andropogoneae and C_4 members of the Paniceae, which produce panicoid phytoliths, reach their apex at intermediate elevations and available soil moistures. The C_3 tribes (Aveneae and Poeae), which contain pooid phytoliths, occur only at soil moisture indices of 60 or greater and at higher elevations where temperatures are lower. The transitional range for the decline of C_4 grasses and the concomitant rise of C_3 grasses occurs at available soil moistures of 50 to 70 and between elevations of 2,000 to 3,000 meters. Few C_3 grasses occur below 2,050 meters and few C_4 grasses occur above 3,050 meters. Tieszen *et al* concluded that open grasslands at low elevations or low available soil moistures consist of C_4 species and open grasslands at high elevations contain C_3 species.

Livingstone and Clayton (1980) obtained similar results in tropical Africa; they reported that temperature controls the ultimate distribution between C_3 and C_4 grasses. They also suggested that pooid grass cuticles, identifiable to genera (Palmer 1976) and recovered from lake sediments, could provide evidence for

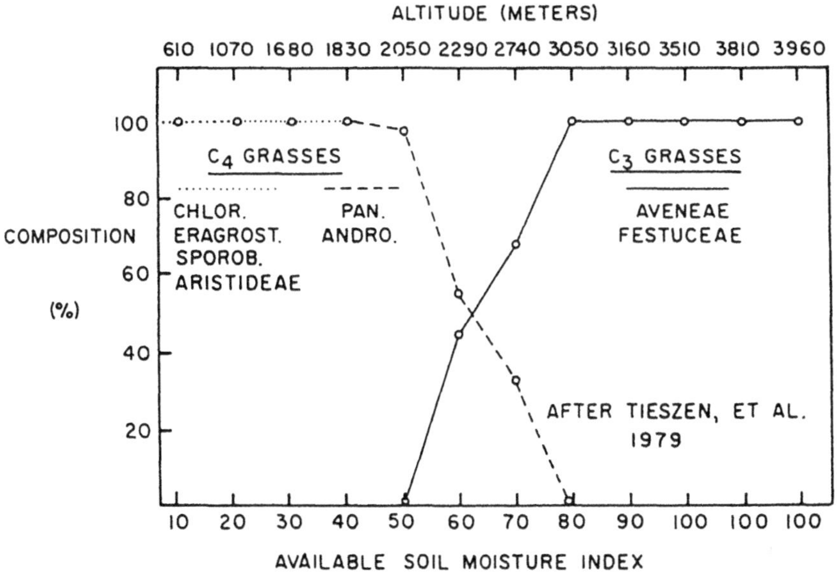

Fig 6.6. Percentage of C_3 and C_4 grass species along an altitudinal transect in Kenya (adapted from Tieszen *et al* 1979). Chloridoid grasses are represented as dots, panicoid grasses as dashes, and pooid grasses as solid lines.

past temperatures in the region. Opal phytoliths can serve a similar function, especially when combined with fragments of grass leaves, pollen, diatoms and other fossils.

Hartley (1964: 31) emphasized that extensive grasslands are widely distributed and are most common in the interior of the continents. They range from about 55° N in Asia and North America to 40° S in South America. "Treeless prairies, steppes, and pampas predominate in the temperate region, and are replaced by savannas, with varying proportions of trees and shrubs, in the tropics and subtropics." He also stressed (1964: 35) that very little relationship exists between the "distribution of grass tribes, as mapped on a floristic basis, and their ecological significance." Mapping the relative species concentration reveals only part of the story. Hartley (1964) used the example of tribe Andropogoneae, which has the greatest number of species in forests, not grasslands. In the tallgrass prairie of the Great Plains of the U.S., only three andropogonoids are dominant: *Andropogon gerardi* (big bluestem), *Schizachyrium scoparium* (little bluestem) and *Sorghastrum nutans* (indiangrass). These three species probably insure that panicoid phytoliths would be the predominant types of regular phytoliths in the soils and sediment in this region.

The quantity and relative abundance of grass species in a grassland can change significantly over a few years. Tomanek and Hulett (1970) documented the changes in species, abundance and production of grass cover in central Kansas (Central Great Plains) during two extreme periods of drought in the 1930s and 1950s. During and immediately following periods of drought, the average rate of evaporation increased 16% during the growing season, and average maximum daily temperatures and wind velocities increased also. Grass cover dropped from more than 85 percent in 1932 to less than 20 percent in 1940. Moreover, in 1932, buffalo grass (*Buchloe dactyloides*) and blue grama (*Bouteloua gracilis*) each formed about 50 percent of the cover, but by 1940 blue grama became dominant. Three years after the drought, the original ratio of 50/50 was restored. Over 27 years, annual production of grass ranged from less than 900 pounds per acre during and immediately after a drought to more than 4,100 pounds per acre during periods of greater than average annual rainfall. Both grasses produced chloridoid phytoliths that were the dominant phytoliths in the sediment of individual dust storms near Hays, Kansas, during this period (Twiss, Suess & Smith 1969).

Although discrete opal phytoliths in sediment are not traceable to individual grass taxa, different ratios of regular phytoliths (pooid, chloridoid and panicoid phytoliths) may serve as climatic indices for interpreting the paleoclimate of a region. Where the three types occur, the ratio of C_3 to C_4 grasses can be approximated by comparing the number of pooid forms to the sum of pooid, chloridoid and panicoid phytoliths as compared to 100. High values suggest a cool climate such as occurs in the high latitudes, or in high altitudes where C_3

grasses are prevalent. Low values suggest a warm and humid to arid climate that would be concentrated at low latitudes and elevations.

Diester-Haass, Schrader and Thiede (1973) used an aridity index in studying the phytolith assemblages in two box cores recovered from the continental slope in 3,000 meters of water about 100 miles west of northern Africa in the North Atlantic. These cores penetrated sediment of Holocene and Pleistocene age with a maximum age of about 140,000 years BP. Along with opal phytoliths were fresh water diatoms and terrigenous silt and sand. They compared the number of chloridoid phytoliths to the total number of chloridoid and panicoid phytoliths based on a total count of 100. A high index, close to 100, indicated arid conditions in nearby Africa, whereas a low index suggested a humid climate in the adjacent land mass. The index expresses the range in abundance of the two subfamilies of C_4 grasses in the source area. An increase in chloridoid phytoliths occurred with an increase in terrigenous sand and content of $CaCO_3$ in the cores. Diester-Haass, Schrader and Thiede assumed that the terrigenous influx came from wind and torrential rivers from the nearby continent. Their interpretations matched the marine paleotemperatures obtained from foraminifera in the same cores.

The distribution of phytoliths throughout the world is dependent on the location of major grasslands, the climate of the region, the location and direction of the major wind belts, major streams and major ocean currents. Phytoliths, like pollen, spores, diatoms and other microfossils, can accumulate in situ, or they can be transported many hundreds of kilometers from their place of origin. Most studies have been directed towards correlating phytoliths in soils, paleosols and sediment with extant grasses, or describing phytoliths from dust deposits in deep sea cores. Regional studies that assess the proportion of phytoliths from local and remote sources are lacking. Only by integrating the global distribution of the grasses with climatic belts, wind belts, drainage patterns and ocean currents can we approach the goal of interpreting ancient environments from the analyses of grass opal phytoliths.

SUMMARY

Opal phytoliths from grasses do occur as discrete particles in soils, paleosols, sediment and sedimentary rock of the Cenozoic Era. Of the several forms, regular phytoliths (pooid, chloridoid and panicoid) derived from the leaf epidermis seem to offer the greatest potential for environmental interpretation.

Using ratios of regular phytoliths that are preserved in sediment as climatic indices, one may infer the photosynthetic pathways of the parent grasses. Pooid phytoliths are produced dominantly in C_3 grasses that flourish in cool seasons and climates in the high latitudes and elevations. Chloridoid and panicoid phytoliths are produced largely by C_4 grasses; chloridoid phytoliths occur in most grasses that are adapted to warm semiarid to arid regions, and panicoid forms occur dominantly in grasses that favor warm and humid habitats.

Ratios of pooid to total regular phytoliths emphasize the relative abundance of C_3 grasses in a region. Ratios of chloridoid to total chloridoid and panicoid phytoliths are measures of the aridity of an area.

Individual grass phytoliths are not effective for taxonomic identification, but when combined with preserved cuticles or leaf fragments some taxa may be recognized. Additional work is necessary in tabulating the distribution and abundance of phytoliths in common extant grasses of the steppes, prairies and savannas of the world. Grass phytoliths are widely distributed in Cenozoic sediments throughout the globe, but their full potential has yet to be realized.

REFERENCES

Barkworth, ME and Everett, J 1987 Evolution in the Stipeae: Identification and relationships of its monophyletic taxa. *In* Soderstrom, TR, Hilu, KW, Campbell, CS and Barkworth, ME, eds, *Grass Systematics and Evolution.* Washington, DC, Smithsonian: 251–264.

Bertoldi de Pomar, H 1971 Ensayo de clasificacion morfologica de los silicofitolitos. *Ameghiniana, Revista de la Asociacion Paleontologica Argentina* 8: 317–328.

Brown, DA 1984 Prospects and limits of a phytolith key for grasses in the central United States. *Journal of Archaeological Science* 11: 345–368.

Brown, WA 1977 The Kranz Syndrome and its subtypes in grass systematics. *In* Kiger, RW, ed, *Memoirs of the Torrey Botanical Club* 23: 1–97.

Conert, HJ 1987 Current concepts in the systematics of the Arundinoideae. *In* Soderstrom, TR, Hilu, KW, Campbell, CS and Barkworth, ME, eds, *Grass Systematics and Evolution.* Washington DC, Smithsonian Institution Press: 239–250.

Deflandre, G 1963 Les phytolithaires (Ehrenberg). *Protoplasma* 57: 234–259.

Diester-Haass, L, Schrader, HJ and Thiede, J 1973 Sedimentological and paleoclimatological investigations of two pelagic-ooze cores off Cape Barbas, Northwest Africa. *Meteor Forschung-Ergebnisse* 16: 19–66.

Ehrenberg, CG 1854 *Mikrogeologie. Das Erden und Felsen schaffende Wirken des unsichtbar kleinen selbststadigen Lebens auf der Erde.* Leipzig, Leopold Voss: 374 p.

Ellis, RP 1987 A review of comparative leaf blade anatomy in the systematics of the Poaceae: The past twenty-five years. *In* Soderstrom, TR, Hilu, KW, Campbell, CS and Barkworth, ME, eds, *Grass Systematics and Evolution.* Washington, DC, Smithsonian Institution Press: 3–10.

Gould, FW and Shaw, RB 1983 *Grass Systematics.* Second edition, College Station, Texas, A&M University Press: 397 p.

Hartley, W. 1958a Studies on the origin, evolution, and distribution of the Gramineae. I. The tribe Andropogoneae. *Australian Journal of Botany* 6: 116–128.

_____1958b Studies on the origin, evolution, and distribution of the Gramineae. II. The tribe Paniceae. *Australian Journal of Botany* 6: 343–357.

_____1964 The distribution of the grasses. *In* Barnard, C, ed, *Grasses and Grasslands.* London, Macmillan: 29–46.

_____1973 Studies on the origin, evolution, and distribution of the Gramineae. V. The subfamily Festucoideae. *Australian Journal of Botany* 21: 201–234.

Hartley, W and Slater, C 1960 Studies on the origin, evolution, and distribution of the Gramineae. III. The tribes of the subfamily Eragrostoideae. *Australian Journal of Botany* 8: 256–276.

Hattersley, PW 1987 Variations in photosynthetic pathway. *In* Soderstrom, TR, Hilu, KW, Campbell, CS and Barkworth, ME, eds, *Grass Systematics and Evolution.* Washington, DC, Smithsonian: 49–64.

Hattersley, PW and Watson, L 1975 Anatomical parameters for predicting photosynthetic pathways of grass leaves: The "maximum lateral cell count" and the "maximum cells distant count." *Phytomorphology* 25: 325–333.

Jacobs, SW 1987 Systematics of the Chloridoid grasses. *In* Soderstrom, TR, Hilu, KW, Campbell, CS and Barkworth, ME, ed, *Grass Systematics and Evolution.* Washington, DC, Smithsonian Institution Press: 277–286.

Livingstone, DA and Clayton, WD 1980 An altitudinal cline in tropical grass floras and its paleoecological significance. *Quaternary Research* 13: 392–402.

Locker, S and Martini, E 1986 Phytoliths from the southwest Pacific, site 591. *In* Blakeslee, JH, ed, *Initial Reports of the Deep Sea Drilling Project* 90(2): 1079–1084. Washington, DC, US Government Printing Office.

Macfarlane, TD 1987 Poaceae subfamily Pooideae. *In* Soderstrom, TR, Hilu, KW, Campbell, CS and Barkworth, ME, eds, *Grass Systematics and Evolution.* Washington, DC, Smithsonian Institution Press: 265–276.

Metcalfe, CR 1960 *Anatomy of the Monocotyledons. I. Gramineae.* Oxford, Clarendon Press: 731 p.

Mulholland, SC and Rapp, G, Jr 1989 Characterization of grass phytoliths for archaeological analysis. *Materials Research Society Bulletin* 14(3): 36–39.

Mulholland, SC, Rapp, G, Jr and Ollendorf, AL 1988 Variation in phytoliths from corn leaves. *Canadian Journal of Botany* 66: 2001–2008.

Ollendorf, AL, Mulholland, SC and Rapp, G, Jr 1988 Phytolith analysis as a measure of plant identification: *Arundo donax* and *Phragmites communis. Annals of Botany* 61: 209–214.

Palmer, PG 1976 Grass cuticles: A new paleoecological tool for East African lake sediments. *Canadian Journal of Botany* 54: 1725–1734.

Piperno, DR 1988 *Phytolith Analysis: An Archaeological and Geological Perspective.* San Diego, California, Academic Press: 280 p.

Rovner, I 1971 Potential of opla phytoliths for use in paleoecological reconstruction. *Quaternary Research* 1: 343–359.

Sase, T and Kondo, R 1974 The study of opal phytoliths in the humus horizon of buried volcanic ash soils in Hokkaido. *Research Bulletin of Obihiro University* 8: 465–483.

Smith, BN and Brown, WV 1973 The Kranz Syndrome in the Gramineae as indicated by carbon isotope ratios. *American Journal of Botany* 60: 505–513.

Smith, RM and Twiss, PC 1965. Extensive gaging of dust deposition rates. *Transactions of the Kansas Academy of Science* 68: 311–321.

Smithson, F 1958 Grass opal in British soils. *Journal of Soil Science* 9: 148–154.

Soderstrom, TR, Hilu, KW, Campbell, CS and Barkworth, ME, eds 1987 *Grass Systematics and Evolution.* Washington, DC, Smithsonian Institution Press: 473 p.

Soderstrom, TR and Ellis, RP 1987 The position of bamboo genera and allies in a system of grass classification. *In* Soderstrom, TR, Hilu, KW, Campbell, CS and Barkworth, ME, eds, *Grass Systematics and Evolution.* Washington, DC, Smithsonian Institution Press: 225–238.

Teeri, JA and Stowe, LG 1976 Climatic patterns and the distribution of C_4 grasses in North America. *Oecologia* 23: 1–12.

Thomasson, JR 1984 Miocene grass (Gramineae: Arundinoideae) leaves showing external micromorphological and internal anatomical features. *Botanical Gazette* 145: 204–209.

Tieszen, LL, Senyimba, MM, Imbamba, SK and Troughton, JH 1979 The distribution of C_3 and C_4 grasses and carbon isotope discrimination along an altitudinal and moisture gradient in Kenya. *Oecologia* 37: 337–350.

Tomanek, GW and Hulett, GK 1970 Effects of historical droughts on grassland vegetation in the Central Great Plains. *In* Dort, W, Jr and Jones, JK, eds, *Pleistocene and Recent Environments of the Central Great Plains.* Lawrence, Kansas, University of Kansas Press: 203–210.

Twiss, PC 1980 Opal phytoliths as indicators of C_3 and C_4 grasses. *Abstracts with Programs, Geological Society of America* 12: 17.

Twiss, PC, Suess, E and Smith, RM 1969 Morphological classification of grass phytoliths. *Soil Science of America, Proceedings* 33: 109–115.

Watson, L 1987 Automated descriptions of grass genera. *In* Soderstrom, TR, Hilu, KW, Campbell, CS and Barkworth, ME, eds, *Grass Systematics and Evolution.* Washington, DC, Smithsonian Institution Press: 343–351.

Watson, L, Clifford, HT and Dallwitz, MJ 1985 The classification of Poaceae: Subfamilies and supertribes. *Australian Journal of Botany* 33: 433–484.

Zuloaga, FO 1987 Systematics of New World species of *Panicum* (Poaceae: Paniceae). *In* Soderstrom, TR, Hilu, KW, Campbell, CS and Barkworth, ME, eds, *Grass Systematics and Evolution.* Washington, DC, Smithsonian Institution Press: 287–306.

7

PRELIMINARY IDENTIFICATION OF SILICA SKELETONS FROM NEAR EASTERN ARCHAEOLOGICAL SITES: AN ANATOMICAL APPROACH

ARLENE MILLER ROSEN

ABSTRACT

Silica skeletons (fossilized sections of epidermal tissue) are frequent components in Near Eastern archaeological sediments. Culm (straw) and husks (grain bracts) produce different cell patterns that allow determination of plant part; this in turn provides information about use of activity areas. Identification of taxa is still preliminary; data is presented for wheat, barley, date palm, and three genera of weedy grasses. The distinctive anatomical characteristics of wheat, barley and date palm have not yet been compared to all possible wild confusers. However, within Levantine archaeological sites, strong suggestions of these cultigens can be proposed based on phytolith evidence.

INTRODUCTION

Phytoliths are abundant in archaeological sediments in the Near East. In Israel, they are known from sites dating from the Mousterian period (Rosen, personal observation) through the Iron Age (Folk & Hoops 1982; Ollendorf 1987; Rosen 1987, 1989, in preparation; Thomson & Rapp 1989). These silica bodies usually originate in monocotyledons (grasses, sedges, rushes, palms), and often consist of complete sections of silicified epidermal tissue in the form of contiguous cells. Composite phytoliths such as these have been previously referred to as silica skeletons (Helbaek 1960: 540; Renfrew 1973: 9).

The formation of silica skeletons may be related to the growing environment. Increased silica deposition is induced by greater evapo-transpiration in hot arid environments (Renfrew 1973: 17; Hutton & Norrish 1974). It has been suggested that this process is enhanced by irrigation farming or in microenvironments with poor drainage that serves to add soluble silica from excess soil water (Rosen 1991).

Phytolith Systematics, Edited by G. Rapp, Jr. and S.C. Mulholland, Plenum Press, New York, 1992

Because silica skeletons are fossilized sections of epidermal tissue, consisting of multiple types of cells, it is not practical to use generally accepted methods of phytolith-shape terminology developed for individual shapes (see Piperno 1988: 52–58). It is much more informative for identification purposes to relate these silica skeletons to their anatomical counterparts in the plant epidermis. For example, instead of referring to rectangular forms as trapezoids, it is more functionally useful and appropriate to call them long-cells. Likewise, cone-shaped phytoliths are called papillae when viewed in their anatomical context.

Theoretically, it should be possible to identify taxa of silica skeletons by comparison with the anatomical features of living plant epidermis. Also, silica skeletons provide a method to differentiate between specific plant parts that contributed the phytoliths, thus providing a potential aid in archaeological studies such as room functions and activity areas.

In the first part of this paper I will illustrate the differences in anatomical features between cereal straw (culms) and husks (inflorescence bracts), because these are among the most abundant phytolith remains in Near Eastern agricultural sites. In the second section I will address the more complicated problem of identifying cereal phytoliths to genus and date palm phytoliths to species.

METHODS

Reference material from living plants was prepared according to a technique modified after Kaplan and Smith (1980). The fresh or dried plant parts were soaked in a solution of 50% household bleach and 50% distilled water for one or two days. The plant material was then placed on a microscope slide and repeatedly washed with pure ethyl alcohol until all the water in the plant cells was replaced with alcohol. Finally the sample was washed in xylene until transparent, and then mounted on the slide with "Permount" in preparation for viewing under an optical microscope.

IDENTIFICATION OF PLANT PARTS

The anatomical features of the culm and husks of cereals are quite distinctive and allow the phytolith researcher to determine which plant parts were present in a given sediment. Room function and activity area analyses can profit from this type of information because, for example, silica skeletons from cereal husks may indicate a storage room or flour processing station, whereas phytoliths from straw may be derived from bedding or animal pens.

The epidermis of the culm is composed of long-cells, short-cells and stomata aligned in longitudinal strips. Although this paper will not address the taxonomic significance of the epidermal cells in the culm, Metcalfe (1960: xxxii) suggests that there are a number of distinctive features. In wheat and barley, the long-cells of the culm are generally narrow (ca 5–10 microns in width), with side walls that are smooth or slightly sinuous (Fig 7.1). The short cells, often silicified along with their related silica bodies, tend to be cubical or cuboid with rounded corners, and are sometimes elliptical. Papillae (small nipple-like projections) are absent. Phytoliths from the culm are generally colorless or contain remnants of brownish organic matter (Fig 7.2).

In this paper, cereal husks include glumes (the basal bracts on a grass spikelet), the lemma (outer bract enclosing the flower or grain) and the palea (inner bract enclosing the flower or grain). These are grouped together because of the similarity of the cell morphology in the epidermal tissue, as well as the likelihood of their co-occurrence in an archaeological context.

The epidermal anatomy of husks is quite different from that of culms, as well as differing from leaves and leaf sheaths (not treated in this paper). Characteristic features include comparatively wide long-cells (ca 15–20 microns on the medial portion of the husk), pronounced sinuosity of the long-cell walls, often very thick long-cell walls and numerous papillae with elaborate ornamentation including pitting and minute surface projections (see Figs 7.3 and 7.4). The silica skeletons are usually clear or bluish in color, and only occasionally brownish from organic inclusions.

IDENTIFICATION OF TAXA

The taxonomic identification of silica skeletons is still at a preliminary stage. Although botanical literature describing the epidermal tissue of grasses with respect to taxonomic identity exists (eg, Lloyd 1921; Percival 1921; Wilkens 1928; Prat 1932, 1933; Metcalfe 1960), only a small proportion of this deals with the husks of the plants, which are the most diagnostic and yield the most silica skeletons in archaeological sites. Even those studies that describe the husk do so in very general terms.

One of the earliest researchers to employ this technique for the identification of plants in archaeological material was Schellenberg (1908). He examined plant impressions in pottery from the Neolithic site of Anau in southwestern Turkistan. His initial identifications of wheat and barley were based on clay casts of the husks. He then described the silica skeletons whose remains appeared in the casts and compared them to modern cereal epidermal tissue with the intent of distinguishing wheat from barley husks. However, judging from his illustrations (Schellenberg 1908: Plate 97), he mistook the papillae in the wheat husk for short cells. Then he compared the different forms of the wheat "short cells" (really papillae) with the true barley short cells. The differences that he described were

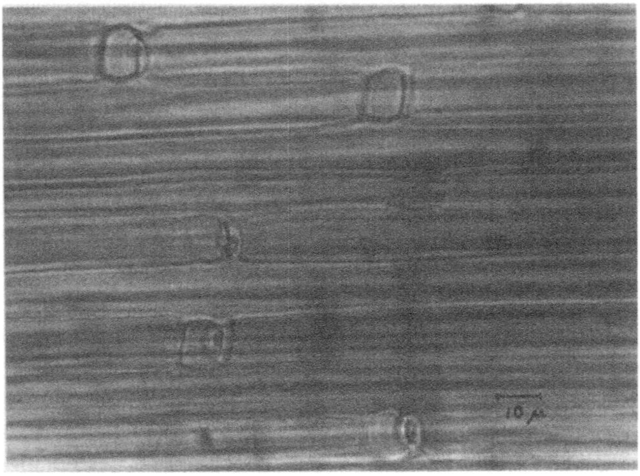

Fig 7.1. Culm epidermis from *Hordeum vulgare*. Bar is 10 microns.

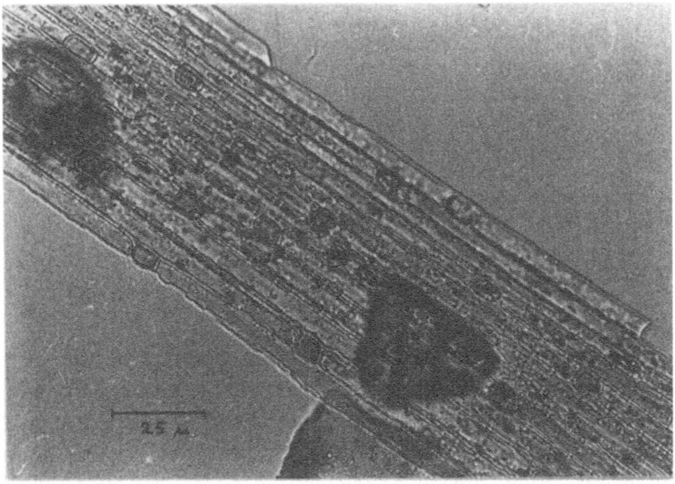

Fig 7.2. Chalcolithic silica skeleton from cereal culm (Shiqmim, Israel). Bar is 25 microns.

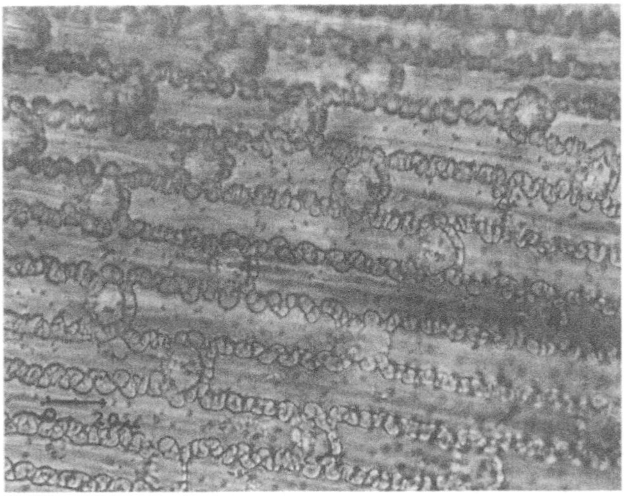

Fig 7.3. Epidermis of the lower portion of domestic emmer (*Triticum dicoccum*) husk showing a thick low wave on long-cell walls. Bar is 20 microns.

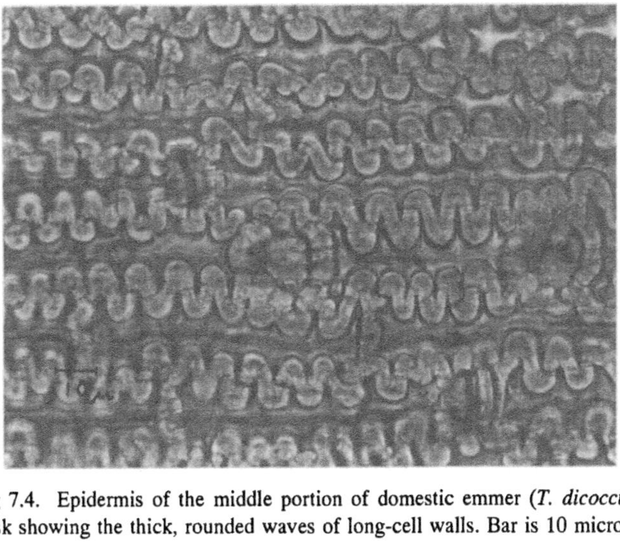

Fig 7.4. Epidermis of the middle portion of domestic emmer (*T. dicoccum*) husk showing the thick, rounded waves of long-cell walls. Bar is 10 microns.

therefore a function of two distinct cell types rather than features indicative of genus.

Helbaek (1960) also examined plant impressions and silica skeletons from Neolithic levels from the 'Amuq sites in Syria. Like Schellenberg, his initial identifications were based on the casts of cereal spikelets in pottery sherds. He then described and photographed the characteristic features of the long-cell walls in wheat and oat grass, and noted the density of papillae (called isodiametric cells) in rye grass. Although he reported the dimensions of the epidermal cells, he cautioned that a certain amount of cell shrinkage should be taken into account in the analysis of the silica skeletons. In more recent studies, Rosen (1987, 1989) analyzed wheat and barley silica skeletons from Neolithic sites in the Galilee and Chalcolithic sites in the Negev desert, Israel, with the aid of the distinguishing criteria outlined below.

The plants selected for the present study are among the most likely to appear as cultigens in Near Eastern archaeological sites. They include wheat (*Triticum* sp), barley (*Hordeum* sp) and date palm (*Phoenix dactylifera*). Selected weed grasses, including goat grass (*Aegilops* sp), oat grass (*Avena* sp) and rye grass (*Lolium* sp), will be discussed for comparative purposes.

The problems of silica skeleton identification are similar to those of single component phytoliths, although to a somewhat lesser degree because of the variety of cell shapes in any one silica skeleton. Many plant species, usually within the same family or subfamily, produce cells that are similar. Some, however, seem to be distinctive to genus, especially when several characteristics are considered together. Furthermore, diagnostic features of husks, such as long-cell widths, degree of sinuosity and density of papillae, sometimes vary depending upon the location on the husk (upper, middle or lower), as well as the side of the husk (inner or outer). Therefore, other researchers are advised to prepare their own reference samples before beginning the analyses of archaeological material. Since this is a preliminary study, anatomical features from domesticated cereals were compared to only a handful of wild grasses. Future studies should be directed towards the examination of more varieties of wild plants.

Characteristics of Epidermal Cells from Cereal Husks

In grass husks, one of the most diagnostic features is the shape of the waves of long-cell walls on the outer epidermis. Although these walls are usually much thicker than the long-cell walls in other parts of the plant, they are sometimes thin, and the thin version can differ in wave shape from the thick. Therefore, both shapes will be described here. Other distinctive criteria include the width of the long-cells and the density and ornamentation of papillae. When all of these factors are considered, it should be possible to determine the identity of the grass in question. In the following sections, descriptions refer to the abaxial (outer) epidermis, unless stated otherwise, and the lower and middle portions of the husks. The uppermost section of the husk, near the awn, is not as diagnostic and therefore will be omitted from the descriptions.

Wheat Husks. The examples described here come primarily from cultivated emmer wheat (*Triticum dicoccum*) with occasional references to differences in einkorn (*T. monococcum*), bread wheat (*T. aestivum*) and wild emmer (*T. dicoccoides*). In the lowermost portion of the domestic emmer husks, the long-cells are ca 18–23 microns in width (Helbaek 1960: 542). In describing bread wheat, Percival (1921: 107) notes that the long-cells of the glume are 80–100 microns long and ca 25 microns wide. Both thin and thick long-cell walls have very low, blunted waves (Fig 7.3) that are often squarish in the thin version. The wave heights are ca 4 microns in the thin waves, and ca 5–8 microns in the thick. According to Percival (1921: 107), the thin long-cell walls occur most often in the inner epidermis of wheat glumes. Hairs and papillae are found on both the inner and outer epidermis, but generally hairs are more common on the inner, and papillae are more numerous on the outer face (Wilkens 1928: 315).

In the central portion of the husks, both the thin and thick long-cell walls have high rounded waves of irregular amplitude (Fig 7.4). Helbaek (1960: 542) describes these waves as "soft winding sinuate walls." The height of the wave is ca 10 microns in the thin version and ca 15.5 microns in the thick. The waves in the long-cell walls of domestic emmer appear to be somewhat more regular in amplitude than the very erratic waves found in einkorn and bread wheat.

Papillae in domestic emmer range from 22–30 microns in diameter and are distributed evenly over the lower and middle portions of the husks. Wild emmer (*T. dicoccoides*) tends to have larger papillae (21–43 microns), although there is considerable overlap in size. Papillae in einkorn can obtain an even larger size, ranging from 25–50 microns in diameter or length in the case of oblong-shaped papillae.

The ornamentation of papillae may be distinctive, but much more research on this aspect is needed. Hodson and Sangster (1988) used scanning electron microscopy to examine the inflorescence bracts of bread wheat (*Triticum aestivum*). They observed that there were consistently 10 to 12 pits around the margins of each papilla. This may be contrasted to the observation by Hayward and Parry (1980) that in one species of barley (*Hordeum sativum*) there were only 7 to 9 surrounding depressions. In this study, the depressions around papillae in domestic emmer wheat corresponded to Hodson's and Sangster's description, but the two-row barley, *Hordeum distichon*, also had papillae with ca 10 to 12 marginal pits. A more significant contrast, however, is the difference in number of papillae pits between domestic and wild emmer, where the wild emmer exhibits more pits (16–18) than the domestic species. This is the only clear difference between the two species as yet identified; however, it could have great potential significance for future studies of early wheat cultivation in the Levant. Archaeologically derived silica skeletons with features similar to those found in the husk of domestic emmer are illustrated in Figure 7.5A and B.

Barley Husks. Reference samples for the description of barley husk epidermis came from both two-row barley (*Hordeum distichon*) and six-row barley (*H. vulgare*). The long-cells of the two-row barley husks are narrower than those in the lower and middle portions of the wheat husk (ca 15–18 microns). The same measurement in six-row barley is ca 12–15 microns. In the lower and middle parts of the husks, the thin cell walls are serrated and end in sharp or knobbed points (Fig 7.6) with a wave height of ca 7 microns. The thick cell walls are often squarish and usually occur in waves of even amplitude (Fig 7.7). The height of these waves is ca 10 microns.

Papillae range in size from ca 18–25 microns. In two-row barley they are more numerous on the lower husk than on the middle portion. The difference in the numbers of marginal pits has already been discussed above. Figure 7.8A and B illustrates archaeological silica skeletons resembling barley husks.

Characteristics of Epidermal Cells from the Husks of Weed Grasses

Goat Grass (*Aegilops* sp). *Aegilops* is a wild grass that is closely related to wheat. In spite of this association, the waves of the long-cell walls are most similar to those of barley, with squared thick waves of even amplitude (Fig 7.9), and thin waves that are also typically squared at the apex. The thin waves differ from the pattern found in the thin waves of barley long-cells, although the pointed serrations also occasionally occur in *Aegilops*. The width of the long cells in *Aegilops searsii* is likewise similar to that of two-row barley, 15–18 microns, although the long-cell width in *Ae. bicornis* is only ca 12 microns. The main difference between *Aegilops* and *Hordeum* seems to be the number of marginal pits around papillae. *Aegilops* has from 16–18 pits, far more than emmer wheat or the species of barley studied.

Oat Grass (*Avena* sp). The wild oat is a common field weed, and is a plant whose phytoliths might turn up in archaeological samples. The long-cell waves of the palea of oat are quite distinctive and have been previously described by Helbaek (1960: 542) from Neolithic contexts at the 'Amuq sites in northwestern Syria. The thick waves are rod-like and straight, with thick rounded knobs at the crests. The papillae vary in size, ranging from small and rounded (ca 15 microns in diameter) to large and elliptical (ca 40 × 50 microns). They are highly ornamented, with lobed borders that extend onto the papilla itself, and numerous hair-like projections on their surfaces (Fig 7.10). The number of marginal pits ranges from 18–20.

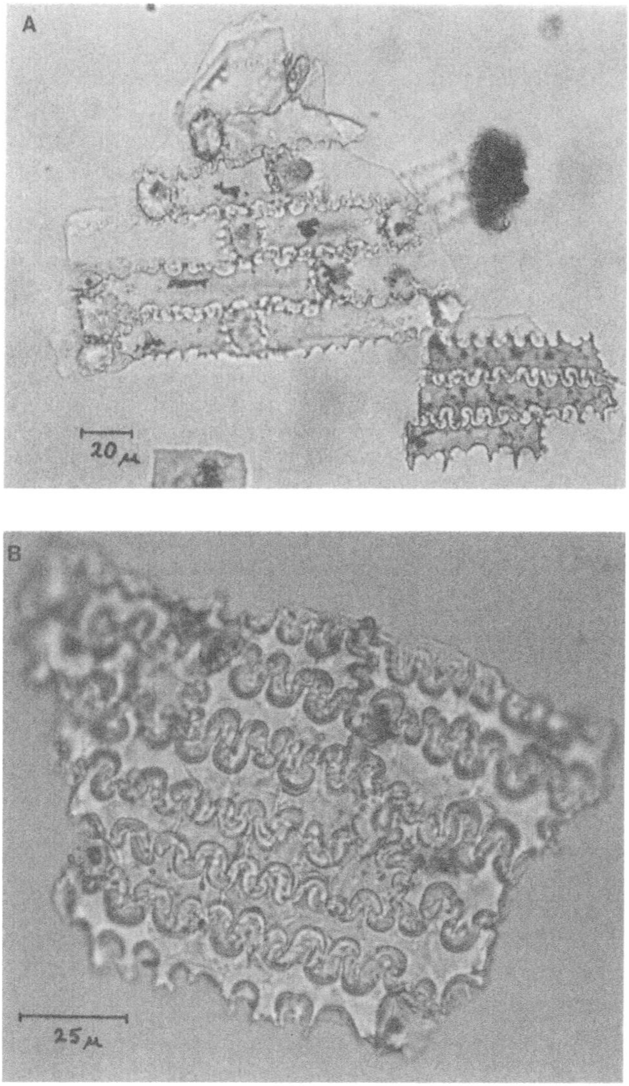

Fig 7.5. Silica skeleton of emmer-like husks. A. Lower husk from Pottery Neolithic level at Tel Teo, Galilee. Bar is 20 microns. B. Middle husk from Chalcolithic Shiqmim. Bar is 25 microns.

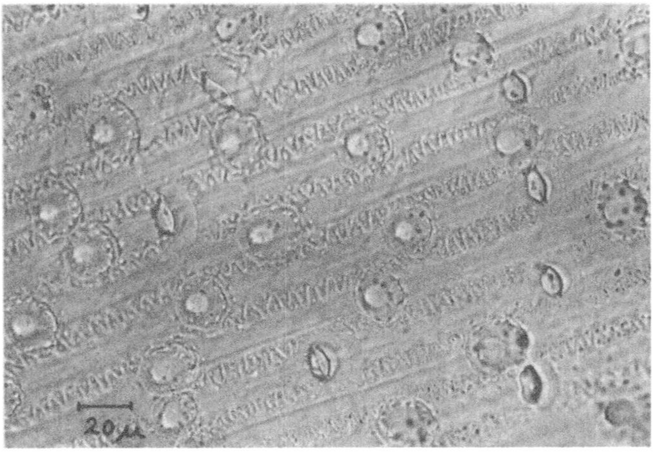

Fig 7.6. Epidermis of the middle portion of a barley husk (*H. vulgare*) with thin, pointed long-cell waves. Bar is 20 microns.

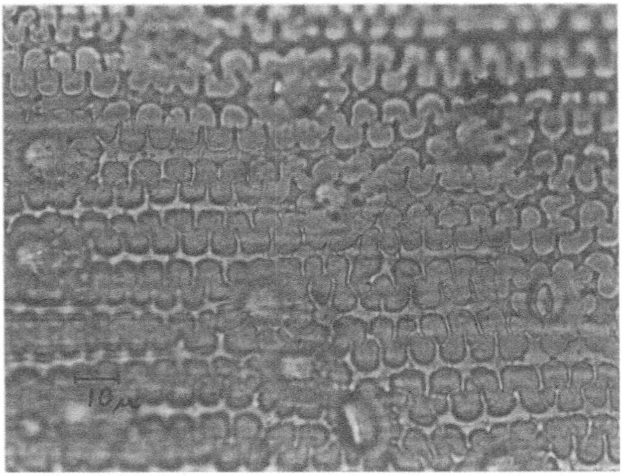

Fig 7.7. Epidermis of the middle section of a barley husk (*H. distichon*) showing thick, squarish long-cell waves. Bar is 10 microns.

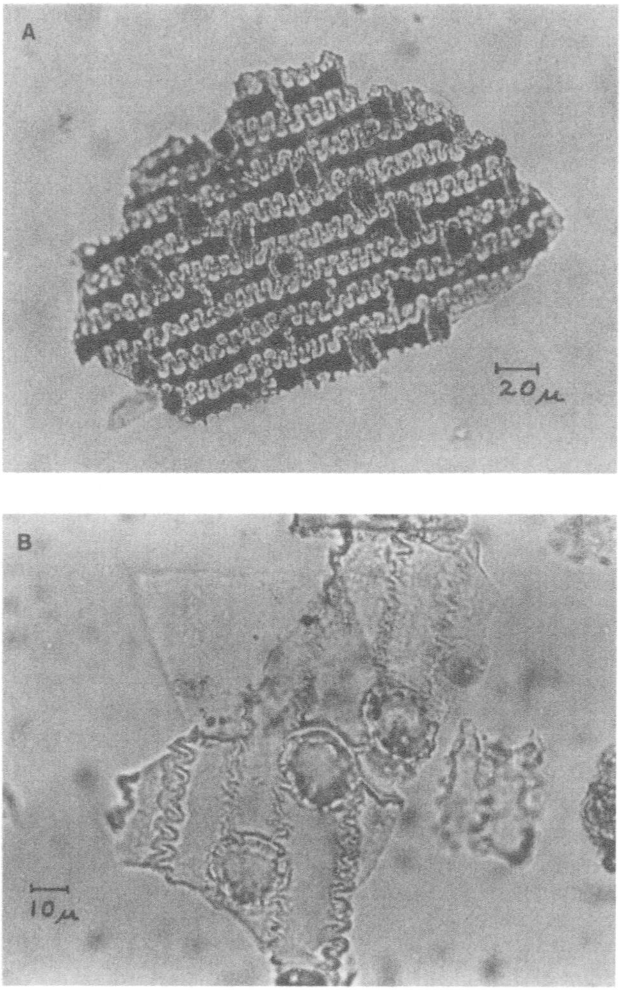

Fig 7.8. Iron Age silica skeletons of barley-like husks (Kadesh Barnea, Sinai). A. With thick long-cell walls. Bar is 20 microns. B. With thin long-cell walls. Bar is 10 microns.

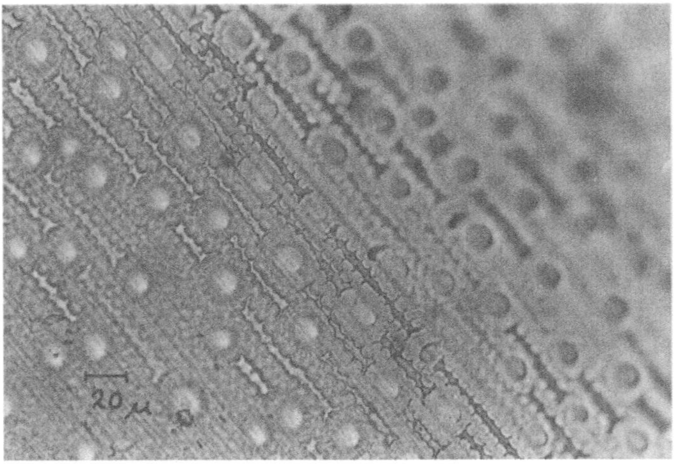

Fig 7.9. Husk epidermis of goat grass (*Aegilops* sp) with long-cell waves resembling those of barley, but larger papillae with more numerous marginal pits. Bar is 20 microns.

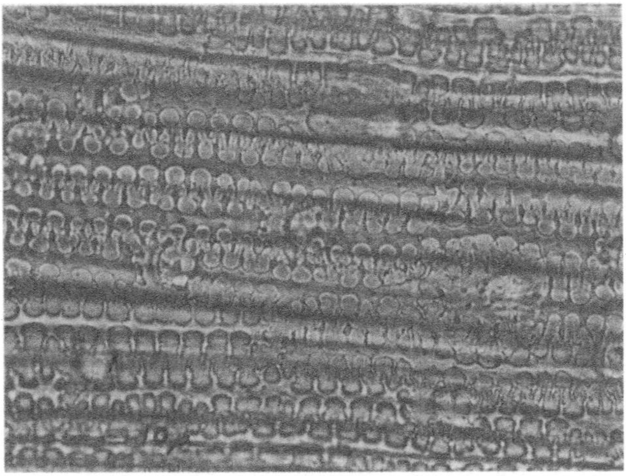

Fig 7.10. Husk epidermis of oat grass (*Avena* sp) showing irregular papillae and lobed long-cell waves. Bar is 10 microns.

Rye Grass (*Lolium* sp). The walls of the long-cells in the husks have waves that are more regular and rounded at the crest than those in wheat (Fig 7.11). The long-cell width is 15 microns in some species and 22–25 microns in others. Papillae are very numerous in the lower portion of the husks and display from 18–20 marginal pits.

The wave patterns of long-cell walls in the husks of wheat, barley, goat grass, oat and rye grass are illustrated schematically in Figure 7.12. Table 7.1 presents a summary of measurements from the grass husks treated in this paper.

Characteristics of Date Palm (*Phoenix dactylifera*) Leaf Epidermis

In the leaf of the date palm, the epidermal cells over the veins are rectangular, vary in length and have slightly sinuous walls. The intercostal cells are less regular, shorter and have smooth walls. Stomata are only found in the intercostal regions and do not appear in regular rows. Silica bodies overlay both costal and intercostal cells. These silica bodies are spherical in shape, aligned in short rows and are usually 5–6 microns (occasionally up to ca 12 microns) in diameter. Although Tomlinson (1961: 308) simply describes these bodies as spherical, they are also spiked around the circumference (Fig 7.13). Their coloring is purplish in both the fresh and archaeological material. A possible Iron Age date palm phytolith can be observed in Figure 7.14.

The phytoliths from date palm differ in size and surface morphology from the silica bodies in the leaves of doum palm (*Hyphaene thebaica*). Those of the latter are generally larger, and fall into two size groups. The "small" group averages about 9 microns in diameter, and the "large" averages about 32 microns. Their surface morphology is markedly more spinulose than the phytoliths from date palm leaves.

In archaeological samples from the Negev and Sinai desert sites of Ein Zik, Hatzeva and Kadesh Barnea, the spherical silica bodies of the date palm are much more abundant than the silica skeletons. They are quite distinctive in size, shape and coloring.

It should be stressed here that the identification of these two types of palms can only apply in the Levant. There are many palm species worldwide, and some undoubtedly produce phytoliths similar in size and shape to those of the date and doum palms. The date tree is the only palm that occurs naturally in Israel, except in the northernmost portion of the Gulf of Aqaba where the doum palm (*Hyphaene thebaica*) grows (Danin 1983: 50). The date palm was also the only palm cultivated in the area in antiquity.

One must also address the question of possible silica bodies that resemble these palm types, but in fact come from completely unrelated plants. Although this is a problem that should always be kept in mind when making an identifica-

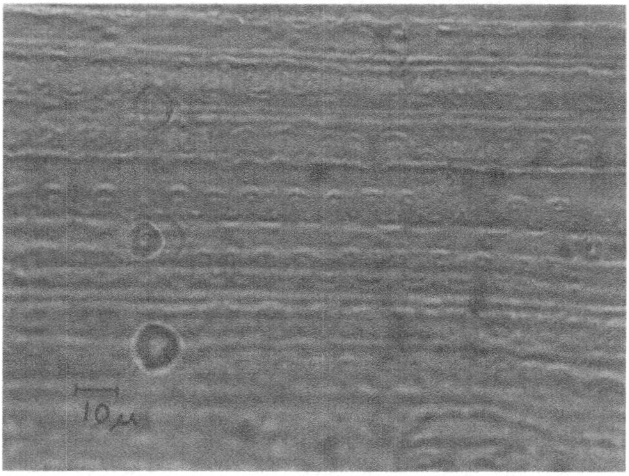

Fig 7.11. Epidermis of husk from rye grass (*Lolium* sp) showing thin, rounded long-cell waves with regular amplitude. Bar is 10 microns.

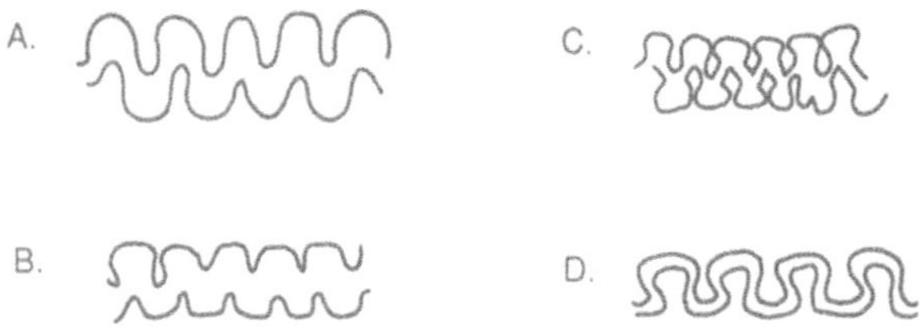

Fig 7.12. Schematic drawings of wave patterns from the long cell walls of cereals and weed grass husks. A. Wheat; B. Barley and goat grass; C. Oat grass; D. Rye grass.

TABLE 7.1. Summary Table of Measurements from the Husk Epidermis of Wheat, Barley, Goat Grass, Oat Grass and Rye Grass

	L-C wall wave height				
	L-C Width	Lower Husk	Middle Husk	Papillae Size	No. of pits
Wheat (*Triticum*)					
Emmer	18–23*	4 thin	10 thin	22–30	10–12
(*T. dicoccum*)		5–8 thick	15 thick		
Wild Emmer	18–23	" "	" "	21–43	16–18
(*T. dicoccoides*)					
Einkorn	18–23	" "	" "	25–50	12–14 (?)
(*T. monococcum*)					
Barley (*Hordeum*)					
H. distichon	15–18	7 thin	7 thin	18–25	10–12
(two-row)		10 thick	10 thick		
H. vulgare	12–15	" "	" "	18–25	7–9
(six-row)					
Goat Grass (*Aegilops*)					
Ae. searsii	15–18		8.5 thin	25–27	16–18
			15 thick		
Ae. bicornis	18		10 thick	25–32	16–18
Oat Grass (*Avena sp*)	17		8.5 thin 16 thick	15, 50	18–20
Rye Grass (*Lolium sp*)	15–25		7 thin 8.5 thick	22	16–18

* Measurements given in microns.

tion, in the case of the Levantine material the probability is slight. This researcher has examined numerous archaeological sediments for their phytolith content, ranging in time from the late Mousterian (ca 55,000 BP) through the Byzantine period, and in geographic locations from the northernmost tip of the Galilee to southern Sinai. These palm-like phytoliths were found only around the natural desert oasis habitat of the wild palm trees. Therefore, within this region, the above criteria for the identification of palm phytoliths should be enough to suggest strongly the presence of date or doum palm.

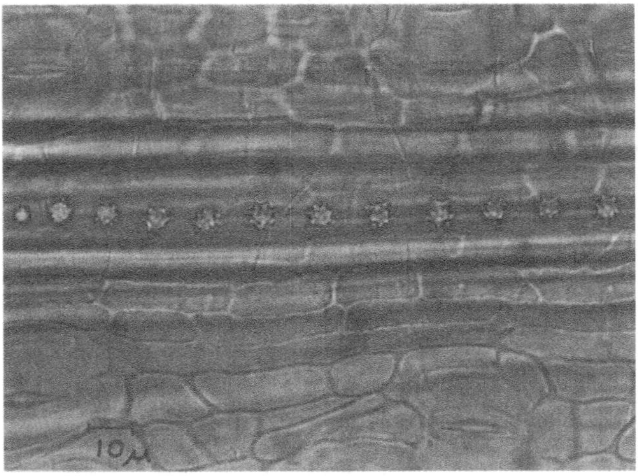

Fig 7.13. Epidermis of date palm (*Phoenix dactylifera*) leaf with spherical, spiked silica bodies. Bar is 10 microns.

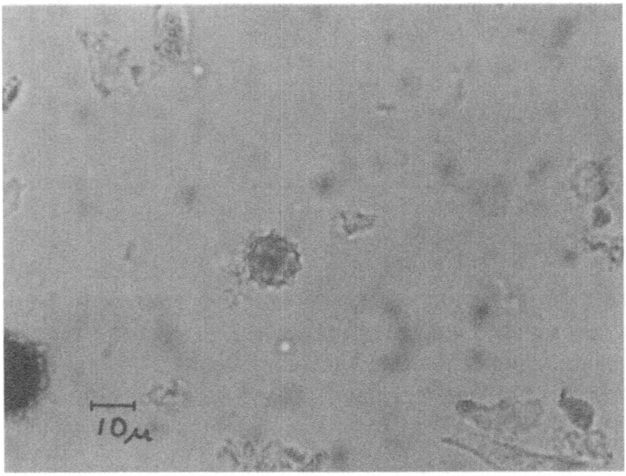

Fig 7.14. Probable date palm phytolith (center) from Iron Age Hatzeva (Israel). Bar is 10 microns.

SUMMARY AND CONCLUSIONS

In this paper I have pointed out distinctive anatomical characteristics of silica skeletons from three Near Eastern domesticates: wheat, barley and date palm. The cereals were compared to several wild grass species. Although phytolith studies at Near Eastern archaeological sites are much in need of basic information on anatomical features of wild plants, the analysis of archaeological materials can and should continue together with the examination of modern wild plants.

Even though the diagnostic features of wheat and barley husks do not necessarily exclude some as yet unexamined grass genus, it is still possible to use this information in studies of archaeological sediments where wheat and barley cultigens are expected to dominate. Analyses of carbonized material from Near Eastern agricultural sites show that the floral portions of wheat and barley far outnumber the floral parts of other grasses. Therefore, in agricultural sites, the number of phytoliths from wheat and barley husks should be much greater than those of wild plants. It is useful, at this early stage of analytical development, to compare ratios obtained through phytolith analysis with those resulting from the classification of macrobotanical remains at the same site. At the Chalcolithic sites of Shiqmim and Horvat Beter in the Negev desert, there was general agreement between the relative frequencies of wheat and barley. Although at Shiqmim the relative percentages of phytoliths did not exactly match those determined from analysis of macrobotanical remains (Kislev 1987; Rosen 1987), both data sets showed significantly more barley than wheat. At Horvat Beter, the phytoliths from wheat were more prevalent than those of barley, as was indicated previously by Zeitchek (1959) in his analysis of the macrobotanical remains.

At sites where the use of cereals is only suspected, the presence of cereal phytoliths can form the basis of working hypotheses to be tested by corroborative evidence. Apart from taxonomic considerations, the use of silica skeletons to identify plant parts, such as husks vs stems, is in itself applicable to a number of archaeological research problems.

Therefore, even at this preliminary stage of phytolith research in the Near East, it is possible to make significant contributions to archaeological problems. This paper is intended as one stone in the foundation of future phytolith research in the Near East and hopefully will encourage other phytolithariens to initiate new projects in this relatively untapped source of paleobotanical knowledge.

ACKNOWLEDGMENTS

I wish to thank Moshe Feldman and Ilana Strauss (Department of Plant Genetics, Weizmann Institute of Science) who kindly provided samples of *Triticum* and *Aegilops*, and identified some of the wild grasses. Paul Goldberg

(Institute of Archaeology, Hebrew University) generously allowed me free access to the facilities of the Archaeometry Laboratory. Steven Rosen (Ben Gurion University), Irwin Rovner (North Carolina State University) and two anonymous reviewers made helpful comments on an earlier version of this paper. This work was completed while the author was a Sir Charles Clore Post-Doctoral Fellow at the Weizmann Institute of Science, Rehovot, Israel.

REFERENCES

Danin, A 1983 *Desert Vegetation of Israel and Sinai*. Jerusalem, Cana: 148 p.

Folk, RL and Hoops, GK 1982 An early Iron-Age layer of glass made from plants at Tel Yin'am, Israel. *Journal of Field Archaeology* 9: 455–466.

Hayward, DM and Parry, DW 1980 Scanning electron microscopy of silica deposits in the culms, floral bracts and awns of barley (*Hordeum sativum* Jess.). *Annals of Botany* 46: 541–548.

Helbaek, H 1960 Cereals and weed grasses in Phase A. *In* Braidwood, RJ and Braidwood, LS, eds, *Excavations in the Plain of Antioch I*. Chicago, University of Chicago Press: 540–543.

Hodson, MJ and Sangster, AG 1988 Silica deposition in the inflorescence bracts of wheat (*Triticum aestivum*). I. Scanning electron microscopy and light microscopy. *Canadian Journal of Botany* 66: 829–838.

Hutton, JT and Norrish, K 1974 Silicon content of wheat husks in relation to water transpired. *Australian Journal of Agricultural Research* 25: 203–212.

Kaplan, L and Smith, MB (ms) 1980 Procedures for phytolith reference materials.

Kislev, M 1987 Chalcolithic plant husbandry and ancient vegetation at Shiqmim. *In* Levy, TE, ed, Shiqmim I: Studies concerning Chalcolithic societies in the Northern Negev Desert, Israel (1982–1984), *British Archaeological Reports, International Series* 356: 251–279.

Lloyd, FE 1921 The structure of cereal straw. *Pulp and Paper Magazine* 19: 953–954, 973–976, 1002–1004, 1025–1026, 1048–1050, and 1071–1075.

Metcalfe, CR 1960 *Anatomy of the Monocotyledons. I. Gramineae*. Oxford, Clarendon: 731 p.

Ollendorf, A 1987 Archaeological implications of a phytolith study at Tel Miqne (Ekron), Israel. *Journal of Field Archaeology* 14: 453–463.

Percival, J 1921 *The Wheat Plant*. London, Duckworth: 463 p.

Piperno, DR 1988 *Phytolith Analysis: An Archaeological and Geological Perspective*. San Diego, Academic Press: 280 p.

Prat, H 1932 L'épiderme des Graminées: Etude anatomique et systématique. *Annales des Sciences Naturelles Botanique et Biologie Végétale Séries* 10, 14: 117–324.

_____1933 Les affinités des blés (*Triticum*) et des Chiendents (*Agropyrum*) d'après leurs caractères épidermiques. *Transactions of the Royal Society of Canada, Sec* 5, 28: 103–105.

Renfrew, JM 1973 *Palaeoethnobotany*. New York, Columbia University: 248 p.

Rosen, AM 1987 Phytolith studies at Shiqmim. *In* Levy, TE, ed, Shiqmim I: Studies concerning Chalcolithic societies in the Northern Negev Desert, Israel (1982–1984), *British Archaeological Reports, International Series* 356: 243–249.

_____1989 Microbotanical evidence for cereals in Neolithic levels at Tel Teo and Yiftahel in the Galilee, Israel. *Mitekufat Haeven (Journal of the Israel Prehistoric Society)* 22: 68–77.

_____1991 Phytoliths as indicators of ancient irrigation farming. *In* Anderson-Gerfaud, P, ed, *Préhistoire de l'Agricul-*

ture: Nouvelles Approches Expérimentales et Ethnographiques. Paris, CNRS: 1–7.

____(ms) Phytolith evidence for early cereal exploitation in the Levant. *In* Pearsall, DM and Piperno, DR, eds, *Current Research in Phytolith Analysis: Applications in Archaeology and Paleoecology*. In preparation.

Schellenberg, HC 1908 The remains of plants from the North Kurgan, Anau. *In* Pumpelly, R, ed, *Explorations in Turkestan* 2: 271–474. Washington, DC, Carnegie Institute.

Thomson, M and Rapp, G, Jr 1989 Paleobotany from phytoliths. *In* Herzog, Z, Rapp, G, Jr and Negbi, O, eds, *Excavations at Tel Michal, Israel*. Minneapolis, University of Minnesota: 223–225.

Tomlinson, PB 1961 *Anatomy of the Monocotyledons II. Palmae*. Oxford, Clarendon: 453 p.

Wilkens, VE 1928 Anatomical studies of certain Gramineae. *Annals of Botany* 42: 305–316.

Zeitchek, DV 1959 Remains of cultivated plants from Horvat Beter (Beersheva). Preliminary notes. *'Atiqot* 2: 48–52.

8

CEREAL GRAIN PHYTOLITHS OF SOUTHWEST ASIA AND EUROPE

LAWRENCE KAPLAN, MARY B. SMITH and
LESLEY ANN SNEDDON

ABSTRACT

Phytoliths from cereal grains are classified into 23 types based on cellular origin and morphological details. Description of the types are illustrated by scanning electron micrographs and line drawings; both spodograms and cell-free preparations were utilized. Distribution of the 20 types recognizable by optical microscopy is given for 43 cereal species. Combinations of these types may be characteristic of specific taxa.

INTRODUCTION

Applications of phytolith analysis in archaeology, beginning with Helbaek (1961) and initiated in their contemporary methodology by Rovner (1971, 1972), have increasingly provided the stimulus for critical studies in phytolith classification, many of which are cited by Piperno (1988).

Our study of cereal grain phytoliths was initiated when one of us (Kaplan), prompted by difficulties with the identification of the carbonized archaeological remains of these plants, began to build a reference collection for use at the Center for Materials Research in Archaeology and Ethnology at the Massachusetts Institute of Technology. Using Renfrew's *Palaeoethnobotany* (1973) as a guide, we drew up a list of species and set out to compile a cereal grain reference collection of microscope slide preparations, photomicrographs and scanning electron micrographs. In our study, we have sought to record the phytolith array of a group of plants selected for their common or overlapping geographic ranges and their significance in human culture. We have organized our findings in a way that we hope can be applied to the analysis of archaeological sediments and other remains.

Phytolith Systematics, Edited by G. Rapp, Jr. and
S.C. Mulholland, Plenum Press, New York, 1992

METHODS AND MATERIALS

Cereal grains and related weed grass seed reported (Renfrew 1973) to occur in archaeological contexts in prehistoric Southwest Asia and Europe were obtained from the Plant Breeding Institute, Trumpington, U.K.; U.S. Department of Agriculture, Pullman, Washington; the North Central Regional Plant Introduction Station, Ames, Iowa; and the Southern Regional Plant Introduction Station, Experiment, Georgia. See Table 8.1 for a complete list of species included in this study. The plant names used throughout this study are those applied to the seed stocks by the agency from which they were obtained. Except for the wheats and allied species, the nomenclature is sufficiently regular in its application that no comment is required in this report. The nomenclature of the wheats, however, warrants some explanation. Kimber and Feldman (1987), in their monographic treatment of wild wheats, point out that *Triticum* has long been applied to the cultivated wheats, *Aegilops* to the wild wheats. Several authors cited by Kimber and Feldman have treated both groups as *Triticum* because of the absence of any genetic barrier separating them. Others have attempted to reorganize *Triticum-Aegilops* phylogenetically on the basis of shared genomes, but Kimber and Feldman find this line of evidence controversial and present a working synonymy of *Triticum-Aegilops*; we record (Appendix 8.1) that portion of their synonymy that includes the binomials of wheats, the phytoliths of which we have studied. Binomials used in our report, especially Table 8.1, are in the left column, synonyms in the right column. Appendix 8.2 lists common names for those species included in our study that have English names which are not merely contrived from the Latin binomials.

Plants were grown in the greenhouse in 6-inch plant pots in a mixture of equal parts peat moss and perlite with added lime. Four seeds per pot were planted, and the seedlings were thinned as required. A 20-20-20 soluble fertilizer in tap water was used as needed to maintain vigorous growth. The addition of sodium silicate (Na_2SiO_3) to replicate plantings resulted in no differences in over-all growth or silicate deposition between these plants and those that received no applications of sodium silicate. There is evidently sufficient silica in the potting mix and in the tap water-fertilizer solution to provide for phytolith formation.

After flowering, the above-ground portions of plants were harvested and divided into their main organ components: stems, leaves and floral parts. Each component was stored separately in polyethylene bags in the freezer. Although silicates are deposited in root structures of grasses, roots were not included in this study since the techniques for extraction would be different from those developed for aerial parts. Root silicates should receive attention in the future.

To obtain cell-free silicates for scanning electron microscopy, whole above-ground cereal grain plants were harvested, coarsely chopped and processed as described below.

Bulk Extraction for Cell-Free Phytolith Preparations

Chopped leaf, stem and inflorescence samples were combined in beakers with distilled water and sonicated to remove most of the surface dust and pollen. The water was decanted and the tissues were dehydrated overnight at approximately 57°C. A total of 0.1 gram of the dehydrated tissues comprising a mixed or composite sample of the major plant parts was added to 40 mL of commercial Fisher chromic-sulfuric acid solution. This solution is a strong oxidizing agent routinely used to remove organic stains and residues from laboratory glassware. The destructive oxidation of organic cell wall and cell content material in the plant tissues should be complete after standing at room temperature overnight. The resulting solution was swirled and divided among four 15-mL tapered glass centrifuge tubes and spun for 5 minutes at a #3 setting in a clinical centrifuge with a swing-out head. The supernatant was discarded and the precipitate, consisting of silicate phytoliths and a very small amount of organic debris, was twice resuspended and washed in distilled water with the aid of a vortex agitator. The distilled water wash was followed by a wash in 95% ethanol, which was decanted. The alcohol washed precipitate was resuspended in 95% ethanol and stored in a glass vial. Heavy liquid flotation using various densities of zinc chloride was tested for separating silicates from organic material but was found to be less effective in both separation and recovery than centrifugation.

Control samples without plant tissue were run initially and repeated periodically to determine whether the process was subject to contamination with opal silicates from glassware, reagents, water or atmosphere.

Scanning Electron Microscopy (SEM)

The phytoliths in ethanol were prepared for study with scanning electron microscopy by agitating the vial and transferring one or more drops, depending on the density of phytoliths, of the ethanol-phytolith suspension onto a clean inspected glass slide in a dust-free area. The ethanol was allowed to evaporate, and an aluminum SEM stub coated with "microstick" was touched to the dried sample. Samples mounted in this way were then coated with carbon followed by gold-palladium in the vacuum evaporator for SEM examination. Micrographs were made of each of the phytolith types initially identified with light microscopy as described in the following section.

Spodograms

Samples of the separated leaf, stem and inflorescence structures were processed to clear cytoplasmic contents, and to detach epidermal layers from underlying tissues while maintaining the integrity of the epidermal layer itself and

Table 8.1. The Distribution of Phytolith Types in Cereal Grains

	Ia	Ib	Ie	Ii	Ih	I2a	I2g	IIa1
Aegilops speltoides	X	X	X		X			X
Aegilops squarrosa		X	X					X
Aegilops umbellulata	X	X					X	X
Agropyron repens	X	X			X		X	X
Avena abyssinica	X	X	X	X				X
Avena barbata	X	X	X	X	X?			X
Avena brevis	X			X	X		X	X
Avena byzantina	X	X	X	X			X	X
Avena fatua	X	X		X	X		X	X
Avena nuda	X	X			X			X
Avena sativa	X		X	X	X			X
Avena sterilis	X	X		X	X?			X
Avena strigosa	X	X	X	X				X
Bromus arvensis			X	X		X	X	X
Bromus mollis		X			X			X
Bromus secalinus				X				
Hordeum distichon		X	X		X		X	X
Hordeum spontaneum	X	X		X				X
Hordeum vulgare	X	X			X			X
Lolium perenne	X	X			X	X		X
Lolium remotum	X	X	X		X			X
Lolium temulentum	X	X	X					X
Secale anatolicum	X	X	X	X	X	X		
Secale cereale	X							
Secale vavilovii		X		X	X			X
Triticum aegilopoides	X	X	X	X	X			X
Triticum aestivum	X			X	X		X	X
Triticum araraticum	X	X	X	X	X			X
Triticum carthlicum	X	X	X					
Triticum compactum	X	X	X	X	X	X	X	X
Triticum dicoccoides	X	X	X		X	X?		X
Triticum dicoccum	X	X	X	X				X
Triticum durum		X	X		X		X	
Triticum macha	X	X						X
Triticum monococcum	X	X	X				X	X
Triticum polonicum	X	X	X	X			X	X
Triticum spelta	X	X		X			X	X
Triticum sphaerococcum	X	X		X		X	X	X
Triticum thaoudar	X	X	X	X				X
Triticum timopheevii	X	X	X	X	X		X	X
Triticum turgidum	X	X				X		X
Triticum urartu		X		X	X	X?		X
Triticum vavilovii	X	X				X	X	X

IIa2	IIa3	IIb1	IIb2	IIc1	IIc2	IIIa	IIIb	Va1	Va2	Vb1	Vb2
X	X	X	X	X	X		X	X		X	X
X		X			X		X	X	X	X	
	X	X					X	X	X	X	
			X	X	X	X	X	X	X		
X		X	X	X	X		X	X			
				X	X	X	X	X	X?	X	X?
		X	X	X			X				
		X		X	X		X	X	X	X	X?
		X		X			X	X	X	X	
		X			X			X		X	
X		X	X					X		X	
X		X			X		X	X		X	X?
		X	X	X	X	X	X				X
		X		X	X		X	X			
X		X		X		X	X	X	X		
X	X	X					X	X			X
		X		X			X	X	X	X	
		X				X	X	X	X		
		X				X	X	X	X	X	
		X	X	X	X	X	X	X		X	
X	X	X	X		X	X		X			
X			X					X			
X		X	X	X		X	X	X	X	X	
		X	X					X			
		X		X		X	X	X?	X	X	X?
X		X?	X	X	X	X	X	X	X	X	X?
X		X		X			X	X	X	X	
			X	X		X	X		X		
		X		X	X	X	X	X	X	X	
X			X?	X	X			X	X	X	X?
X				X	X?	X?	X	X	X	X	
X?				X	X		X	X?	X		
X	X	X						X	X	X	
		X	X	X	X	X	X	X	X		X?
X	X	X		X		X	X	X	X	X	
X		X		X	X	X	X	X	X	X	
		X		X	X	X	X	X	X	X	X
		X	X	X	X		X	X	X	X	
		X	X	X	X	X	X	X	X	X	X?
		X	X	X		X	X		X		
	X	X		X	X		X	X?		X	
		X		X		X	X		X		
X		X		X	X		X	X	X	X	

of the associated phytoliths. Such a preparation is a spodogram, and consists of the cleared cellulose framework of the epidermis with the phytoliths in situ.

Plant parts were trimmed with a razor blade into approximately 0.5-cm square pieces with certain special treatments: leaf squares include one edge that is the leaf margin; the stem epidermis was trimmed from larger plants, whereas an entire stem segment was cut from smaller plants; flowers were separated from the inflorescence, and grains were removed. The tissue blocks were placed in petri dishes containing a 50% solution of commercial 5.35% sodium hypochlorite (commercial Clorox) and allowed to stand overnight or longer until the tissues cleared.

After sufficient clearing, the tissue blocks were transferred from the sodium hypochlorite solution to dishes of distilled water to remove traces of the oxidizing agent from the tissues. At this stage, the tissues were stored in water for several weeks under refrigeration while awaiting further processing. The upper and lower epidermal layers were separated from the underlying tissues by a combination of agitating the solution, tapping the dish on the bench top and gently prodding with a glass rod or forceps. Very thin leaves do not require separation. Floral structures were opened so that the glumes, lemmas and paleas lay flat.

Structures so prepared were then floated onto slides where they were dehydrated and infiltrated by pipetting and replacing in series: distilled H_20; ethanol 10%, 50%, 75%—2 changes, 95%—3 changes, 100%—3 or more changes; xylene—2 changes; dilute Canada balsam in xylene.

When tissues are thoroughly dehydrated in 100% ethanol, the subsequent addition of xylene should not result in turbidity. When turbidity did occur in our preparations, treatment with fresh 100% ethanol was repeated. After addition of the mounting medium, slides were covered with No. 1 cover glasses and placed on the slide warming table until the medium was set. Excess Canada balsam was removed with xylene. Six replicate slides were prepared for each of the 3–4 structural divisions of each plant, resulting in 18–24 slides for each taxon.

Photomicrographs

Light photomicrographs at 250 and 400 magnifications were prepared to record all types of phytoliths found in each plant tissue. The optimum contrast was obtained by using a green filter in the microscope (Leitz Ortholux) light source and 35-mm Panatomic X film ASA 25. Contrast was further increased by slight underexposure of the negatives and a 20% increase in developing (Eastman-Kodak D76) time. Prints were exposed using a #4 polycontrast filter and standard developing times in Dektol.

Figures

Figures 8.1–8.37 illustrate the phytolith types observed. Electron micrographs (A) are paired with line drawings (B, C) in most cases.

RESULTS

We recognize 23 phytolith types (Figs 8.1–8.37) in our preparations. Three are difficult to distinguish in optical microscopy (see trapezoids below) and are not included in Table 8.1. All are included in our identification key (Appendix 8.3). Our descriptive names follow Metcalfe (1960) and Palmer (1976), and we have adhered as closely as we could to Brown (1984) in the use of alphanumeric symbols to designate types in Table 8.1, and in our key (Appendix 8.3).

Our record of phytolith occurrence is a presence:absence distribution. From the outset, we avoided any attempt to develop a frequency distribution of phytolith types in our sample because of the possible effects of variations in plant maturity, development, and possible sampling errors in the extraction and microscopic analysis.

Among the 43 plant taxa (Table 8.1) included in our study, the occurrence of phytolith types varies considerably. For example, type IIa3 (Figs 8.12A, B, C), a small prickle with a conical tip, is present in 7 taxa in contrast with IIa1 (Figs 8.8A, B; 8.9A, B, C), a small prickle with an acute tip, present in 38 of the 43 taxa.

DISCUSSION

Because our emphasis in this study is to delineate the phytolith types of Southwest Asian cereal grains, our discussion of the possible utility of phytoliths in archaeological analysis will consist only of some suggestions of how our data might be used by sediment analysts. We caution those who might wish to use our data in this way that: (1) the absence of any given phytolith type may be the result of a sampling error and therefore must be less significant than the presence of that type, and (2) the present grass flora and vegetation history of the region in which the analysis occurs must be taken into account because of the possibility that a particular phytolith type may be produced by species other than those we have investigated. In an early stage of our project, soil sediments from Wadi Kubbaniyah (Egypt) were examined in an effort to detect the presence of cereal grain remains. No phytoliths resembling those of cereal grains were detected: a negative finding corroborating other evidence from that site (Kaplan 1989).

Sheet Elements

Five of our types are classified as sheet elements. They are formed on epidermal surfaces as the result of depositions from watery exudates, or they are "multicelled sections of silica-replaced epidermal tissue" as detected by Rosen (1987: 244) at Shiqmim in the northern Negev Desert, Israel. We distinguish sheet elements on the basis of their margins or marginal protuberances. Among the sheet elements, those with entire margins, Ib (Fig 8.1A, B), sinuous margins, Ia (Fig 8.2A, B), and those with dendritically branched, Ii (Fig 8.4A, B), and unbranched crateriform, Ih (Fig 8.5A, B), protuberances occur so widely that it is unlikely that they will be of use either in archaeological analysis or in systematic studies in the grasses. Sheet elements having clavate protuberances over the entire surface, I2g (Fig 8.7A, B), are interesting in that they appear to distinguish the wheats—diploid, tetraploid and hexaploid—from common cultivated 6-row barley, rye and commonly cultivated hexaploid oats (*Avena sativa*). Such a distinction might be significant in those regions outside the Near East where any combinations of these grains might be grown—Europe and North America for example.

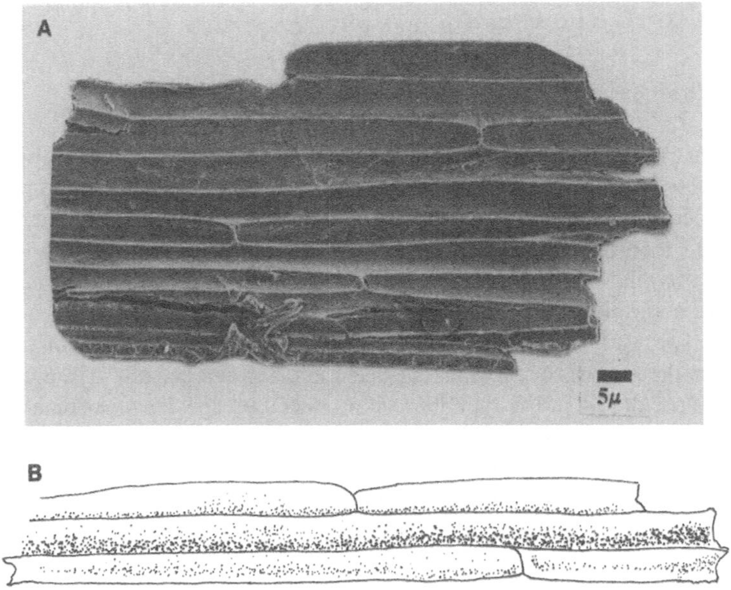

Fig 8.1. Sheet element, margin entire, Ib, *Lolium perenne*. A. Scanning electron micrograph. Bar is 5 microns. B. Line drawing.

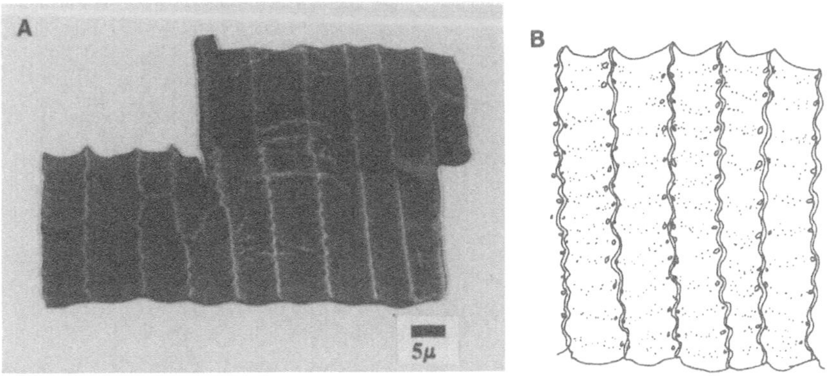

Fig 8.2. Sheet element, margin sinuous, Ia, *Avena sativa*. A. Scanning electron micrograph. Bar is 5 microns. B. Line drawing.

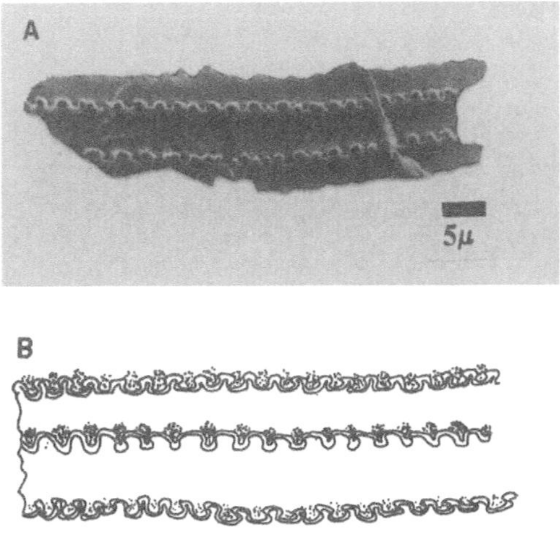

Fig 8.3. Sheet element, margin undulate-interlocking, Ie, *Avena sativa*. A. Scanning electron micrograph. Bar is 5 microns. B. Line drawing.

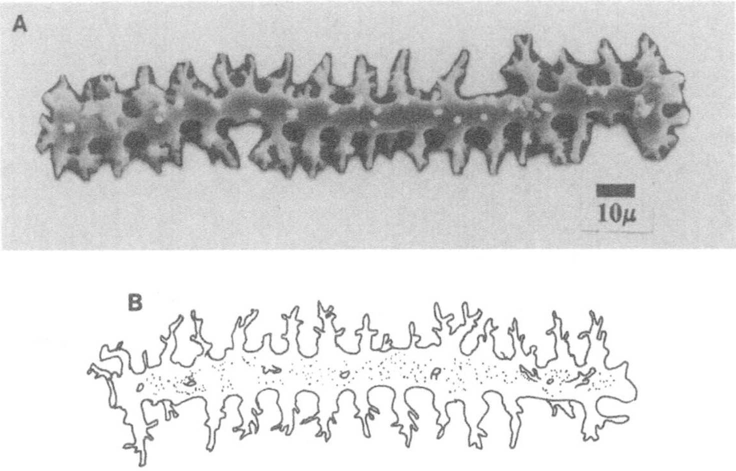

Fig 8.4. Sheet element, protuberances branched, dendritic, Ii, *Avena sativa*. A. Scanning electron micrograph. Bar is 10 microns. B. Line drawing.

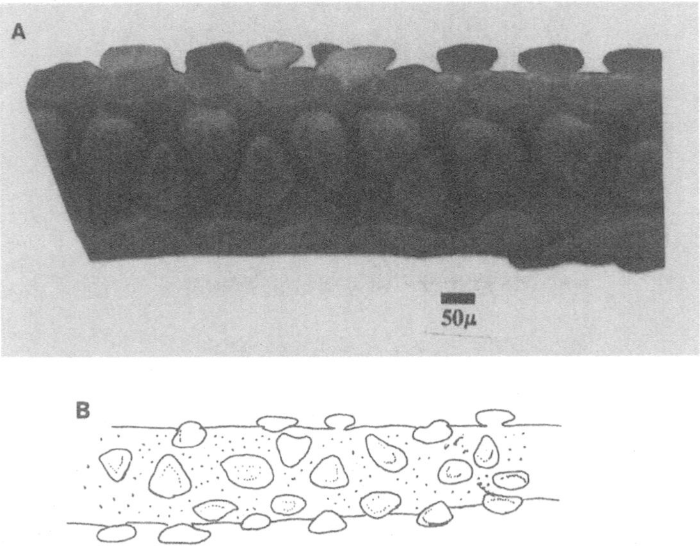

Fig 8.5. Sheet element, protuberances crateriform, Ih, *Avena sativa*. A. Scanning electron micrograph. Bar is 50 microns. B. Line drawing.

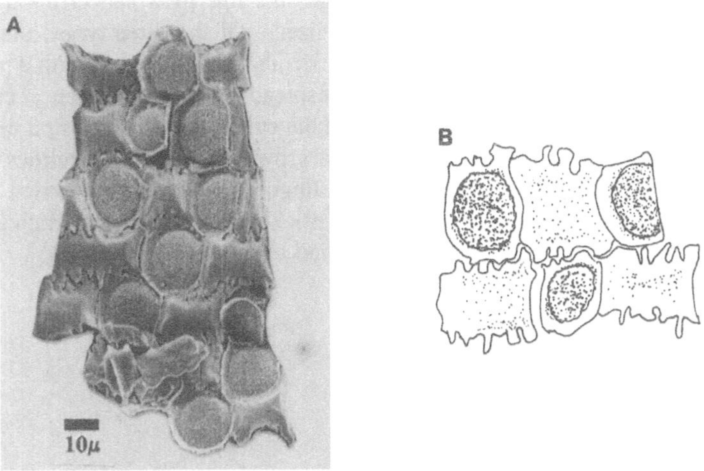

Fig 8.6. Sheet element, margins irregularly dentate, I2a, *Lolium perenne*. A. Scanning electron micrograph. Bar is 10 microns. B. Line drawing.

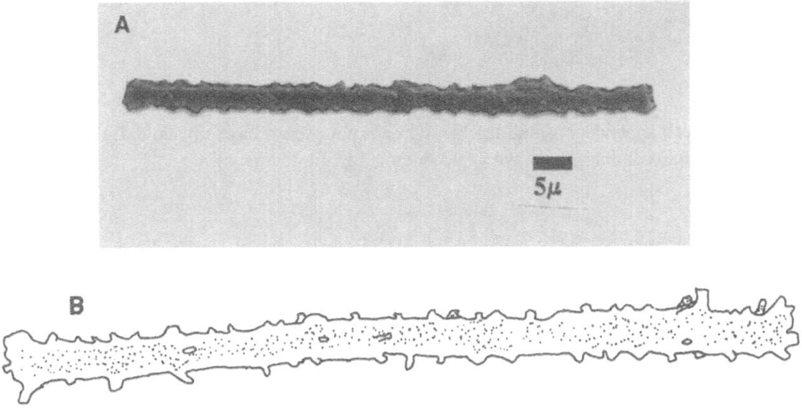

Fig 8.7. Sheet element, protuberances clavate, I2g, *Triticum durum*. A. Scanning electron micrograph. Bar is 5 microns. B. Line drawing.

Long-Cell Appendages

Small Prickles are long-cell appendages the tips of which differ from one another so that three types may be recognized. Of the three types, those with acute tips, IIa1 (Figs 8.8A, B; 8.9A, B, C), are ubiquitous. Those with a papillate tip, IIa2 (Figs 8.10A, B; 8.11), are widespread in their occurrence but their absence in 6-rowed barley and presence in the cultivated wheats tested may help to distinguish these crops, which are often grown in the same localities. Similarly, this papillate-tipped small prickle distinguishes between cultivated rye and cultivated oats. The conical-tipped small prickle seems to occur erratically and we feel that further study of its structure and occurrence is needed.

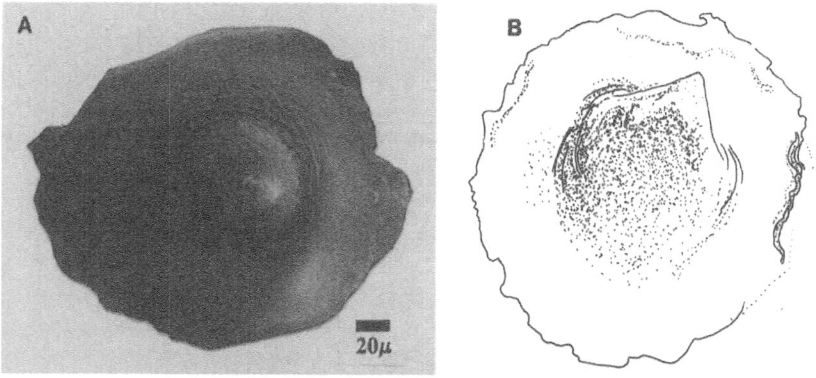

Fig 8.8. Long cell appendage, small prickle, tip does not exceed base, tip acute, IIa1, *Avena fatua*. A. Scanning electron micrograph. Bar is 20 microns. B. Line drawing.

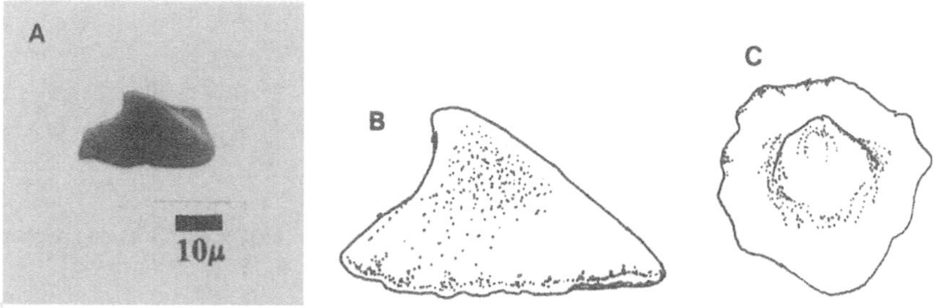

Fig 8.9. Long cell appendage, small prickle, tip does not exceed base, tip acute, IIa1, *Aegilops squarrosa*. A. Scanning electron micrograph. Bar is 10 microns. B. Line drawing, side view. C. Line drawing, top view.

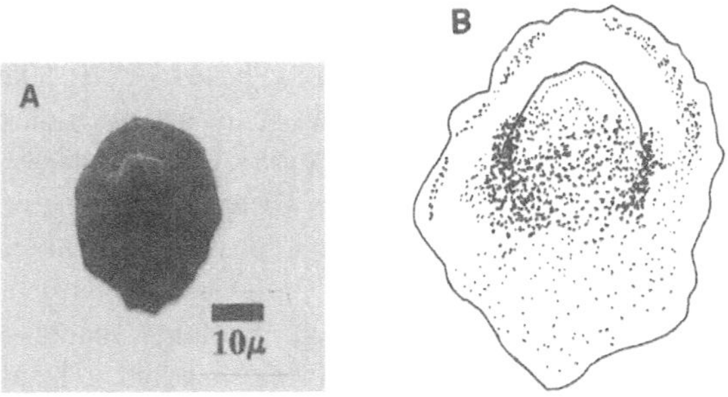

Fig 8.10. Long cell appendage, small prickle, tip does not exceed base, tip papillate, IIa2, *Triticum monococcum*. A. Scanning electron micrograph. Bar is 10 microns. B. Line drawing.

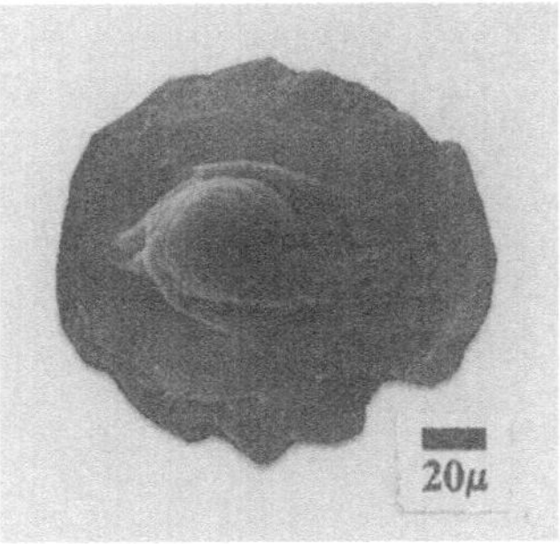

Fig 8.11. Long cell appendage, small prickle, tip does not exceed base, tip papillate, IIa2, *Triticum aestivum*. Scanning electron micrograph. Bar is 20 microns.

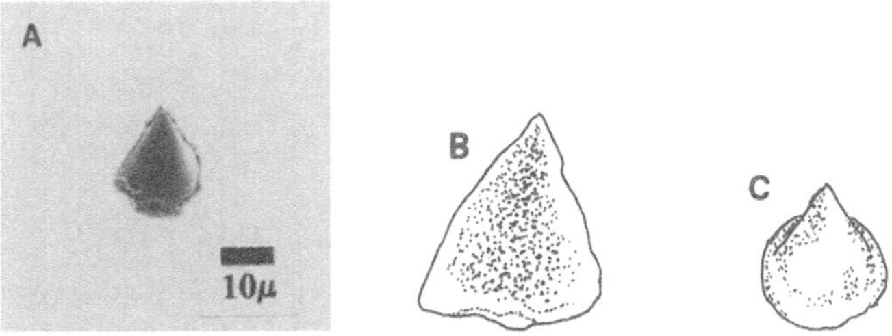

Fig 8.12. Long cell appendage, small prickle, tip does not exceed base, tip conical, IIa3, *Aegilops speltoides*. A. Scanning electron micrograph. Bar is 10 microns. B. Line drawing, side view. C. Line drawing, top view.

Large Prickles are long-cell appendages having reflexed tips that extend beyond the one-celled swollen base. The swollen base may be elongate, IIb1 (Figs 8.13A, B; 8.14A, B), or circular, IIb2 (Fig 8.15A, B). The swollen-base large prickle, IIb1, occurring in all but seven of the species studied, is one of the most widespread of phytoliths. The distribution of the circular-base large prickle is more restricted in its distribution, and its absence from most of the *Triticum-Aegilops* group and the barleys provides a possible useful contrast with its presence in cultivated rye and oats.

Macrohairs, another long-cell appendage, are large, macroscopically visible and widespread, occurring in 31 of the 43 species studied. The apparent absence of this type from 6-row barley and oats and its presence in bread wheat and rye may be a useful contrast in some analyses.

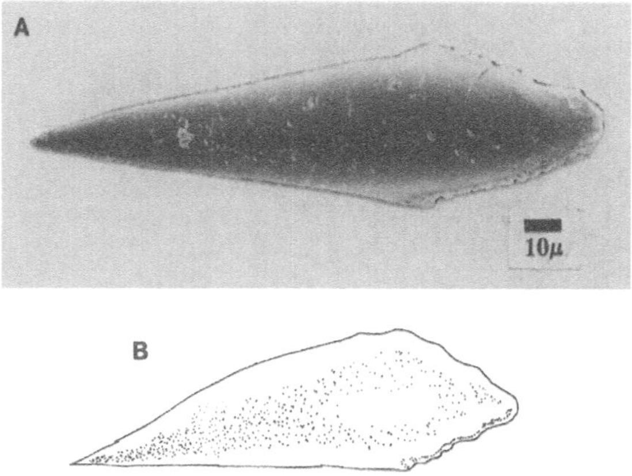

Fig 8.13. Long cell appendage, large prickle, tip exceeds base, base elongate, IIb1, *Secale anatolicum*. A. Scanning electron micrograph. Bar is 10 microns. B. Line drawing.

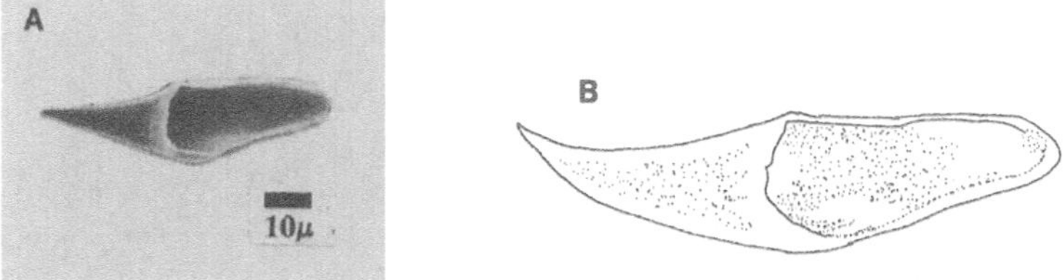

Fig 8.14. Long cell appendage, large prickle, tip exceeds base, base elongate, IIb1, *Hordeum distichon* var *'trifurcatum'*. A. Scanning electron micrograph. Bar is 10 microns. B. Line drawing.

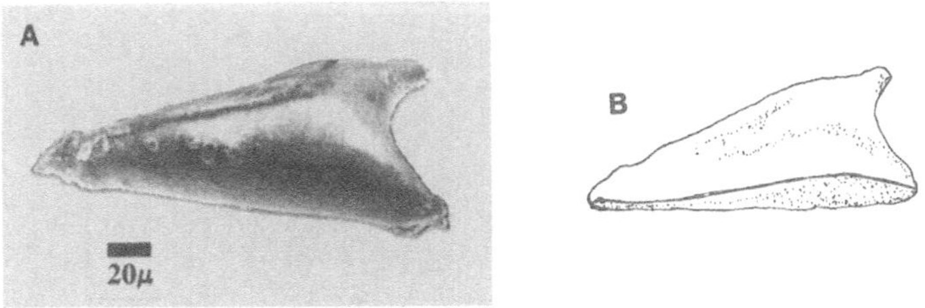

Fig 8.15. Long cell appendage, large prickle, tip exceeds base, base circular, IIb2, *Avena brevis*. A. Scanning electron micrograph. Bar is 20 microns. B. Line drawing.

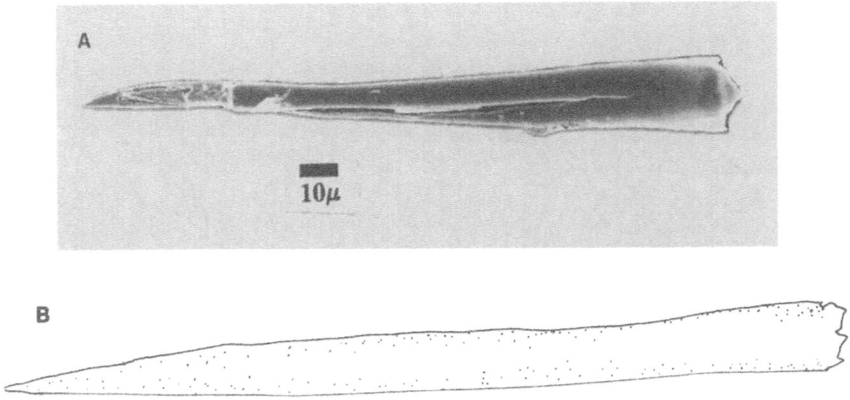

Fig 8.16. Long cell appendage, large macrohair, IIc1, *Bromus mollis*. A. Scanning electron micrograph. Bar is 10 microns. B. Line drawing.

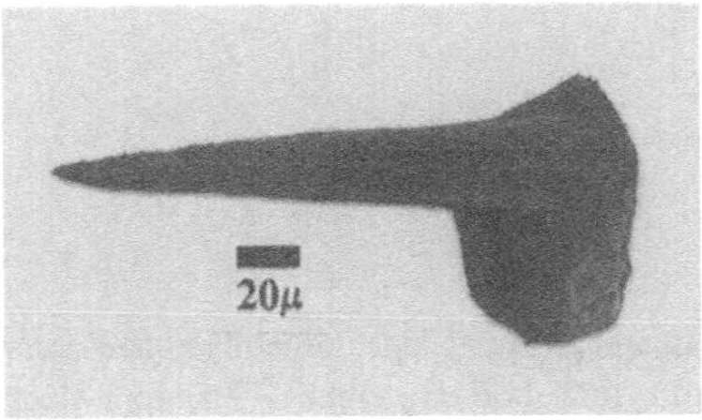

Fig 8.17. Long cell appendage, large macrohair, IIc1, *Triticum monococcum*. Scanning electron micrograph. Bar is 20 microns.

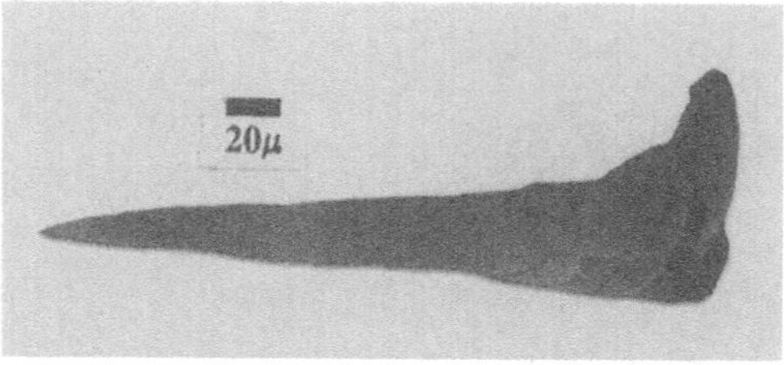

Fig 8.18. Long cell appendage, large macrohair, IIc1, *Triticum monococcum*. Scanning electron micrograph. Bar is 20 microns.

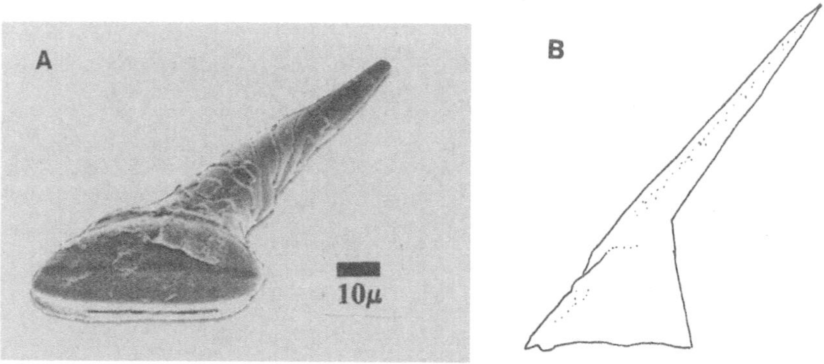

Fig 8.19. Long cell appendage, large macrohair, (smaller than IIc1), IIc2, *Aegilops speltoides*. A. Scanning electron micrograph. Bar is 10 microns. B. Line drawing.

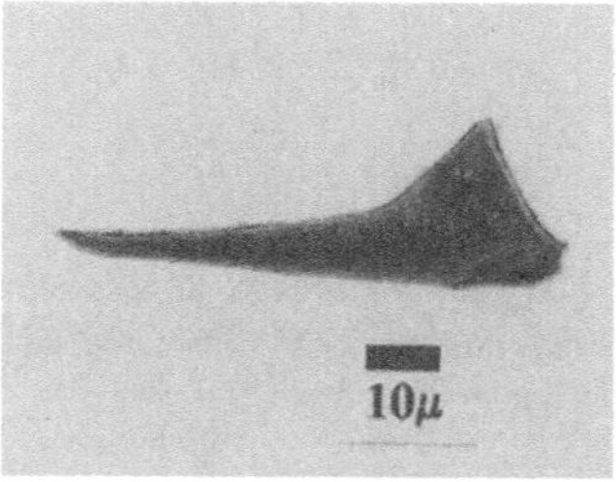

Fig 8.20. Long cell appendage, large macrohair, (smaller than IIc1), IIc2, *Lolium perenne*. Scanning electron micrograph. Bar is 10 microns.

Trichome Bases

These are epidermal cellular elements from which trichomes diverge. They may be annular, with an entire margin along at least two quadrants, IIIa (Fig 8.21A, B), or they may have a cupped or lobed margin, IIIb (Figs 8.22A, B; 8.23). The first type is sporadically distributed in our study sample but its absence from the commonly cultivated grains except for 6-row barley may, with further study, prove to be useful in sediment analysis. Those trichome bases having cupped or lobed margins, IIIb (Figs 8.22 A, B; 8.23), are so widely distributed that their utility in analysis is doubtful.

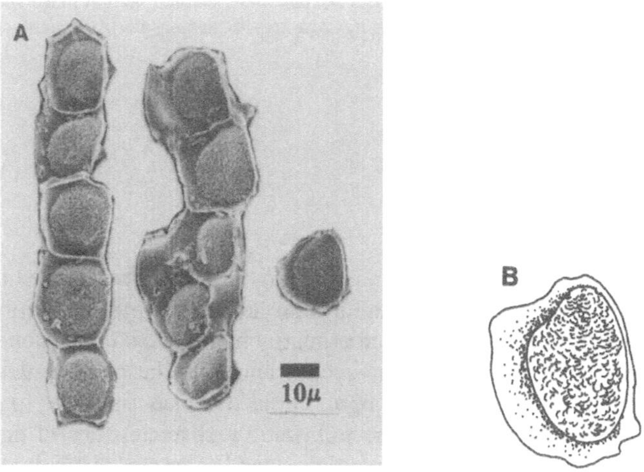

Fig 8.21. Trichome base, annular margin entire along at least two quadrants, IIIa, *Lolium perenne*. A. Scanning electron micrograph. Bar is 10 microns. B. Line drawing.

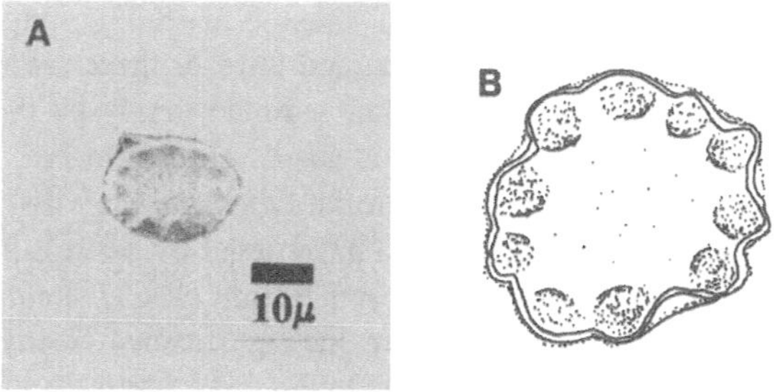

Fig 8.22. Trichome base, annular margin lobed and cupped, IIIb, *Aegilops speltoides*. A. Scanning electron micrograph. Bar is 10 microns. B. Line drawing.

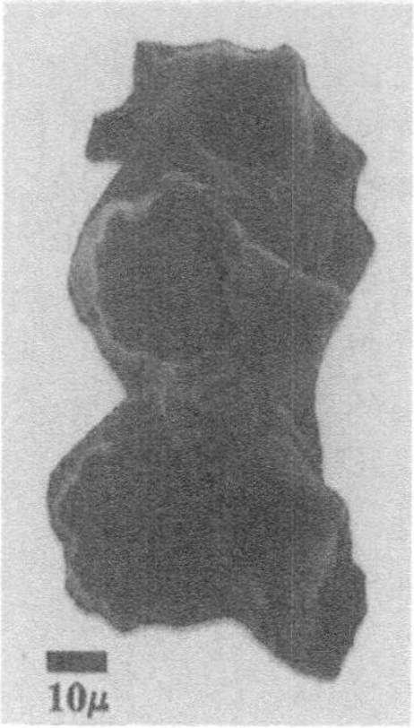

Fig 8.23. Trichome base, annular margin lobed and cupped, IIIb, *Triticum monococcum*. Scanning electron micrograph. Bar is 10 microns.

Trapezoids

These forms comprise the remaining phytolith types recognized in our study, and their morphological traits exhibit more levels of organization than do other types. Trapezoids are distinguished primarily by the three-dimensional form from which their name derives. Their overall length is the primary determinant of identity; characteristics of the margin, entire or lobed and alate or not provide secondary and tertiary determiners followed by characteristics of the "top," that facet of least area. Characteristics of the "top" often are difficult to distinguish both in the optical and scanning electron microscope images. For this reason we have described these types and included them in our key (Appendix 8.3), but in Table 8.1 we combine types having different tips but similar margins and lengths: (1) Vb1 (Fig 8.24A, B), short with margins entire; (2) Va2 (Figs 8.27–8.28), long with margins alate and lobed; (3) Val (Figs 8.33–8.37), long with margins sinuous, not alate, concave top. The first and third of these are widespread and likely to be of little use in archaeological sediment analysis; the second, Va2, among the commonly cultivated grains, is absent only from rye.

The survey and classification of phytolith types and their distribution among plant groups has been carried on at the family level by several investigators, of whom Metcalfe (1960), Twiss, Suess and Smith (1969), Palmer (1976) and Piperno (1988) are illustrative. However, the significance of phytoliths in plant systematics has been minor when compared with applications in soils analysis and archaeology. Probably the best explanation for this observation is the great multiplicity and low specificity of phytoliths in those plant groups that are at a level

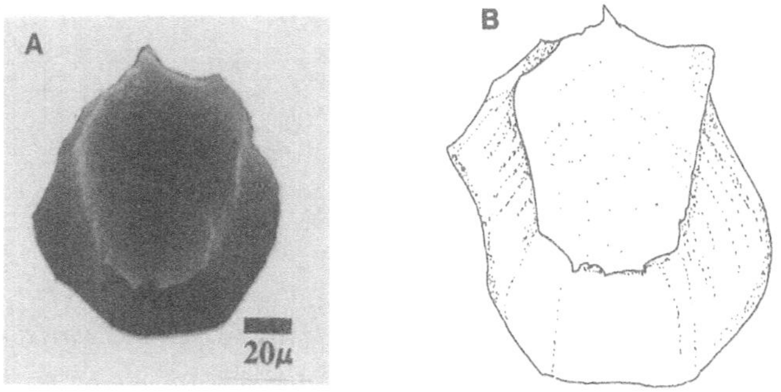

Fig 8.24. Trapezoid, short, margins entire, Vb1, *Triticum durum*. A. Scanning electron micrograph. Bar is 20 microns. B. Line drawing.

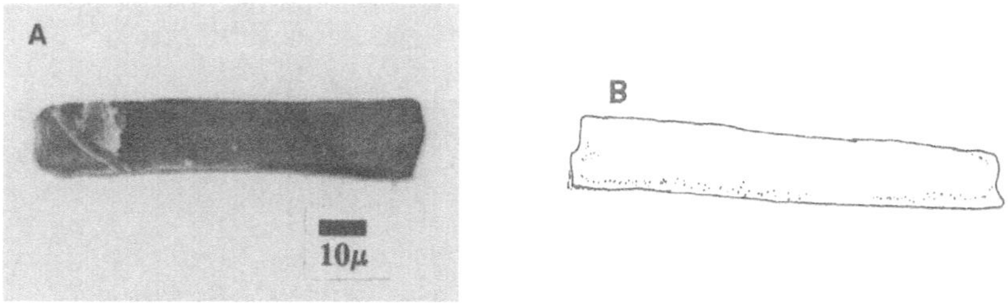

Fig 8.25. Trapezoid, long, margins entire, Vb2, *Bromus secalinus*. A. Scanning electron micrograph. Bar is 10 microns. B. Line drawing.

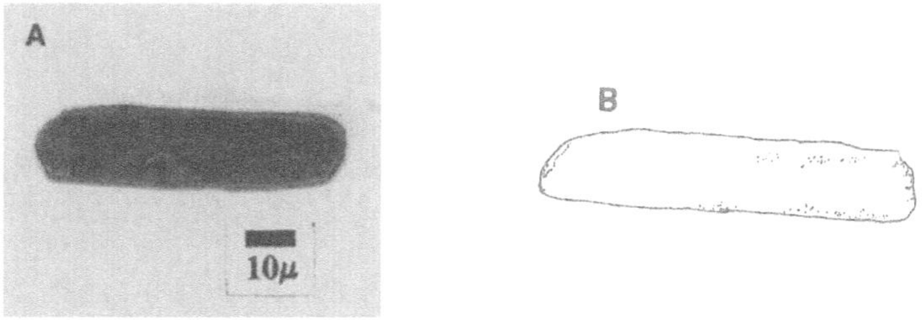

Fig 8.26. Trapezoid, long, margins entire, Vb2, *Aegilops speltoides*. A. Scanning electron micrograph. Bar is 10 microns. B. Line drawing.

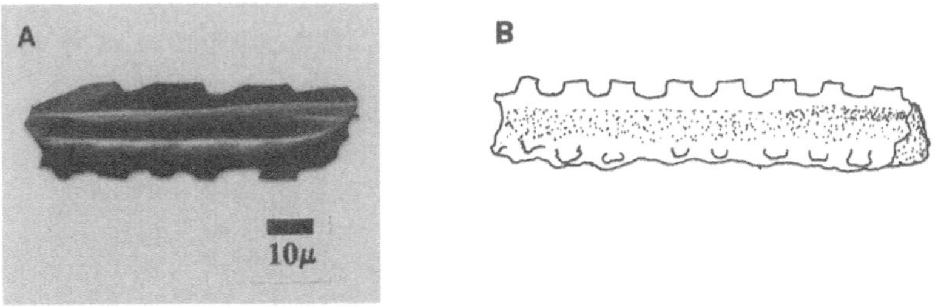

Fig 8.27. Trapezoid, margins alate and lobed, Va2, *Triticum aestivum* var *'Maris Nimrod'*. A. Scanning electron micrograph. Bar is 10 microns. B. Line drawing.

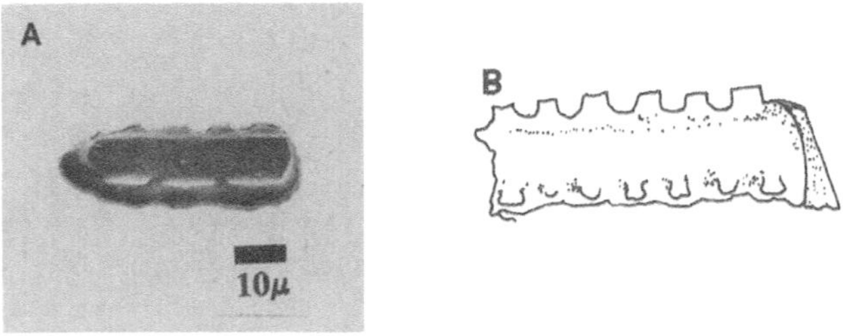

Fig 8.28. Trapezoid, margins alate and lobed, flat top, Va2a, *Hordeum distichon* var *'trifurcatum'*. A. Scanning electron micrograph. Bar is 10 microns. B. Line drawing.

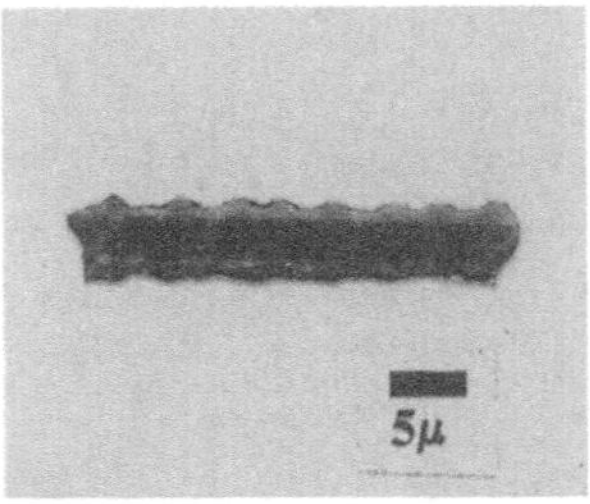

Fig 8.29. Trapezoid, margins alate and lobed, ridged top, Va2b, *Triticum aestivum* var *'Maris Nimrod'*. Scanning electron micrograph. Bar is 5 microns.

Fig 8.30. Trapezoid, margins alate and lobed, ridged top, Va2b, *Triticum aestivum* var *'Maris Nimrod'*. Scanning electron micrograph. Bar is 5 microns.

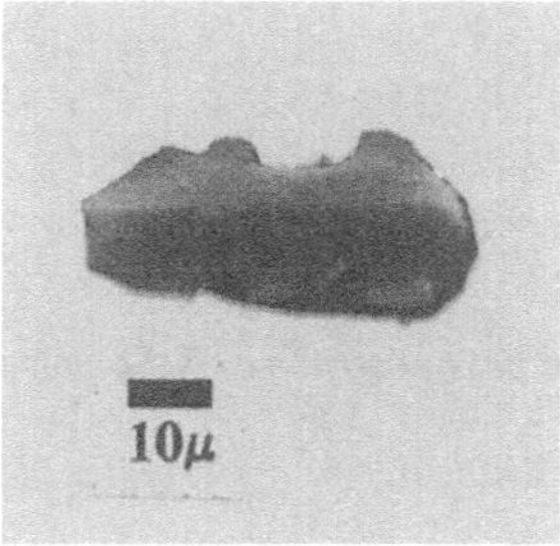

Fig 8.31. Trapezoid, margins alate and lobed, ridged top, Va2b, *Triticum aestivum* var *'Maris Nimrod'*. Scanning electron micrograph. Bar is 10 microns.

Fig 8.32. Trapezoid, margins alate and lobed, ridged top, Va2b, *Hordeum vulgare* var *'hexastichum'*. Scanning electron micrograph. Bar is 5 microns.

Fig 8.33. Trapezoid, margins sinuous, flat or concave top, Va1a, *Secale anatolicum*. A. Scanning electron micrograph. Bar is 10 microns. B. Line drawing.

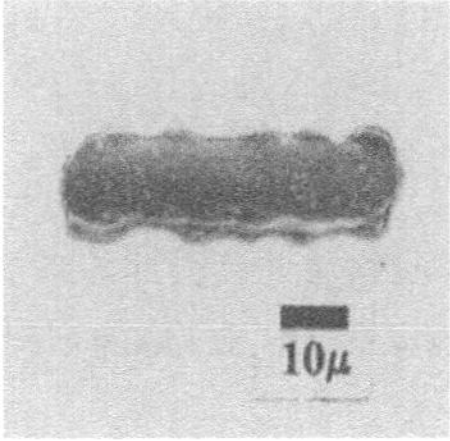

Fig 8.34. Trapezoid, margins sinuous, flat or concave top, Va1a, *Aegilops umbellulata*. Scanning electron micrograph. Bar is 10 microns.

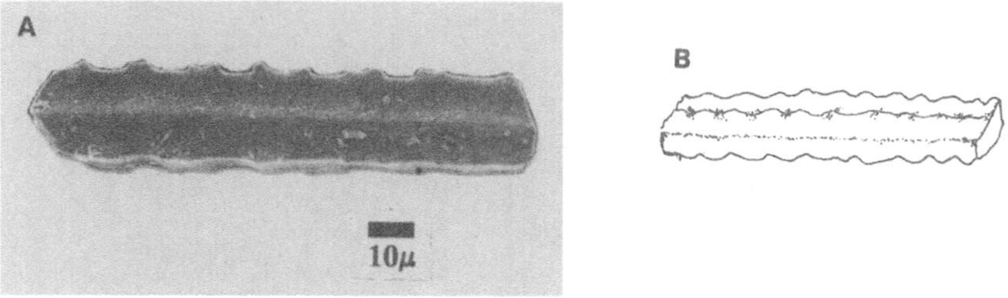

Fig 8.35. Trapezoid, margins sinuous, 1-ridged top, Va1b, *Bromus secalinus*. A. Scanning electron micrograph. Bar is 10 microns. B. Line drawing.

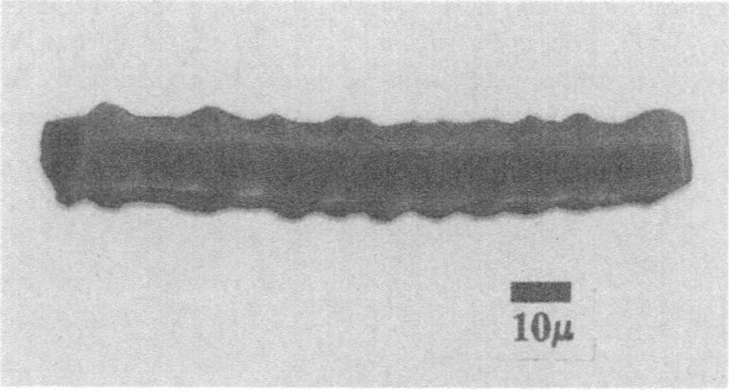

Fig 8.36. Trapezoid, margins sinuous, 1-ridged top, Va1b, *Triticum monococcum*. Scanning electron micrograph. Bar is 10 microns.

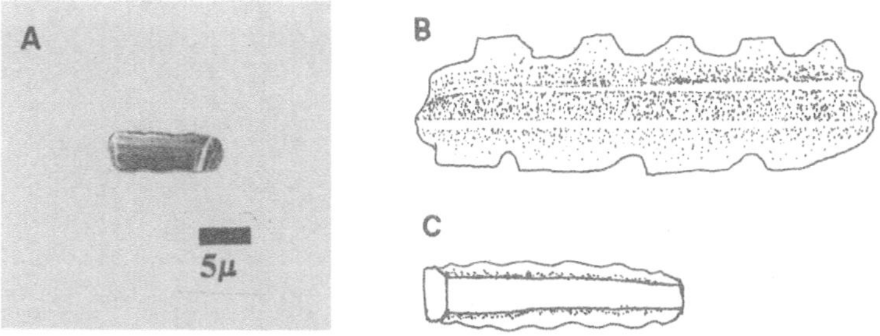

Fig 8.37. Trapezoid, margins sinuous, 2-ridged top, Va1c, *Lolium perenne*. A. Scanning electron micrograph. Bar is 5 microns. B. Line drawing. C. Line drawing.

of concern in contemporary plant sciences. Plant biologists have long been more concerned with the systematics and evolutionary mechanisms at the specific and infraspecific levels than at the family level. Major structural and biochemical features that reflect phylogeny have been the subject of investigation at the family and higher levels, while at the species and population levels the powerful methods of genetics, biochemistry and population biology have been the techniques of choice. Morphology of floral structure continues to be used in plant systematics, and micromorphology in recent decades has emphasized selected structures such as pollen, epidermal hairs and the like.

Our data may find application in plant systematics, but a preliminary phenetic cluster analysis shows no apparent relationship between phytolith distribution and the taxonomic relationships among the plant groups included in our study. In such an analysis congeneric species are not shown to be more closely related than species of distinct genera, and polyploid levels within the wheats (*Triticum*) are not closely related. However, the polyploid series in wheats are of diverse origin (Zohary & Hopf 1988: 24) and as such may not be expected to share the same phytolith types.

SUMMARY

We have presented the results of a survey of biogenic opaline phytoliths of cereal grains and related graminaceous species native to southwest Asia and Europe. Twenty-three types are recognized; 20 of these are sufficiently distinguishable using optical microscopy so that in certain combinations they may be characteristic of the species of grains from which they derive. We delineate some of these combinations and propose that they be used in the analysis of archaeological sediments to determine the presence of cereal grain species where the macrofossil record is inconclusive. A table of phytolith distributions among 43 species, a key for identification and a full set of scanning electron micrographs and line drawings of cereal grain phytolith types are included.

ACKNOWLEDGMENT

Our work was supported in part by a National Endowment for the Humanities grant to the Massachusetts Institute of Technology, and by the College of Arts and Science, University of Massachusetts/Boston. Our thanks are extended to Mary Ellen Ruane who devoted many hours to the preparation of slides, and to Kathleen Dooley, who executed the line drawings.

REFERENCES

Brown, D 1984 Prospects and limits of a phytolith key for grasses in the central United States. *Journal of Archaeological Science* 11: 221–243.

Helbaek, H 1961 Studying the diet of ancient man. *Archaeology* 14: 95–101.

Kaplan, L 1989 Phytolith analysis at Wadi Kubbaniyah (Egypt). *In* Wendorf, F and Schild, R, eds, *The Prehistory of Wadi Kubbaniyah. 2. Stratigraphy, Paleoeconomy and Environment.* Dallas, Southern Methodist University Press: 267–269.

Kimber, G and Feldman, M 1987 Wild wheat, an introduction. Special Report 353. College of Agriculture, Columbia, University of Missouri.

Mabberley, DJ 1989 *The Plant Book.* Cambridge, Cambridge University Press: 706 p.

Metcalfe, CR 1960 *Anatomy of the Monocotyledons. I. Gramineae.* Oxford, Clarendon Press: 731 p.

Palmer, PG 1976 Grass cuticles: A new paleoecological tool for East African lake sediments. *Canadian Journal of Botany* 54: 1725–1734.

Piperno, DR 1985 Phytolith analysis and tropical paleo-ecology: Production and taxonomic significance of siliceous forms in New World plant domesticates and wild species. *Review of Paleobotany and Palynology* 45: 185–228.

_____1988 *Phytolith Analysis: An Archaeological and Geological Perspective.* San Diego, Academic Press: 280 p.

Renfrew, JM 1973 *Palaeoethnobotany.* New York, Columbia University Press: 248 p.

Rosen, AM 1987 Phytolith studies at Shiqmim. *In* Levy, TE, ed, Shiqmim I: Studies concerning Chalcolithic societies in the Northern Negev Desert, Israel (1982–1984), *British Archaeological Reports, International Series* 356: 243–249.

Rovner, I 1971 Potential of opal phytoliths for use in paleoecological reconstruction. *Quaternary Research* 1: 343–359.

_____1972 Notes on a safer procedure for opal phytolith extraction. *Quaternary Research* 2: 591.

Twiss, PC, Suess, E and Smith, RM 1969 Morphological classification of grass phytoliths. *Soil Science Society of America, Proceedings* 33: 109–115.

Zohary, D and Hopf, M 1988 *Domestication of Plants in the Old World.* Oxford, Oxford University Press: 249 p.

APPENDIX 8.1. Synonymy of Wild Wheats Included in Phytolith Study*

Triticum monococcum	=	*T. aegilopoides, T. urartu*
T. araraticum	=	*T. timopheevii*
T. turgidum	=	*T. carthlicum, T. dicoccoides, T. durum, T. polonicum*
Aegilops speltoides	=	*T. speltoides (ligustica)*
Ae. squarrosa	=	*T. tauschii*
Ae. umbellulata	=	*T. umbellulatum*

*Nomenclature according to Kimber and Feldman (1987: 78–79).

APPENDIX 8.2. Cereal Grains and Grasses—Common Names and Polyploidy*

Aegilops speltoides	Goat-grasses = *Triticum*
Ae. squarrosa	"
Ae. umbellulata	"
Agropyron repens	Couch-grass
Avena abyssinica	Abyssinian oat, (limited cult.) 4n
A. brevis	Short oat, 2n
A. fatua	Wild oat, 6n
A. sativa	White oat, cult., 6n
Bromus arvensis	Bromegrasses
B. mollis	"
B. secalinus	"
Hordeum distichon = *H. distichum*	2-row barley
H. spontaneum	2-row wild barley
H. vulgare	6-row barley
Lolium perenne	Rye-grass, perennial
L. remotum	Annual weed rye-grass
L. temulentum	Darnel
Secale anatolicum	Rye, wild
S. cereale	Rye, cult.
Triticum aestivum	Bread wheat, 6n
T. durum	Pasta wheat, 4n
T. monococcum	Einkorn, 2n

*According to sources of seed grain; Zohary and Hopf (1988); and Mabberley (1989).

Appendix 8.3. Key to Phytoliths of Cereal Grains of Southwestern Asia and Europe

- Plate or Sheet Elements
  - Surfaces without protuberances
    - Margin entire ... Ib
    - Margin not entire
      - Sinuous ... Ia
      - Undulate interlocking ... Ie
  - Surface with protuberances
    - Protuberances branched dendritically ... Ii
    - Protuberances unbranched
      - Crateriform ... Ih
      - Not crateriform
        - Irregularly dentate on margins ... I2a
        - Clavate over entire surface ... I2g
- Not a plate or sheet element
  - Long cell appendage, small
    - Prickle: 1-celled, robust, swollen base
      - Tip does not exceed base, Small prickle
        - Tip acute ... IIa1
        - Tip papillate ... IIa2
        - Tip conical ... IIa3
      - Tip exceeds base, Large prickle
        - Base elongate ... IIb1
        - Base circular ... IIb2
  - Long cell appendage, Large
    - Macrohair:
      - Visible with unaided eye or × 5–10 ... IIc1
      - Smaller, not visible with unaided eye ... IIc2
  - Trichome base
    - Annular margin entire along at least two quadrants ... IIIa
    - Annular margin lobed and cupped ... IIIb
  - Trapezoid
    - Short; Margins entire ... Vb1
    - Length at least 2× width
      - Margins entire ... Vb2
      - Margins not entire
        - Margins alate, lobed
          - Flat top ... Va2a
          - Ridged top ... Va2b
        - Margins not alate
          - Sinuous with flat or concave top ... Va1a
          - Sinuous with 1-ridged top ... Va1b
          - Sinuous with 2-ridged top ... Va1c

9

ILLUSTRATED PHYTOLITHS FROM ASSORTED FOOD PLANTS

LINDA SCOTT CUMMINGS

ABSTRACT

Various food plants were examined for phytoliths. Identification of cultivation or availability of food plants may rest with recovery of phytoliths from the edible or even inedible portions of the plants. Recovery of phytoliths from coprolites represents the edible portions, and may be used to identify plants that contributed to the diet. Occasionally another portion of the plant produces yet another type of phytolith, which might be recovered from plant parts that decay in place when examining sites such as garden plots.

INTRODUCTION

Identification of phytoliths from food items has significant potential to contribute to our understanding of prehistoric diets. Studies of prehistoric and historic coprolites may benefit from phytolith analysis, provided that the diet of the individuals or population studied included silica and/or calcium oxalate crystals. In this study, the term "phytolith" is used to include opal phytoliths as well as crystalline structures produced by plants. Analysis of ground stone artifacts, features or floor samples for phytoliths may also contribute to our understanding of subsistence systems in areas where diets contained plants that produce silica or calcium oxalate crystals. A recent study of prehistoric diet (Cummings 1989) not only relied upon phytoliths to identify a portion of the plants consumed, but also demonstrated the necessity of examining pollen, phytoliths and macrofloral remains to reconstruct the vegetal portion of the diet.

This description of phytoliths is an outgrowth of the study of a group of coprolites from the Sudan in Africa (Cummings 1989). Modern plants examined for phytoliths include items of the diet expected to occur for that population, other related plants and a few miscellaneous food and disturbance plants (Table 9.1). This study addresses silica and/or crystalline structures that may be recovered

Phytolith Systematics, Edited by G. Rapp, Jr. and
S.C. Mulholland, Plenum Press, New York, 1992

TABLE 9.1. Families and Plants for which Phytoliths Were Examined

Family	Scientific Name	Common Name
Araceae	*Alocasia macrorhiza*	Ape, taro
	Colocasia antiquorum esculenta	Taro
Cactaceae	*Opuntia*	Prickly pear
Capparidaceae	*Cleome*	Beeweed
Chenopodiaceae	*Beta vulgaris*	Beet
	Spinacia oleracea	Spinach
Brassicaceae (Cruciferae)	*Brassica oleracea botrytis cymosa*	Broccoli
	Nasturtium officinale	Watercress
Cucurbitaceae	*Luffa* sp	Luffa
Euphorbiaceae	*Ricinus*	Castor bean (not a food item but a common disturbance plant in the tropics)
Fabaceae	*Vigna repens*	Cow pea, black-eyed pea
(Leguminosae)	*Vigna luteola*	Cow pea, black-eyed pea
	Pisum sp	Sweet pea
	Phaseolus	Bean
	Prosopis	Mesquite
	Tamarindus	Tamarind
Liliaceae	*Allium*	Onion
	Asparagus	Asparagus
Malvaceae	*Hibiscus esculentus*	Okra
Palmae	*Cocos nucifera*	Coconut
	Phoenix	Date
Apiaceae	*Apium*	Celery
(Umbelliferae)	*Daucus*	Carrot
	Petroselinum	Parsley
Vitaceae	*Vitis* sp	Grape

from a diverse group of plants that may contribute to the diet in such geographically widely spaced populations as the Sudan in Africa, the historic Southeastern United States, the prehistoric Southwestern United States and the Pacific Islands, among others.

Plant families have not been methodically examined for their contribution to the phytolith record. Instead, suspected dietary elements were selected individually for examination of phytolith remains. This paper does not address exclusivity of recovery of phytoliths from any genera or families of plants.

EXTRACTION METHODS

Extraction of phytoliths and crystals from plant reference specimens was based on the use of hydrogen peroxide and an occasional catalyst. Extraction of phytoliths and crystals from coprolites did not utilize any acids, but instead relied upon the use of hydrogen peroxide, often with a catalyst and heavy liquid separation (sodium polytungstate or zinc bromide mixed without HCl). Removal of phytoliths and crystals from sediments must use no acids if the crystals are to be preserved. Instead, extraction should use a combination of clay removal (sodium pyrophosphate, calgon, etc.), organic destruction (hydrogen peroxide with or without a catalyst) and heavy liquid separation (zinc bromide mixed without HCl or sodium pyrophosphate, etc.). If calcium carbonates are present, use of acids may be necessary to release phytoliths from the sediment matrix. In this case, all crystals will be destroyed. This is, however, a conscious decision to be made in cases where calcium carbonates are present.

FOOD GROUP PHYTOLITHS

Araceae

Alocasia and *Colocasia* (both taro and members of the Araceae; see Figs 9.1 and 9.2) produce an underground stem or tuberous root, which has been an important vegetable in India, Asia, Australia and the islands of the Pacific, as well as in tropical America (Hedrick 1972: 40–41). In Hawaii, *Alocasia* is referred to as Chinese taro, while *Colocasia* carries the common name taro. The leaves of both plants produce calcium oxalate druses in varying sizes. The druses are pointed, but irregular. Raphids are also noted. No roots or tubers were available for study at the time the leaves were collected, so presence or absence of these forms in the edible portion cannot be addressed. Piperno (1988: 21) notes that silica phytoliths have never been observed in members of the Araceae family. Genua and Hillson (1985: 352–360) note the presence of druses and bundles of raphids in *Colocasia*.

Cactaceae

Opuntia (prickly pear cacti) were also examined for phytoliths and found to contain relatively abundant spheroid spiny forms. A detailed description of *Opuntia* phytoliths may be found in Jones and Bryant (this volume).

Capparidaceae

Cleome (beeweed) is distributed throughout the American Southwest, where it was used prehistorically as a food plant. No opal phytoliths were recovered from either leaves or stems. *Cleome* was used both as a food and a pottery paint. The leaves, as well as stems and flowers, may be gathered when young or mature,

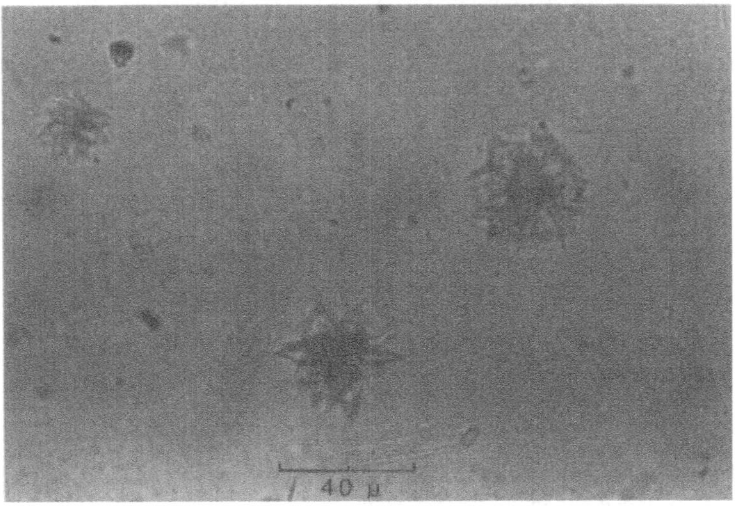

Fig 9.1. *Alocasia* (Chinese taro) leaf phytoliths. Bar is 40 microns.

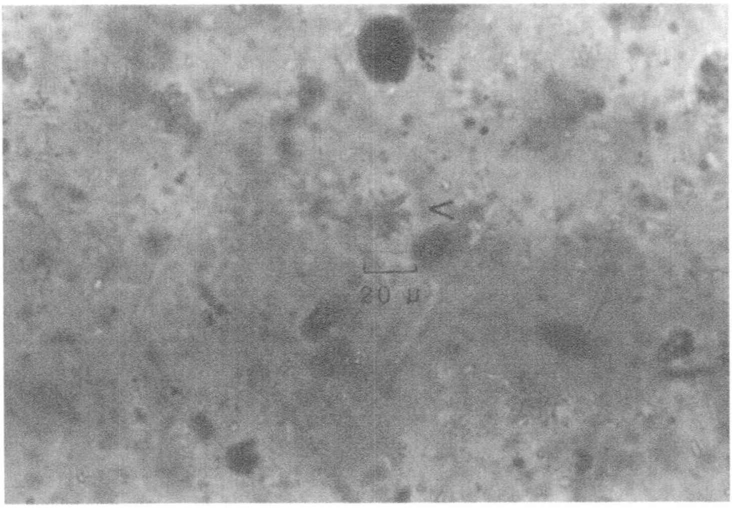

Fig 9.2. *Colocasia* (taro) leaf phytoliths. Bar is 20 microns.

and boiled. The fluid resulting from cooking may be used as a pottery paint, or the greens eaten or dried and reconstituted later for use as either food or paint. The seeds may also be gathered and ground into meal, although utilization as a potherb appears to have been more common (Stevenson 1915: 69, 82; Robbins, Harrington and Freire-Marreco 1916: 58–59; Whiting 1939: 77–78; Harrington 1967: 72). *Cleome* has been allowed to grow in gardens with cultivated plants.

Chenopodiaceae

Both *Beta* (beet) and *Spinacia* (spinach) leaves were examined from the family Chenopodiaceae. Beet leaves contained sclereids, but no phytoliths were recorded for this genus. Spinach leaves, however, contained both the sclereid form and druses of varying sizes (Fig 9.3). The spinach druses varied from approximately 20 microns to more than 60 microns in size and tended to look very irregular. The outer edges of the druses were frequently rounded or blunted, rather than highly pointed. Piperno (1988: 21) notes that the Chenopodiaceae may produce phytoliths, which may be common or rare, and that phytoliths are absent from some members of this family.

Brassicaceae (Cruciferae)

Two members of this family were examined: *Brassica oleracea botrytis cymosa* (broccoli) and *Nasturtium officinale* (watercress). Broccoli leaves contained sclereids and an occasional small druse (Fig 9.4). Some apparent druses exhibited four to six points, were very faint, and appeared not to be three-dimensional. Others were very similar to spinach druses. Watercress leaves were also examined for phytoliths. No druses were observed in this sample.

Cucurbitaceae

A single member of the Cucurbitaceae family (*Luffa*) was selected for examination (Fig 9.5). *Luffa* is an Old World plant cultivated in India as a food frequently used in curry. Two species are noted: *L. acutangula* (strainer vine) and *L. aegyptiaca* (bonnet gourd, dish-cloth gourd and loof). This is the same luffa whose fruit, when ripe, is peeled, dried and used as a "fleshrubber." Luffa sponges are sold in many U.S. boutiques to use as bath sponges. When young, however, the fruit may be boiled, buttered and seasoned with salt and pepper (Hedrick 1972: 341). The fibers of a mature luffa fruit were examined. *Luffa* produces a pyramid crystal with a square base. This phytolith is very similar to and easily confused with that produced by *Hibiscus* leaves. The phytoliths measured approximately 7 to 15 microns on a side. Piperno (1988: 21) notes that members of the Cucurbitaceae vary in their production of phytoliths. Bozarth

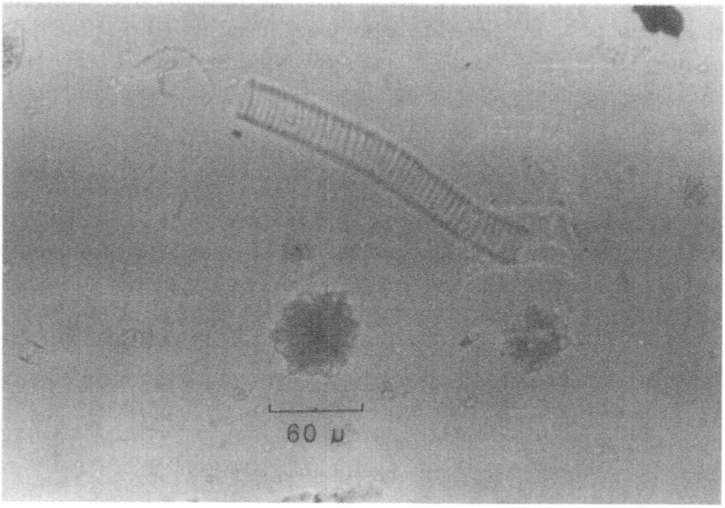

Fig 9.3. *Spinacia* (spinach) leaf phytoliths. Bar is 60 microns.

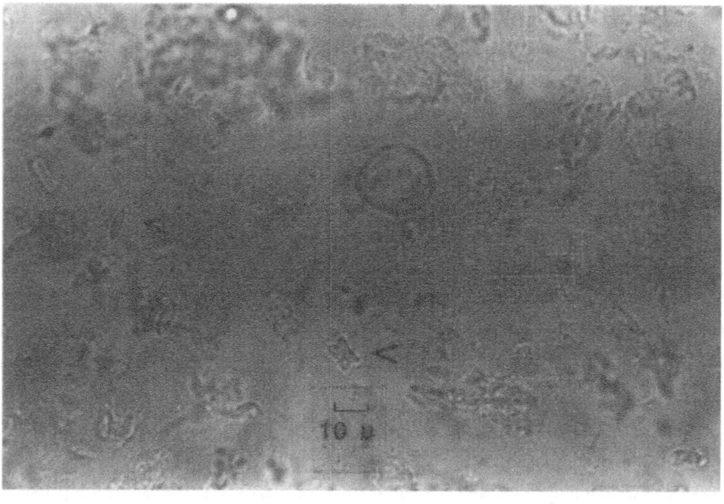

Fig 9.4. *Brassica oleracea botrytis cymosa* (broccoli) phytoliths. Bar is 10 microns.

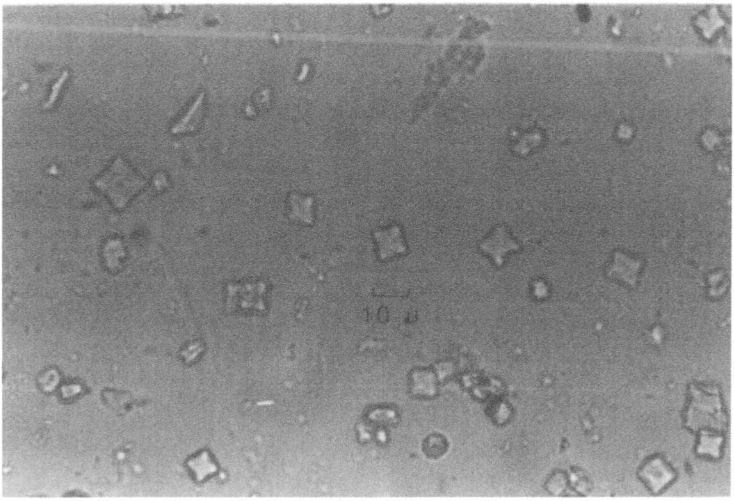

Fig 9.5. *Luffa* (luffa) fruit interior phytoliths. Bar is 10 microns.

(this volume) discusses phytoliths in *Cucurbita* rinds. Segmented hairs produced by *Luffa* are illustrated in Piperno (1988: 227).

Euphorbiaceae

A single member of the Euphorbiaceae was examined for phytoliths (Fig 9.6): *Ricinus* (castor bean). Hedrick (1972: 503) notes that *Ricinus* grows in the tropics. Although this plant produces an oil used in cooking in various areas, all portions of the plant are poisonous. It is, however, a common disturbance plant in the tropics. *Ricinus* leaves were processed to release phytoliths, and were noted to produce a druse, slightly irregular in form and similar to spinach, except more pointed. These druses were noted to vary from less than 10 microns to approximately 60 microns in diameter. Some of the larger druses exhibited more irregularity in form. Production of phytoliths in Euphorbiaceae is recorded as often common to abundant (Piperno 1988: 21).

Fabaceae (Leguminosae)

Numerous specimens from the family Fabaceae (Leguminosae) were examined for their contribution to the phytolith record. These include: *Vigna repens* and *Vigna luteola* (cow pea/black-eyed pea), *Phaseolus vulgaris* (common bean), *Pi-*

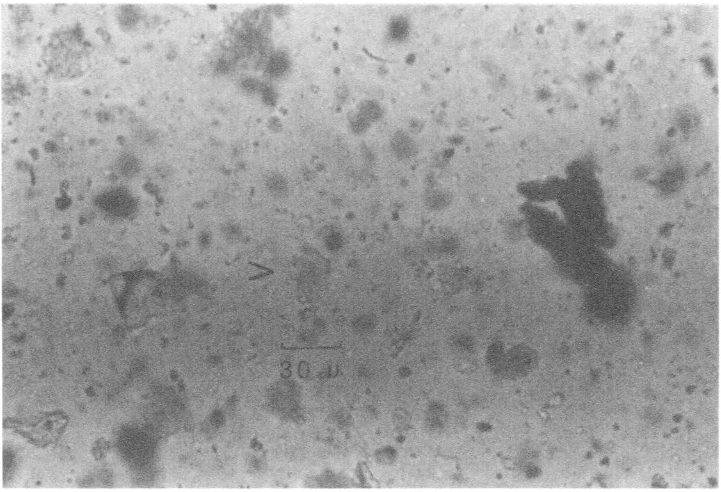

Fig 9.6. *Ricinus* (castor bean) leaf phytoliths. Bar is 30 microns.

sum (sweet pea) and *Tamarindus* (tamarind tree) (Figs 9.7, 9.8, 9.9). Some of these plants are represented by both leaf and pod samples, whereas others are represented by either leaf or pod samples, being limited by availability of part at the time of extraction. Recovery of legume-type crystals was consistent between leaf and pod samples when both were available from the same species. However, crystal recovery was not consistent within the family. For instance, crystals were recovered from *Vigna* leaves and pods, but were absent from both *Phaseolus* leaves and pods.

Hedrick (1972: 597) notes that the genus *Vigna* grows in the East Indies, India, Portugal, Italy, is a native of tropical Africa and is cultivated on the upper Nile. Both the pods and leaves are noted by various authors to have been eaten (Corkhill 1949; Culwick 1951; Hedrick 1972: 597). *Pisum* is noted to have been grown in a variety of locations including Eurasia, Egypt, India, China, England, Scotland and northern Asia (Hedrick 1972: 441–442). *Phaseolus* is noted to have grown in the East Indies, India, Chile, Malay, the tropics (American tropics, tropical Asia, tropical Africa), South America, Mexico, western North America and Asia. At present, *Phaseolus vulgaris* is distributed nearly worldwide (Hedrick 1972: 418–427). *Prosopis* (Fig 9.10) is noted to grow in tropical America, Chile, Argentina, Texas, Mexico, California, Persia and the East Indies (Hedrick 1972: 454–455). *Tamarindus* is an Asian and tropical African tree, the fruit of which is edible (Hedrick 1972: 562).

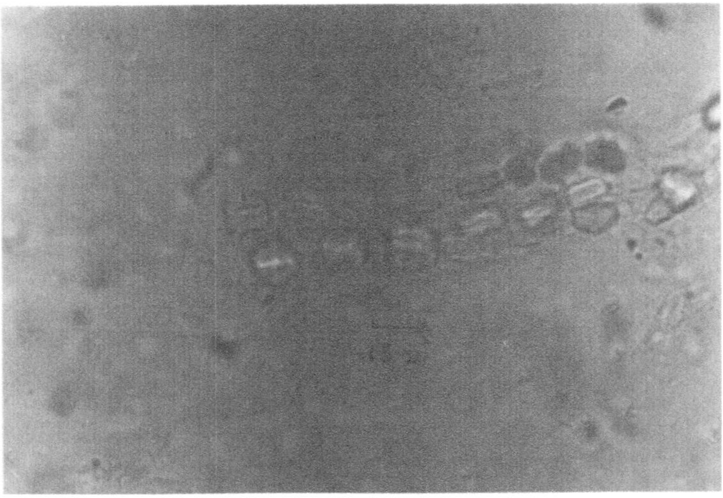

Fig 9.7. *Vigna luteola* (cow pea, black-eyed pea) leaf phytoliths. Bar is 15 microns.

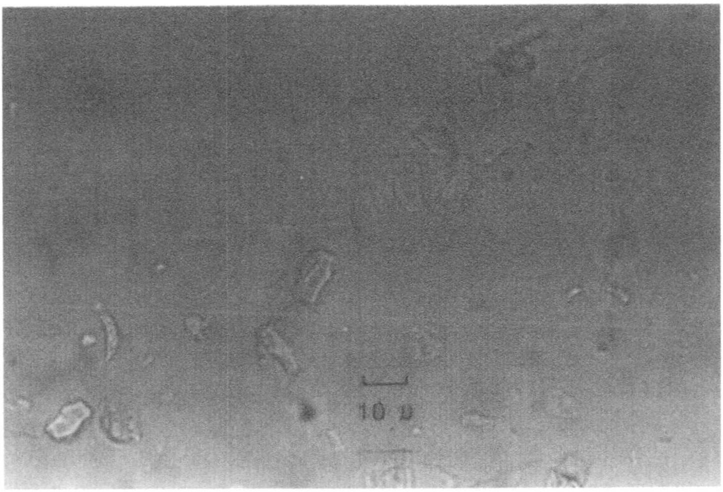

Fig 9.8. *Vigna repens* (cow pea, black-eyed pea) leaf phytoliths. Bar is 10 microns.

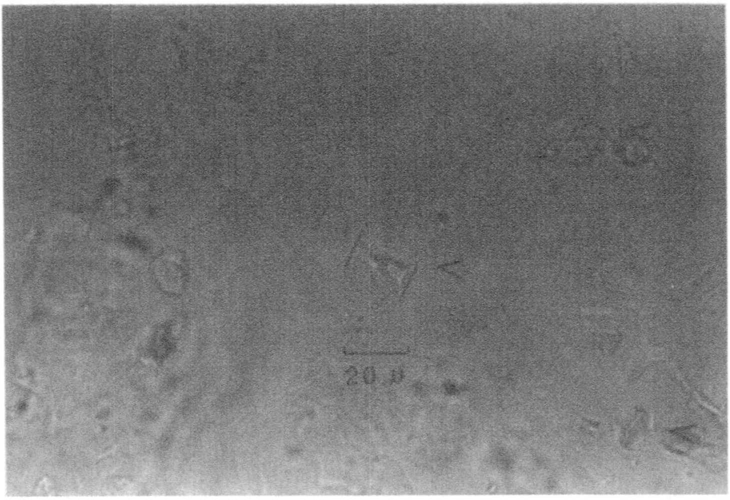

Fig 9.9. *Tamarindus* (tamarind) leaf phytoliths. Bar is 20 microns.

Fig 9.10. *Prosopis* (mesquite) pod phytoliths. Bar is 15 microns.

Crystals recovered from the legume leaves and pods analyzed are similar to one another in shape, and thus have an identification potential at the family level. The phytoliths are rhomboidal. All forms are three-dimensional, and have squared ends, although some resemble cubes superficially. There is some variation in the shape of these phytoliths that may be dependent, in part, on orientation of the phytolith within the medium. Rarely do the phytoliths appear to resemble perfect cubes. More often than not the long sides of the approximately cube-shaped cells appear to have angles in them, frequently two angles, forming what appears to be a three-segmented side.

These phytoliths are easily recognized and distinguishable from pyramid-shaped phytoliths with square bases, which have been noted in *Luffa* and *Hibiscus* (okra). Legume phytoliths may vary in size from 5 or 6 microns to nearly 20 microns, according to observed specimens. The phytoliths may be observed in linear groups within plant cells when reference specimens have been extracted using acids, and some organic matter remains.

Vigna leaves and pods, *Prosopis* pods and *Tamarindus* leaves all exhibit typical legume crystals. The *Phaseolus* and *Pisum* leaves and pods examined did not contain any crystals. Parry and Winslow (1977: 276) note: "It is evident that peas normally have no mechanism comparable with grasses to allow silicon to be absorbed."

Other forms noted as present include sclereids, bases of hair cells, and entire hair cells. Hair cells vary in shape, and are not thoroughly described and illustrated here. All are hollow. Hair cell bases vary from apparently circular to ovoid, with radiating spikes or attachments at the base or point of attachment of the hair cell to the plant. The number and distribution of these spikes apparently vary with genus, and possibly with species. No attempt was made to classify hair cell bases to ascertain identification potential.

Considerable work within this family is recommended to assess interpretability of phytolith identification within the family; *ie*, can genera be separated based on morphological characters? At present, all identifications have been made at the family level.

Prehistoric or archaeological evidence for the presence of Fabaceae (Leguminosae) phytoliths is noted by Cummings (1989). A report on identifiability of hook-shaped hairs attributed to the genus *Phaseolus* should be consulted for identification of these hairs (Bozarth, this volume). For interpretations in the American Southwest, the importance of the recovery of legume phytoliths from mesquite pods (and probably pods of other desert legumes) should not be underestimated. At present, recovery of macrofloral evidence for the use of mesquite in the Hohokam area is far more abundant than recovery of pollen evidence. Although it has not been tested, it is probable that if phytolith analysis were employed in this culture area, use and processing of mesquite pod could be more accurately identified. Southwestern American Indians of the Sonoran Desert valued mesquite pods highly, and ground them into flour.

Liliaceae

Asparagus (asparagus) and *Allium* (onion) were both examined for phytoliths. Phytoliths are reported as absent from this family (Piperno 1988; Pearsall 1989). No opal phytoliths were recovered from *Allium*, although raphids were present in *Asparagus* stem. Piperno (1988: 21) records no phytoliths in the Liliaceae family.

Malvaceae

Hibiscus esculentus (okra) was examined for phytoliths (Figs 9.11, 9.12), because Bock, Lane and Norris (1988) note their presence in okra. Okra pods were examined and found to produce a rectangular crystal. The base is very broad and the pyramid top is low. This form is slightly similar to that of *Luffa* and might occasionally be confused with it. Both *Hibiscus* and *Luffa* grow in the Old World tropics. Okra leaves, however, produce a druse rather than a rectangular crystal. Malvaceae are variable in their production of phytoliths, which may be absent, rare or common (Piperno 1988: 21).

Palmae

Palmae produce a spherical, spiny phytolith in all members of the family examined. Piperno (1988: 21) notes phytolith production as common to abundant in this family. Kondo (1977: 200) also illustrates Palmae phytoliths. Not all members of this family produce phytoliths. Frison (1948: 39–40) used presence or absence of Palmae phytoliths to distinguish between *Palmyra* and Talipot palm used to make writing paper. Several elements of *Cocos nucifera* (coconut) were examined (Fig 9.13) to observe distribution of the phytolith throughout the tree and fruit, including the outer husk, inner husk, fronds and frond bases. All portions of the plant examined produced the same type of phytoliths, which are also described in Piperno (1988). *Phoenix* (date) phytoliths are also spiny, spherical bodies that vary in size from less than 10 microns to approximately 25 microns (Fig 9.14). They are abundantly produced by edible dates, and contribute large quantities to the subsistence record in coprolites from the Sudan (Cummings 1989).

Apiaceae (Umbelliferae)

Plants in the Apiaceae (Umbelliferae) examined for phytoliths include: *Apium* (celery), *Daucus* (carrot) and *Petroselinum* (parsley). *Apium* (celery) grows from

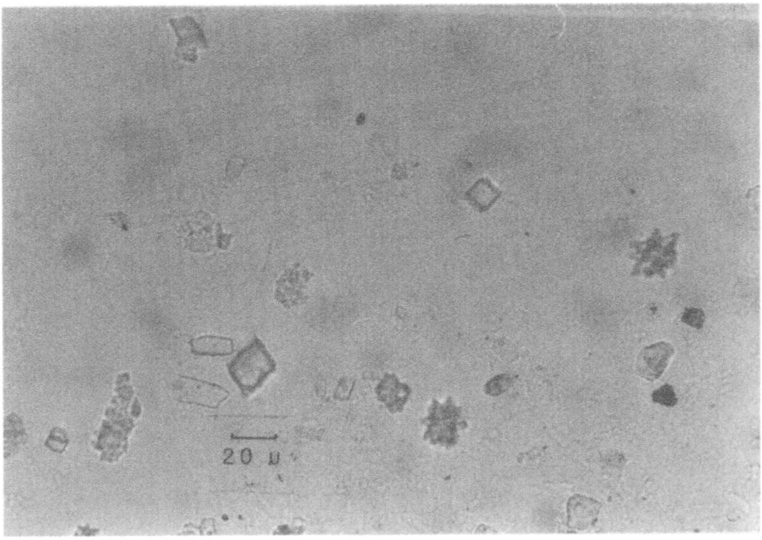

Fig 9.11. *Hibiscus esculentus* (okra) fruit phytoliths. Bar is 20 microns.

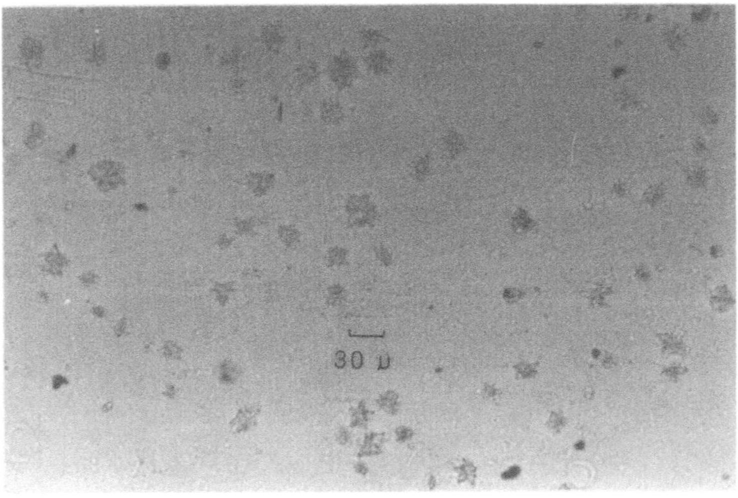

Fig 9.12. *Hibiscus esculentus* (okra) leaf phytoliths. Bar is 30 microns.

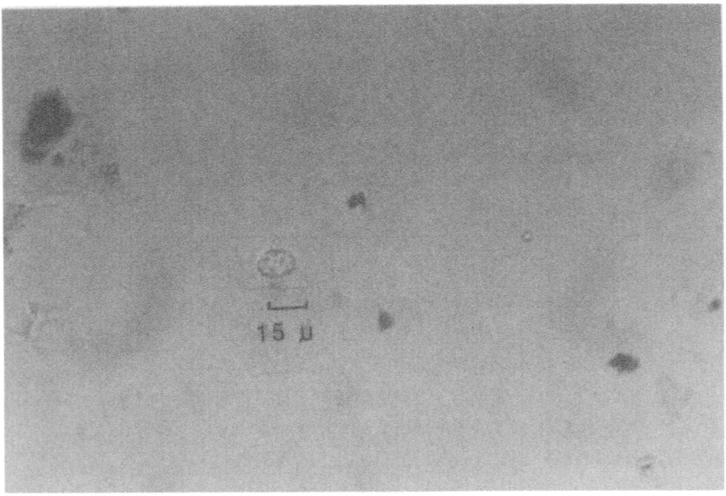

Fig 9.13. *Cocos nucifera* (coconut) husk phytoliths. Bar is 15 microns.

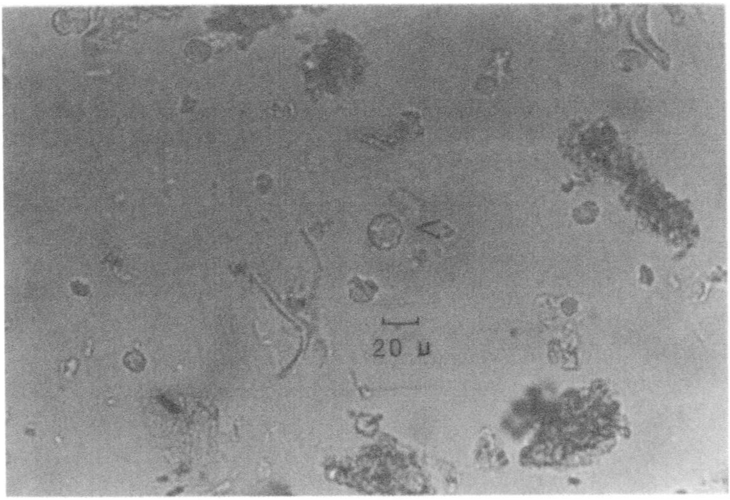

Fig 9.14. *Phoenix* (date) fruit phytoliths. Bar is 20 microns.

Sweden to Algeria, Egypt, Abyssinia, in Asia to the Caucasus, Baluchistan, the Mountains of British India, and also in Tierra del Fuego, California and New Zealand. Celery has a long history as a medicinal plant, but ancient references to use as food appear to be rare (Hedrick 1972: 55–57). Both the leaves and stalk were examined for phytoliths. Sclereids were noted in the leaves as well as the stalk. No opal phytoliths were noted and no forms were present that appeared to be unique to celery or the umbell family.

Daucus is noted to have originated in Europe and the adjoining portions of Asia. The root has long been reported as an article of food (Hedrick 1972: 232). Large sclereid forms were observed in the root examined. No identifiable forms were noted in the leaf sample.

Parsley (*Petroselinum*) originated in the Mediterranean area. Parsley leaves were examined for phytoliths, but no identifiable forms were noted. No opal silica was observed.

Vitaceae

Vitis is the genus containing cultivated and wild grapes. This genus appears to be distributed widely throughout the world, being noted in Australia, Africa, North America, the West Indies, the East Indies, Java, China, Japan, Sumatra, Borneo, India, Asia, Arabia, the Himalayan regions, tropical areas and Europe. The berries are of course edible. In addition, the leaves are eaten in several areas, and are frequently described as acid. The leaves are noted to have proven valuable in cases of scurvy (Hedrick 1972: 598–603).

Crystals recovered from *Vitis* leaves include druses and raphids (Fig 9.15). Raphids tend to form large bundles and may be observed in bundles or singly. Druses are irregular in shape, resembling spinach, but more pointed, and vary in size from less than 10 microns to approximately 20 microns.

Examination of edible plants for phytoliths will assist in building a data base that will allow identification and interpretation of subsistence bases in the archaeological record. Identification of edible plants as phytolith producers or non-producers will contribute significantly to research design and sampling design preparation—the stages at which archaeologists should determine which analyses are most likely to provide interpretable results. A thorough understanding of the modern plants available in a geographic area, and whether or not those plants produce distinct phytoliths, contributes to the decisions about which studies to employ. At present, it is extremely likely that phytolith analysis is under-utilized as a tool for identifying subsistence items from archaeological sites in many areas. Preliminary reference work to identify phytoliths that may be recovered from potential food plants is necessary to identify situations in which phytolith analysis could assist in the identification of a subsistence base. The brief descriptions and

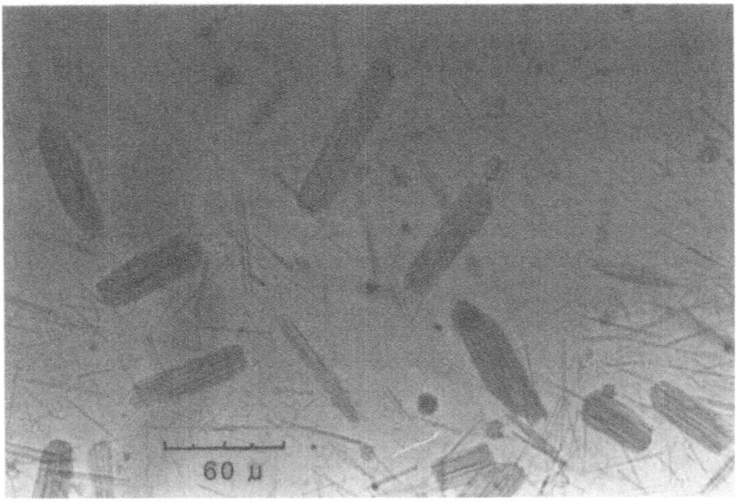

Fig 9.15. *Vitis* (grape) leaf phytoliths. Bar is 60 microns.

illustrations of phytoliths of this study provide information regarding a wide variety of foods that produce at least some remains in the phytolith record.

CONCLUSIONS

Phytoliths, including crystals, were invaluable in reconstructing the diet of early and late Christian Nubians in the Sudan (Cummings 1989). For instance, they verified the importance of dates in the daily diet during both the early and late Christian occupations at Kulubnarti (Figs 9.16, 9.17). Phytoliths also provided evidence that legumes were consumed more often than indicated by the pollen record alone. No macrofloral legume remains were recovered, as expected. Identification of legume phytoliths, however, was unexpected, and lead to identification of consumption of legume leaves and/or pods. Legume leaves are noted to contribute to recent diets in that area (Corkhill 1949; Culwick 1951). That portion of the diet certainly could not have been identified from other data bases examined.

Many of the food plants examined contained crystals, druses and/or raphids rather than opal phytoliths. This recovery of crystals, which are destroyed if acids are used in processing, points to the importance of exploring the use of methods that retain as much as possible of the crystal as well as opal phytolith data base.

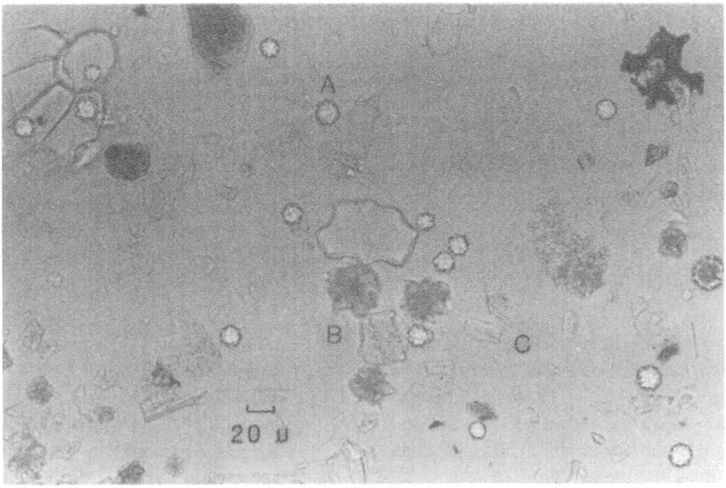

Fig 9.16. Coprolite photograph. Bar is 20 microns. A. *Phoenix*; B. cf *Spinacia*; C. *Vigna*.

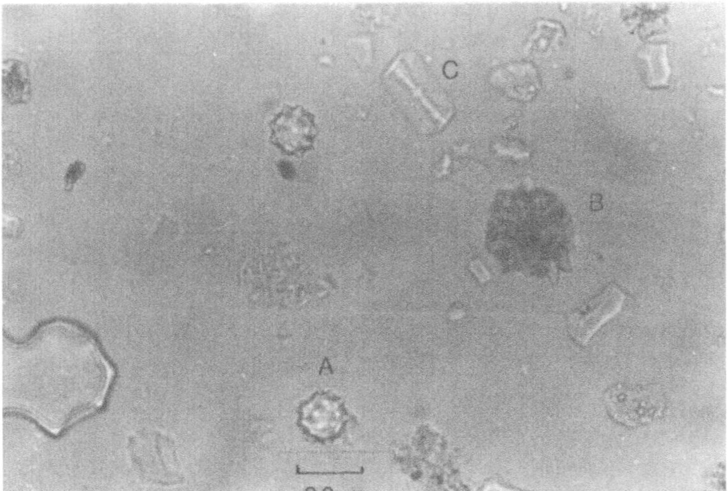

Fig 9.17. Coprolite photograph. Bar is 20 microns. A. *Phoenix*; B. cf *Spinacia*; C. *Vigna*.

Use of these methods will not be possible in all studies nor with all types of sediments or matrices.

ACKNOWLEDGMENTS

I would like to acknowledge Gary McVicker for the use of his darkroom to develop and print the photographs, and for his valuable advice concerning photography.

REFERENCES

Bock, JH, Lane, MA and Norris, DO 1988 *Identifying Plant Food Cells in Gastric Contents for Use in Forensic Investigations: A Laboratory Manual.* Washington, DC, US Department of Justice, National Institute of Justice: 130 p.

Corkhill, NL 1949 Scorbutic diet in a Nile Cataract community. *Transactions of the Royal Society of Tropical Medicine and Hygiene* 43: 293–302.

Culwick, GM 1951 *Diet in the Gezira Irrigated Area, Sudan.* Sudan Survey Department, No 304.

Cummings, L Scott (ms) 1989 Coprolites from medieval Christian Nubia: An interpretation of diet and nutritional stress. PhD dissertation, University of Colorado: 204 p.

Frison, E 1948 On some typical forms of silica found in plants. *Microscope* 7: 38–45.

Genua, JM and Hillson, CJ 1985 The occurrence, type and location of calcium oxalate crystals in the leaves of fourteen species of Araceae. *Annals of Botany* 56: 351–361.

Harrington, HD 1967 *Edible Native Plants of the Rocky Mountains.* Albuquerque, University of New Mexico Press: 392 p.

Hedrick, UP, ed 1972 *Sturtevant's Edible Plants of the World.* New York, Dover Publications, Inc: 686 p.

Kondo, R 1977 Opal phytoliths, inorganic, biogenic particles in plants and soils. *Japan Agricultural Research Quarterly* 11: 198–203.

Parry, DW and Winslow, A 1977 Electron-probe microanalysis of silicon accumulation in the leaves and tendrils of *Pisum sativum* (L.) following root severance. *Annals of Botany* 41: 275–278.

Pearsall, DM 1989 *Paleoethnobotany: A Handbook of Procedures.* San Diego, Academic Press: 470 p.

Piperno, DR 1988 *Phytolith Analysis: An Archaeological and Geological Perspective.* San Diego, Academic Press: 280 p.

Robbins, WW, Harrington, JP and Freire-Marreco, B 1916 Ethnobotany of the Tewa Indians. *Bureau of American Ethnology Bulletin* 55: 124 p.

Stevenson, MC 1915 Ethnobotany of the Zuni Indians. *Thirtieth Annual Report of the Bureau of American Ethnology.* Washington, DC, US Government Printing Office: 31–102.

Whiting, AF 1939 Ethnobotany of the Hopi. *Museum of Northern Arizona Bulletin* 15: 120 p.

10

CLASSIFICATION OF OPAL PHYTOLITHS FORMED IN SELECTED DICOTYLEDONS NATIVE TO THE GREAT PLAINS

STEVEN R. BOZARTH

ABSTRACT

This study describes diagnostic phytoliths from dicotyledons native to the Great Plains. The study is based on analysis of an extensive phytolith reference collection that includes 25 grass species, 11 non-grass monocot species, 65 herbaceous dicot species, 20 woody dicot species and 14 gymnosperm species. Non-dicot species are included in the study for comparison with dicot phytoliths to determine at what level dicot phytoliths are taxonomically useful. Analysis of these reference species indicates that there are at least nine major phytolith types formed only in dicots. Four of the phytolith types identified are formed in both herbaceous and woody dicots. The other types, and several subtypes, appear to be specific to certain dicot families or genera. Only phytoliths produced in dicots are reported in this study.

INTRODUCTION

Phytolith analysis of dicotyledons, particularly herbaceous dicotyledons, has largely been ignored by many phytolith researchers in favor of grass phytoliths, thereby creating a void in phytolith classification. This study is an effort to address that void by describing certain types of diagnostic dicot phytoliths. It is based on analysis of an extensive reference collection that consists of the following: (1) 25 species from 20 genera representing 11 tribes of 6 subfamilies of the Poaceae (Gramineae); (2) 11 species from 4 genera of 4 non-grass monocot families; (3) 61 species from 55 genera representing 10 families of herbaceous dicots; (4) 20 species from 18 genera representing 13 families of woody (mostly arboreal) dicots; (5) 14 species from 7 genera representing 5 families of Gymnosperms; and (6) 2 species of *Equisetum*. All of these species are native to the Great Plains.

Phytolith Systematics, Edited by G. Rapp, Jr. and S.C. Mulholland, Plenum Press, New York, 1992

Reference materials were processed using a modified version of a wet-digestion procedure reported by Piperno (1988). The first step involves thorough cleaning of the plant materials, which were then oxidized with Schulze's solution (three parts concentrated nitric acid to one part saturated potassium chlorate) in a hot water bath. Complete or representative aerial portions (leaves with petioles and stem sections from various parts of the plant plus part of the inflorescence) were processed for each reference species.

PHYTOLITH TYPES

Analysis of the various reference species demonstrates that there are a number of different types of phytoliths with taxonomic value produced in dicots. They are described below in order of their frequency of occurrence. The scientific and colloquial names of the plants that produce these phytoliths are given in Table 10.1. Only phytoliths formed in dicots are reported in this study, and typical examples of the various phytolith types are illustrated in photomicrographs.

POLYHEDRONS

Silicified polyhedral (4–8 sided) epidermal cells are the most common type of dicot phytolith. Polyhedral phytoliths are formed in leaves in many deciduous trees (Rovner 1971; Wilding & Drees 1971; Geis 1973; Wilding, Smeck & Drees 1977). The amount of silicification in the epidermal cell varies, depending on the species, from the walls being weakly silicified to the lumen being completely filled. Length of the long axis varies significantly among certain species (Geis 1973). Polyhedral phytoliths similar to those reported by other researchers are formed in leaves of most of the woody dicots studied. They include: *Acer saccharum* (Aceraceae); *Betula papyrifera* and *Corylus cornuta* (Betulaceae); *Cercis canadensis* (Caesalpiniaceae); *Quercus alba* and *Q. palustris* (Fagaceae); *Carya glabra* (Juglandaceae); *Rosa blanda* (Rosaceae); *Populus deltoides* (Salicaceae) (Fig 10.1A); *Celtis occidentalis* (Ulmaceae) (Fig 10.1B); and *Ulmus americana* (Ulmaceae). Most of these phytoliths are formed in groups and consist only of silicified cell walls. A few completely silicified polyhedral phytoliths are also produced in most of the above species. Silicification of adjacent polyhedral cells produces a reticulate pattern.

It is important to realize that polyhedral phytoliths are not unique to arboreal dicots, as they are also produced in many of the herbaceous dicots studied. In the Asteraceae (Compositae), polyhedral phytoliths are formed in *Ambrosia psilostachya*, *A. trifida*, *Artemisia ludoviciana* (Fig 10.1C), *Eupatorium perfoliatum*, *Helianthus annuus*, *H. grosseseratus*, *H. maximilianii*, *Silphium laciniatum*, and

TABLE 10.1. Dicotyledonous Reference Species

Scientific name	Colloquial name
WOODY DICOTS	
Aceraceae	
Acer saccharum	sugar maple
Aristolochiaceae	
Aristolochia tomentosa	wooly pipevine
Betulaceae	
Betula papyrifera	paper birch
Corylus cornuta	beaked hazelnut
Caesalpiniaceae	
Cercis canadensis	redbud
Gymnocladus dioica	Kentucky coffee-tree
Fagaceae	
Quercus alba	white oak
Q. palustris	pin oak
Juglandaceae	
Carya glabra	pignut hickory
Juglans nigra	black walnut
Menispermaceae	
Menispermum canadense	moonseed
Mimosaceae	
Mimosa borealis	pink mimosa
Moraceae	
Morus rubra	red mulberry
Platanaceae	
Platanus occidentalis	sycamore
Rosaceae	
Rosa blanda	smooth wild rose
Salicaceae	
Populus deltoides	cottonwood
P. tremuloides	quaking aspen
Salix discolor	pussy willow
Ulmaceae	
Celtis occidentalis	hackberry
Ulmus americana	American elm

(Continued)

TABLE 10.1. (Continued)

Scientific name	Colloquial name
HERBACEOUS DICOTS	
Amaranthaceae	
Amaranthus retroflexus	rough pigweed
Asteraceae (Compositae)	
Ambrosia psilostachya	western ragweed
A. trifida	giant ragweed
Artemisia carruthii	wormwood
A. ludoviciana	white sage
Cirsium ochrocentrum	yellowspine thistle
Eupatorium perfoliatum	boneset
Helianthus annuus	common sunflower
H. annuus var *macrocarpa*	domesticated sunflower
H. grosseserratus	sawtooth sunflower
H. maximilianii	Maximilian sunflower
Iva annua	marsh elder
Silphium laciniatum	compass plant
Solidago rigida	rigid goldenrod
Xanthium strumarium	cockelbur
Boraginaceae	
Cryptantha minima	no known colloquial name
Cynoglossum officinale	hound's tongue
Hackelia virginiana	stickseed
Heliotropium convolvulaceum	heliotrope
Lappula redowskii	no known colloquial name
Lithospermum incisum	no known colloquial name
Myosotis verna	forget-me-not
Onosmodium molle	false Gromwell
Caesalpiniaceae	
Caesalpinia jamesii	James rush-pea
Cassia fasciculata	showy partridge pea
Hoffmanseggia glauca	Indian rush-pea
Cucurbitaceae	
Cucurbita foetidissima	buffalo-gourd
Sicyos angulatus	bur cucumber
Euphorbiaceae	
Cnidoscolus texanus	bull nettle

(Continued)

TABLE 10.1. (Continued)

Scientific name	Colloquial name
Fabaceae (Leguminosae)	
Amorpha canescens	lead plant
Amphicarpaea bracteata	hog peanut
Apios americana	groundnut
Astragalus crassicarpus	ground-plum
Baptisia leucantha	white wild indigo
Crotalaria sagittalis	rattlebox
Dalea aurea	golden prairie-clover
Desmodium illinoense	Illinois tickclover
Glycyrrhiza lepidota	wild licorice
Lathyrus polymorphus	hoary vetchling
Lespedeza capitata	round-head lespedeza
Lupinus pusillus	small or rusty lupine
Oxytropis lambertii	purple locoweed
Phaseolus polystachios	hog peanut
Psoralea tenuiflora	wild alfalfa
Sesbania macrocarpa	bequilla
Sophora nuttalliana	white loco
Strophostyles helvola	wild bean
S. leiosperma	piper
Stylosanthes biflora	pencil-flower
Tephrosia virginiana	goat's rue
Vicia americana	American vetch
Mimosaceae	
Acacia angustissima	prairie acacia
Desmanthus illinoensis	Illinois bundleflower
Mimosa borealis	pink mimosa
Prosopis glandulosa	honey mesquite
Schrankia nuttallii	sensitive brier
Polygonaceae	
Polygonum pensylvanicum	Pennsylvania smartweed
Urticaceae	
Boehmeria cylindrica	false nettle
Laportea canadensis	wood nettle
Parietaria pensylvanica	Pennsylvania pellitory
Pilea pumila	clear weed
Urtica dioica	stinging nettle

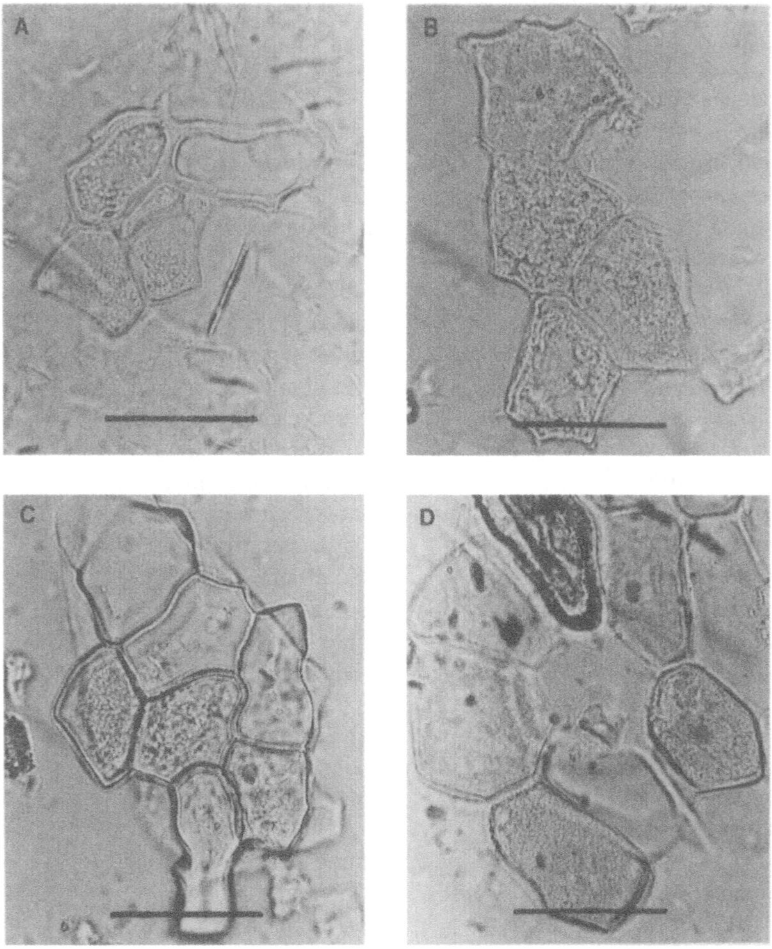

Fig 10.1. Silicified polyhedral epidermal cells. Bar is 40 microns. A. *Populus deltoides*; B. *Celtis occidentalis*; C. *Artemisia ludoviciana*; D. *Strophostyles helvola*.

Xanthium strumarium. In the Boraginaceae, they are produced in *Cynoglossum officinale*, *Cryptantha minima*, *Hackelia virginiana*, *Myosotis verna*, and *Onosmodium molle*. In the Fabaceae (Leguminosae), they are produced in *Amorpha canescens*, *Apios americana*, *Crotalaria sagittalis*, *Desmanthus illinoensis*, *Lespedeza capitata*, *Lupinus pusillus*, *Strophostyles helvola* (Fig 10.1D), *S. leiosperma* and *Tephrosia virginiana*. They are also produced in *Aristolochia tomentosa* (Aristolochiaceae), *Cassia fasciulata* (Caesalpiniaceae), *Cnidoscolus texanus* (Euphorbiaceae), *Menispermum canadense* (Menispermaceae), *Desmodium illinoense* (Mimosaceae) and *Boehmeria cylindrica* and *Urtica dioica* in the Urticaceae. As with the arboreal dicots, most of the polyhedral phytoliths consisted only of silicified cell walls. However, solid polyhedral phytoliths are produced in some of the above species.

Polyhedral phytoliths are not unique to dicots. They are also produced in certain conifers (Rovner 1971; Klein & Geis 1978; Bozarth 1988) in the form of multi-sided, irregularly shaped, silicified endodermal cells (Klein & Geis 1978). These usually have more than eight sides and do not have flat, parallel tops and bottoms, as do polyhedral dicot phytoliths.

Polyhedral phytoliths are also formed in grasses with the silicification of smooth-sided long cells and bulliform (fan-shaped) cells. The vast majority of the shorter long cells are rectangular and can be easily distinguished from polyhedral dicot phytoliths, which have more than four sides.

JIGSAW-PUZZLE PIECES

Phytoliths that have the shape of jigsaw-puzzle pieces, with wavy, undulating walls, are another common type of dicot phytolith. They are produced in many deciduous tree leaves (Rovner 1971; Wilding & Drees 1971; Geis 1973; Wilding, Smeck & Drees 1977). This is the same type of phytolith described by Piperno (1988) as anticlinal. The degree of silicification varies from silicification of the cell wall only to filling of the lumen with silica, thus producing a solid phytolith. Slightly wavy cell walls are common in some species, whereas phytoliths with more deeply undulating walls are produced in other species (Geis 1973).

Jigsaw-puzzle piece phytoliths are produced in leaves of many of the woody dicots studied, including *Acer saccharum* (Aceraceae), *Gymnocladus dioica* (Caesalpiniaceae), *Carya glabra* (Juglandaceae) (Fig 10.2A), *Morus rubra* (Moraceae), *Platanus occidentalis* (Platanaceae) (Fig 10.2B), *Rosa blanda* (Rosaceae), *Populus tremuloides* and *Salix discolor* (Salicaceae). These phytoliths most commonly occur in groups, and normally have only the cell wall silicified. However, in most of the species a few cells are completely silicified, *ie*, the lumina are filled with silica.

As with the polyhedral phytoliths, this type of phytolith is not unique to deciduous trees, because it is also produced in the aerial portion of many of the herbaceous dicots studied. In the Asteraceae (Compositae), jigsaw-puzzle piece phytoliths are formed in *Ambrosia psilostachya*, *Artemisia carruthii*, *A. ludoviciana*, *Eupatorium perfoliatum* (Fig 10.2C), *Helianthus grosseserratus*, *Solidago rigida* and *Xanthium strumarium*. In the Fabaceae (Leguminosae), they are formed in the aerial portion of *Amphicarpaea bracteata* (Fig 10.2D), *Apios americana*,

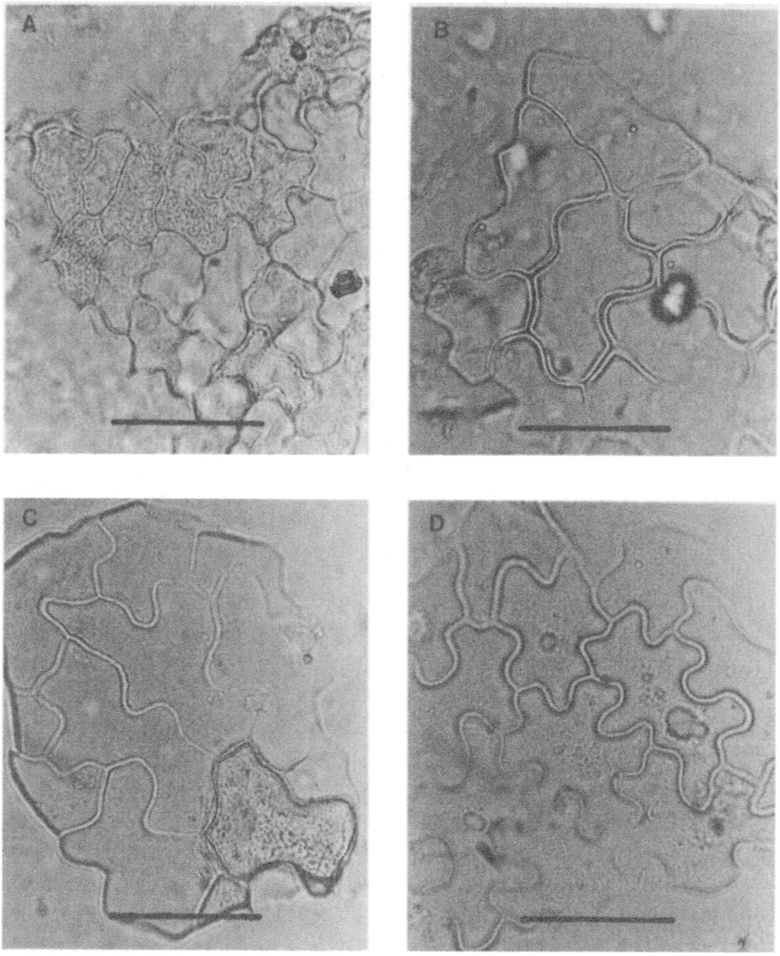

Fig 10.2. Silicified epidermal cells shaped like jigsaw-puzzle piece. Bar is 40 microns. A. *Carya glabra*; B. *Platanus occidentalis*; C. *Eupatorium perfoliatum*; D. *Amphicarpaea bracteata*.

Phaseolus polystachios, *Strophostyles helvola* and *S. leiosperma*. In the Urticaceae, they are formed in *Boehmeria cylindrica*, *Laportea canadensis*, *Parietaria pensylvanica* and *Urtica dioica*. They are also formed in *Cryptantha minima*, *Lappula redowskii*, *Lithospermum incisum* and *Myosotis verna* (Boraginaceae), *Cnidoscolus texanus* (Euphorbiaceae) and in *Acacia angustissima* (Mimosaceae). This type of phytolith most commonly occurs in groups with only the cell wall silicified. In some species, however, the lumina are completely silicified.

There were significant differences in this type of phytolith among certain species. For example, many of the silicified cell walls are much thicker in *Artemisia ludoviciana* (Fig 10.3A) than in the other reference species. The jigsaw-puzzle piece phytoliths produced in *Cryptantha minima* (Fig 10.3B) have an unusually irregular and narrow outline. Those in *Acacia angustissima* have squared processes.

Several members of Boraginaceae (*Cynoglossum officinale, Cryptantha minima* and *Lappula redowskii*) are distinguished from the other reference species (with the exception of *Scirpus pallidus*) by distinctive truncated cones having 1–5 "spires" centered in "star-shaped" jigsaw-puzzle pieces (Fig 10.3C). The multi-spired truncated cones formed in *S. pallidus* (Cyperaceae) differ from those in Boraginaceae in that the cell walls have many more individual waves. Piperno (1988) reported pointed cones with single to multi-crowned bodies in Cyperaceae in Panama. Further study will likely demonstrate other morphological differences in jigsaw-puzzle piece type phytoliths among the various species that produce it.

BRANCHED ELEMENTS WITH SPIRAL THICKENINGS

Silicified branched (anastomosing) tracheary elements are formed in leaves of certain species of deciduous angiosperms (Geis 1973). These silicified tracheids have spiral thickenings and a "branched" structure. They are formed in leaves of many of the arboreal dicots studied. These species include the following: *Acer saccharum* (Aceraceae); *Betula papyrifera* and *Corylus cornuta* (Betulaceae); *Carya glabra* (Juglandaceae); *Mimosa borealis* (Mimosaceae); *Platanus occidentalis* (Platanaceae); *Populus deltoides* and *P. tremuloides* (Salicaceae); and *Ulmus americana* (Ulmaceae).

The same type of phytolith is also formed in many herbaceous dicots. In the Asteraceae (Compositae), it occurs in *Ambrosia psilostachya, A. trifida, Artemisia ludoviciana, Helianthus grosseserratus* (Fig 10.3D) and *Solidago rigida*. In the Fabaceae (Leguminosae), it occurs in *Amorpha canescens, Apios americana, Astragalus crassicarpus, Crotalaria sagittalis, Dalea aurea, Desmodium illinoense, Lathyrus polymorphus, Lespedeza capitata, Lupinus pusillus, Oxytropis lambertii, Psoralea tenuiflora, Sophora nuttalliana, Strophostyles leiosperma, Stylosanthus biflora* and *Vicia americana*. It also occurs in *Amaranthus retroflexus* (Amaranthaceae), *Hackelia virginiana* (Boraginaceae), *Caesalpinia jamesii* (Caesalpinaceae), *Cnidoscolus texanus* (Euphorbiaceae), *Desmanthus illinoensis, Mimosa borealis, Prosopis glandulosa, Schrankia nuttallii* (Mimosaceae) and in *Pilea pumila* (Urticaceae).

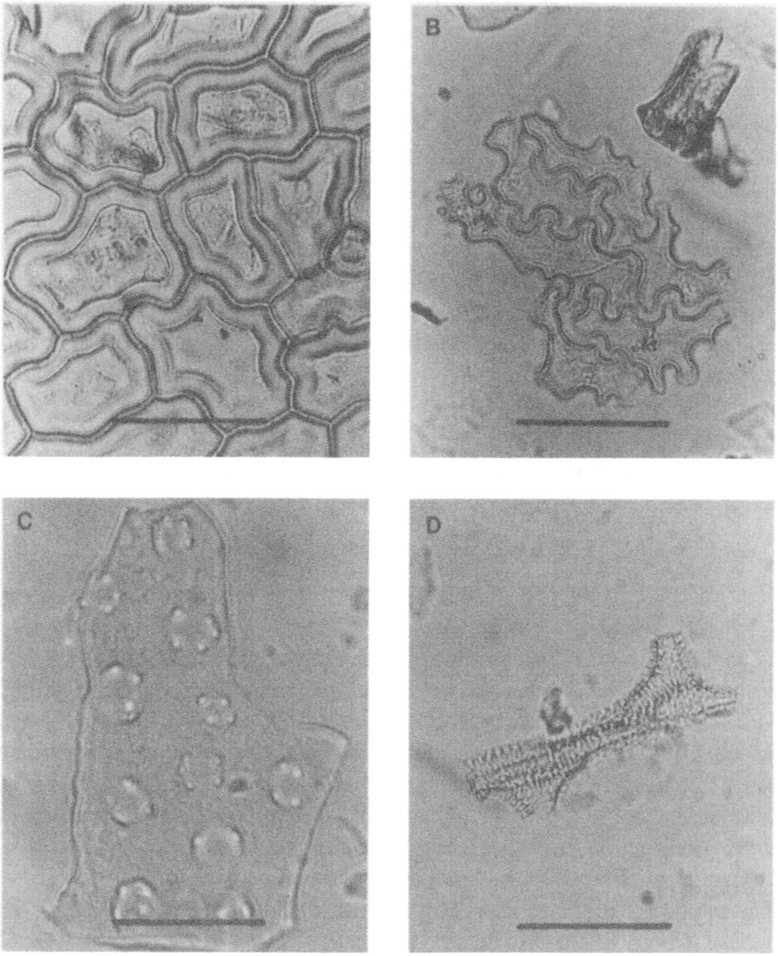

Fig 10.3. Silicified jigsaw-puzzle piece type phytoliths. Bar is 40 microns. A. *Artemisia ludoviciana*; B. *Cryptantha minima*; C. Truncated cones with 2–5 "spires" centered in "star-shaped" jigsaw-puzzle pieces in *Lappula redowskii*. Bar is 40 microns; D. Silicified branched tracheary elements with spiral thickenings in *Helianthus grosseserràtus*. Bar is 40 microns.

HONEYCOMB ASSEMBLAGES

Silicification of clusters of palisade mesophyll cells form elongate honeycomb assemblages in leaves of many deciduous trees, eg, *Acer saccharum* (Aceraceae), *Quercus alba* (Fagaceae) and *Fraxinus americana* (Oleaceae) (Geis 1973; Wilding

& Drees 1973). Elongate honeycomb assemblages are produced in leaves of most of the woody dicots studied. These species include the following: *A. saccharum* (Aceraceae); *Betula papyrifera* and *Corylus cornuta* (Betulaceae); *Cercis canadensis*, *Gymnocladus dioica* (Caesapliniaceae); *Quercus alba* and *Q. palustris* (Fagaceae); *Carya glabra* and *Juglans nigra* (Juglandaceae); *Morus rubra* (Moraceae); *Platanus occidentalis* (Plantanaceae) (Fig 10.4A); *Rosa blanda* (Rosaceae); *Populus deltoides* and *P. tremuloides* (Salicaeae); *Celtis occidentalis* and *Ulmus americana* (Ulmaceae). This type of phytolith is also produced in a few of the herbaceous dicots studied, eg, *Ambrosia trifida* (Asteraceae or Compositae), *Lespedeza capitata* and *Strophostyles helvola* (Fabaceae or Leguminosae). Elongate honeycomb assemblages are not formed on the other reference species studied, strongly suggesting that they are diagnostic of dicots.

Shallow honeycomb assemblages are formed on the bottom of epidermal cells when only the end walls of palisade cells are silicified (Geis 1973; Wilding & Drees 1973). Shallow honeycomb assemblages of small (8 to 13 microns), circular to polyhedral cell walls are formed in all of the arboreal and herbaceous dicots studied that produced elongate honeycomb assemblages. They are also formed in *Heliotropium convolvulaceum* (Boraginaceae) and *Strophostyles leiosperma* (Fabaceae or Leguminosae). The cells do not occur in any regular pattern. This type of phytolith is illustrated in Figure 10.4B. The outlines of the epidermal cells on which these raised silica patterns form are not always obvious in either the arboreal or herbaceous dicots. Similar phytoliths are very rarely formed in certain grass species, eg, *Agropyron smithii*, *Elymus virginicus* and *Stipa viridula*.

Rows of shallow, circular to oval, raised cell walls are occasionally formed on epidermal phytoliths produced in certain grass species, eg, *Buchloe dactyloides*, *Panicum virgatum*, *Sorghastrum nutans*. These assemblages can be differentiated from those formed in dicots because the cell walls occur in adjacent straight rows.

OPAQUE PERFORATED PLATELETS

Wilding and Drees (1973) reported unidentified fossil opaque platelets with systematic perforations. Bozarth (1985a) reported that these distinctive phytoliths are produced in the aerial portion of *Ambrosia psilostachya* (Fig 10.4C), *A. trifida* (Fig 10.4D), *Eupatorium perfoliatum* (Fig 10.5A) and *Xanthium strumarium* (Fig 10.5B). These four species are members of the Asteraceae (Compositae). This type of phytolith can be unusually long. Its maximum length is 200–300 microns in *Ambrosia psilostachya* and *A. trifida*, 500 microns in *Xanthium strumarium* and 1600 microns in *Eupatorium perfoliatum*. The larger ones are visible to the unaided eye. Opaque platelets with systematic perforations are not produced in any of the other reference materials, including other members of the Asteraceae

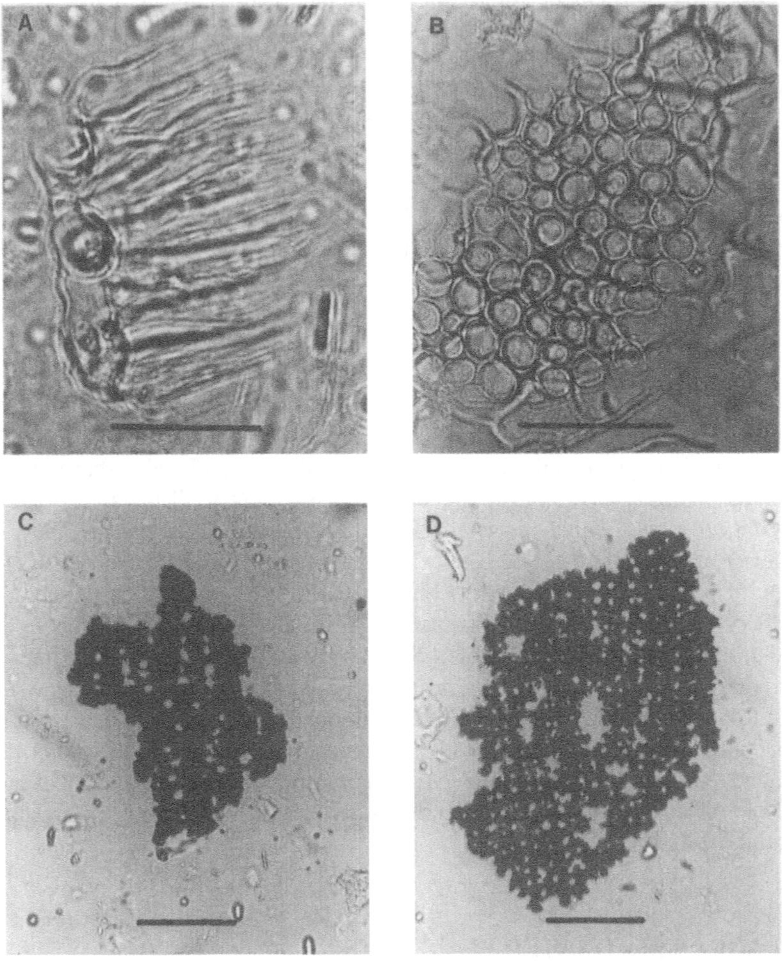

Fig 10.4. A. Elongate honeycomb assemblage of cluster of silicified palisade mesophyll cells produced in *Platanus occidentalis*. Bar is 40 microns; B. Shallow honeycomb assemblage of silicified end walls of palisade mesophyll cells on bottom of epidermal cells formed in *P. occidentalis*. Bar is 40 microns; C. Opaque platelets with systematic perforations formed in *Ambrosia psilostachya* and D. *A. trifida*. Bar is 40 microns.

(Compositae), indicating that they are unique to selected species in the Asteraceae (Compositae). An analysis of an extensive reference collection of plants native to Panama where opaque, perforated platelets were found only in the inflorescence of Asteraceae (Compositae) (Piperno 1988) corroborates this conclusion.

Large, rectangular, opaque platelets with irregularly spaced holes are formed in the achenes of *Helianthus annuus* (Fig 10.5C). They range from 60 to 95 microns in width and several hundred microns in length. Opaque platelets with irregularly spaced holes are also produced in the achenes of the domesticated Seneca sunflower (*H. annuus* var *macrocarpa*), but do not have the distinctive rectangular shape (Bozarth 1985a).

Solid opaque phytoliths are common in many dicots and conifers. They rarely form in grasses. Phytolith opaqueness results from pigmented organic material occluded within the phytolith (Jones & Beavers 1963; Wilding, Brown & Holowaychuk 1967; Wilding, Smeck & Drees 1977).

ECHINATE PLATELETS

Numerous platelets with irregular edges, ranging in length from 15 to 35 microns and in thickness from 6 to 8 microns, are produced in the fruit (presumably the stone) of *Celtis occidentalis* (Bozarth 1987b). These phytoliths are distinguished by echinate (spiny) sculpturing that covers one side, the other side being smooth (Fig 10.5D). Their irregular outline results in no two platelets having exactly the same shape. *Celtis* fruit was identified in a prehistoric archaeological site in northeastern Kansas based on the presence of echinate platelets. *Celtis schippi*, a New World tropical species, produced similar phytoliths (Piperno 1988).

VERRUCATE PHYTOLITHS

Verrucate (bumpy) phytoliths are usually cystoliths or cystolith-like bodies. Cystoliths are silica/calcium carbonate bodies enclosed in specialized cells called lithocysts. Cystoliths have distinct stalks with which they are attached to the cell wall (Solereder 1908). They occur only in a limited number of families in the New World tropics (Piperno 1988).

Spherical phytoliths with contiguous low verrucate sculpturing and stalk-like projections are produced on *Celtis occidentalis* (Fig 10.6A) This type of phytolith was described as mushroom-shaped by Bozarth (1985a, 1985b). There is usually a plate (part of a silicified epidermal cell wall) attached to the base of the stalk. These phytoliths are also formed in the center of groups of silicified epidermal cells lined or filled with coarse verrucae (Fig 10.6B). The mushroom-shaped phytoliths appear to serve as "foundations," because they occur directly below the point where large silicified hairs attach. The surrounding cells probably act as additional structural support for the hairs.

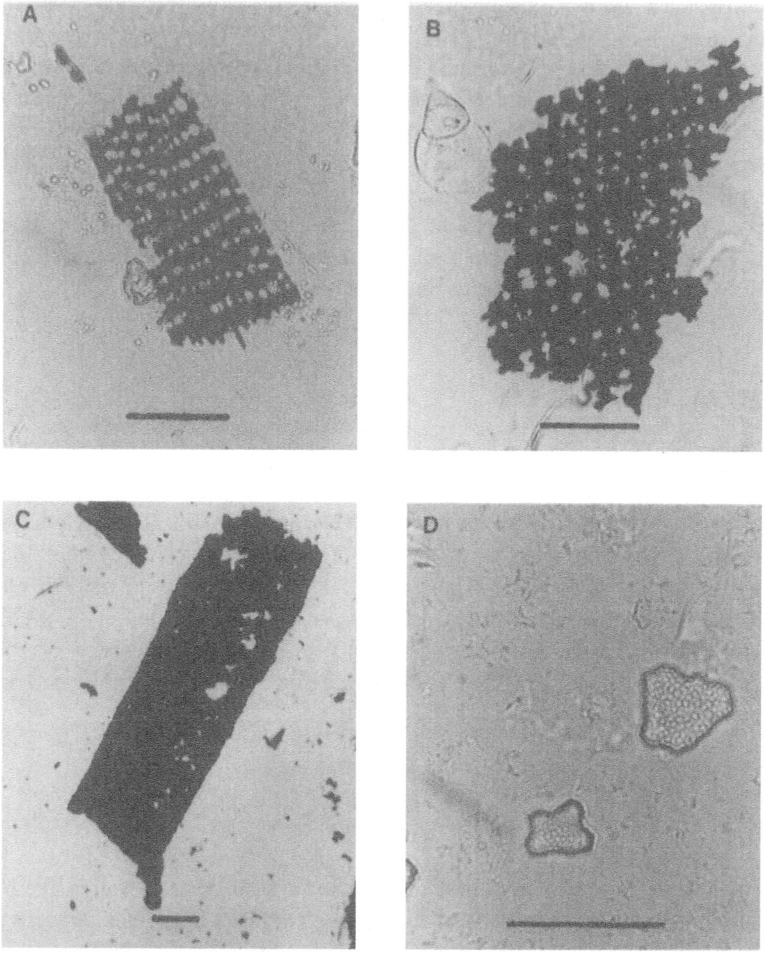

Fig 10.5. Opaque platelets with systematic perforations. Bar is 40 microns. A. *Eupatorium perfoliatum*; B. *Xanthium strumarium*; C. Rectangular opaque platelet with irregularly spaced holes formed in *Helianthus annuus*. Bar is 40 microns; D. Echinate platelets produced in *Celtis occidentalis*. Bar is 40 microns.

Similar groups of silicified epidermal cells containing coarse verrucae surrounding a central hair base are also formed in leaves of *Ulmus americana* (Bozarth 1985a, 1985b). Groups or individual epidermal cells with silicified walls and coarse verrucae in the lumen appear to be unique to Ulmaceae, as they are not found in any of the other taxa studied.

Stalked phytoliths are also formed in leaves of *Morus rubra* (Fig 10.6C), another arboreal dicot. However, they differ from stalked cystoliths formed in *Celtis* in the following ways: they are much smaller in length and width, the body is also much smaller in relation to the stalk, the surface has a grainy appearance instead of being verrucate and the bottom of the stalk is attached to a curved plate instead of a flat plate as in *Celtis*. Another difference is that many of the phytoliths formed in *Morus rubra* consist only of a stalk attached to a curved plate without a body.

Numerous spherical to spheroidal silica cystoliths with contiguous high verrucate sculpturing are produced in the aerial portion of *Boehmeria cylindrica* (Fig 10.6D). Many have no obvious point of attachment to the cell wall. Others are flattened on part of their circumference where they presumably attached to the cell wall. A few have a small, short stalk similar to the mushroom-shaped phytoliths produced in *Celtis*, but without an attached plate-like remnant of silicified cell wall. Virtually identical phytoliths are produced in *Boehmeria aspera*, a New World tropical species (Piperno 1988).

Numerous silica bodies with fine contiguous verrucate sculpturing are formed in the aerial portion of *Urtica dioica* (Fig 10.7A). Many are roughly triangular in shape with rounded corners and small stalks. Elongate verrucate phytoliths with one or both ends tapering to a point are formed in *Pilea pumila* (Fig 10.7B). Similar phytoliths are produced in *P. acuminata*, a New World tropical species (Piperno 1988). Verrucate curvilinear silica plates and spheroidal to globular silica bodies are formed in the aerial portion of *Laportea canadensis* (Fig 10.7C) and *Parietaria pensylvanica*. The curvilinear plates are by far the most common type of phytolith in both species. They are formed by uneven deposition of silica on the inside of the cell wall and do not completely fill the cell lumen. The aerial portion of *Menispermum canadense* produces phytoliths with verrucate sculpturing so fine as to appear grainy (Fig 10.7D). These range in shape from round to globular to irregular and are occasionally formed in clusters. Most of the discrete phytoliths are flat in at least one area where they are attached to the cell wall. Occasionally part of the cell wall remains attached. Most do not have a stalk.

SEGMENTED HAIRS

Various types of silicified segmented hairs are formed in six families in the New World tropics. Several types produced in Asteraceae (Compositae) and Cucurbitaceae appear to be specific to family, genus and species (Piperno 1988). Of the six families reported to produce segmented hairs in the New World tropics, only Acanthaceae, Asteraceae (Compositae), Boraginaceae and Cucurbitaceae are native to the Great Plains.

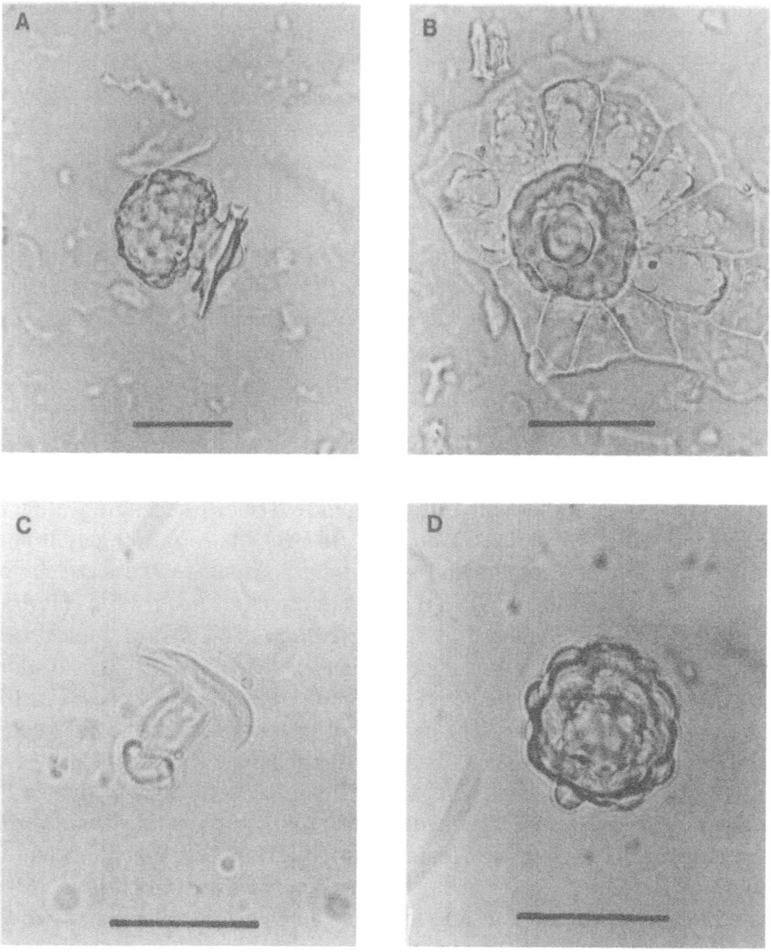

Fig 10.6. A. Spherical silica cystolith with low contiguous verrucate sculpturing and stalk-like projection formed in *Celtis occidentalis*. Bar is 40 microns; B. Same type of phytolith surrounded by silicified epidermal cells in *C. occidentalis*. Bar is 40 microns; C. Silica cystolith with grainy sculpturing and stalk-like projection formed in *Morus rubra*. Bar is 40 microns; D. Spherical silica cystolith with high verrucate sculpturing in *Boehmeria cylindrica*. Bar is 40 microns.

Several types of segmented hairs are common in many, but not all, of the species studied in the Asteraceae (Compositae). Segmented hairs are also common in two members of the Cucurbitaceae. They are only rarely produced in members of the Fabaceae (Leguminosae) and Boraginaceae.

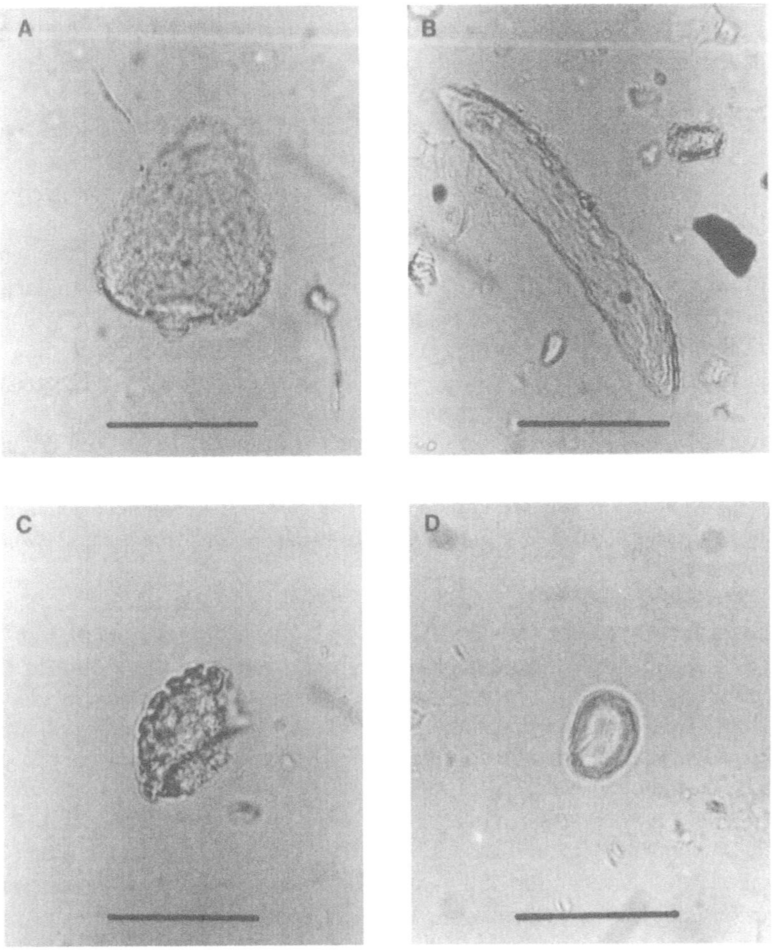

Fig 10.7. A. Triangular silica cystolith with fine verrucate sculpturing formed in *Urtica dioica*. Bar is 40 microns; B. Elongate verrucate phytolith with tapering ends formed in *Pilea pumila*. Bar is 40 microns; C. Curvilinear verrucate silica plate formed in *Laportea canadensis*. Bar is 40 microns; D. Globular phytolith with fine verrucate sculpturing produced in *Menispermum canadense*. Bar is 40 microns.

Silicified hairs composed of 2–4 unarmed (smooth), elongated, thick-walled segments are produced in *Solidago rigida* (Fig 10.8A). The last (terminal) segment is usually notched by transverse striations, especially on the concave side if curved. Most of the hairs are attached to the plant at or near a 90-degree angle.

Eupatorium perfoliatum also produces unarmed silicified hairs of 2–4 elongated segments, a few of which have subtle parallel striations. They are not as thick-walled as those in *Solidago rigida* and are attached to the plant at or near a 90-degree angle.

Helianthus grosseserratus produces silicified hairs of 2–4 segments, most of which are armed (spiny), with the exception of the basal segment. The hairs are attached to the plant at an acute angle, *ie*, one side of the basal cell is considerably shorter than the other.

H. annuus produces silicified hairs of 2–3 segments. The middle segment and lower part of the apex may be armed or unarmed. The apex is attached to an unarmed rounded base at or near a 90-degree angle.

Xanthium strumarium produces unarmed silicified hairs consisting of 2 (occasionally 3) segments. They are distinguished by a short apex, usually attached to a rounded or oval-shaped base at an acute angle.

Cucurbita foetidissima and *Sicyos angulatus*, both members of Cucurbitaceae, produce silicified hairs of 2–3 elongated unarmed segments that attach to the plant at or near a 90-degree angle. The elongated basal segment differentiates these phytoliths from 2–3 segmented hairs in *Helianthus annuus* and *Xanthium strumarium*.

Silicified segmented hairs in *Iva annua* consist of 2 unarmed cells. They are distinguished by a base, usually oval shaped, attached to the plant at an acute angle. Unarmed silicified hairs of 2 segments are rarely produced in *Cynoglossum officinale*, a member of the Boraginaceae, and *Baptisia leucantha*, a member of the Fabaceae (Leguminosae). Those produced in *Baptisia* have very subtle double spiral striations on the terminal segment that result in very small "teeth" on the outer wall.

Cirsium ochrocentrum produces phytoliths that appear to be silicified hairs of 1–2 segments (Fig 10.8B). The cell wall of one segment is completely silicified, with an open top and a solid flat base. The wall flares out near the base, forming a very distinctive phytolith. The other segment, which is usually present, is only partially silicified, forming a stem-like appendage attached to the bottom of the base.

SCALLOPED PHYTOLITHS

Phytoliths with deeply scalloped surfaces of contiguous concavities appear to be unique to *Cucurbita* (Fig 10.8C). *C. foetidissima* is the only *Cucurbita* species native to the Great Plains. Scalloped phytoliths are also produced in selected squash and pumpkins, but these are not native to the Great Plains (Bozarth 1987a).

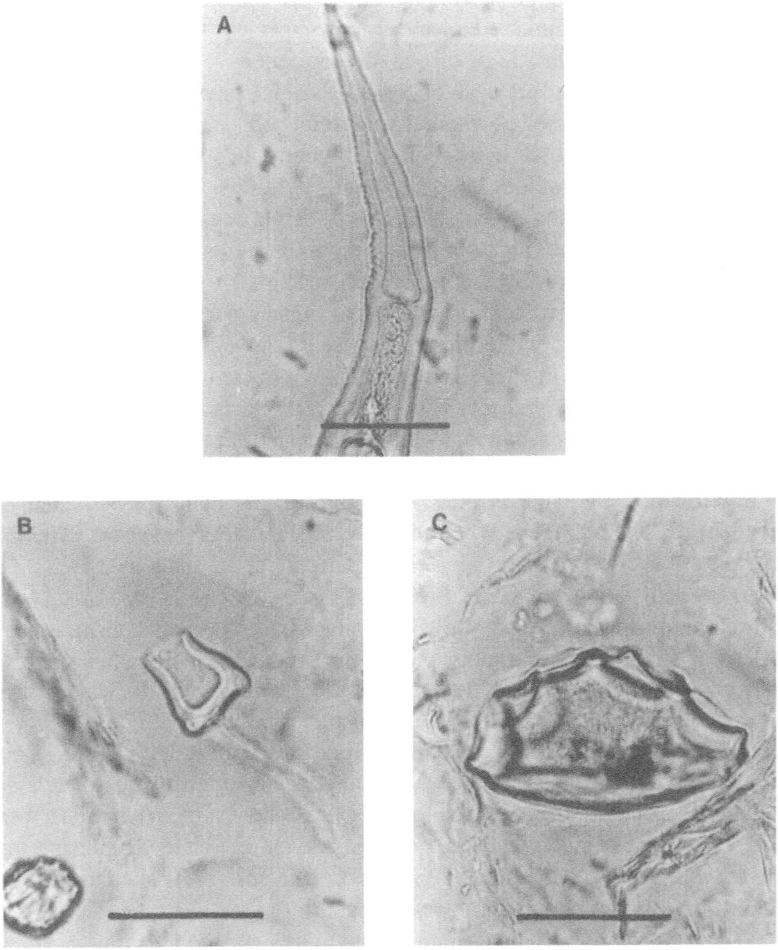

Fig 10.8. A. Three segmented thick-walled silicified hair in *Solidago rigida*. Note transverse notches in terminal segment. Bar is 40 microns; B. Phytolith with open top and solid flared base produced in *Cirsium ochrocentrum*. Bar is 40 microns; C. Phytolith with deeply scalloped surface of contiguous concavities formed in *Cucurbita foetidissima*. Bar is 40 microns.

SUMMARY

This study demonstrates that many dicots produce diagnostic phytoliths. Of the 82 dicot species studied, only 4 (*Sesbania macrocarpa* and *Glycyrrhiza lepidota* in the Fabaceae (Leguminosae), *Hoffmanseggia glauca* in the Caesalpiniaceae and *Polygonum pensylvanicum* in the Polygonaceae) did not produce diagnostic dicot phytoliths. However, these species did produce microscopic opaque bodies characteristic of dicots and conifers.

There are many different taxonomically useful phytolith types produced in dicots. Of these, silicified polyhedral epidermal cells are the most common type. They are produced in both herbaceous and woody dicots, as well as grasses and certain conifers. Non-elongate polyhedral phytoliths with 4–8 sides and flat parallel tops and bottoms are characteristic of dicots because they are common in many dicots but very rarely form in non-dicots. The length of the long axis for this type of phytolith differs among certain species.

Silicified epidermal cells shaped like jigsaw-puzzle pieces are another common type of dicot phytolith. They are found only in dicots. There are significant differences in shape and length along the axis among certain species.

Silicified branched elements with spiral thickenings (anastomosing tracheary elements) are also common in many arboreal and herbaceous dicots. They are not found in non-dicots.

Silicification of palisade mesophyll cells forms two types of honeycomb assemblages. One of these types consists of elongate honeycomb assemblages formed by the silicification of clusters of palisade cells. This type appears to be unique to dicots and is characteristic of deciduous trees and shrubs, as it is much more common in many woody dicots than herbaceous dicots. The other type consists of shallow honeycomb assemblages produced by the silicification only of the ends of palisade cells on the bottom of epidermal cells. It is most common in dicots and is characteristic of woody dicots.

Opaque platelets with systematic perforations are produced only in certain members of Asteraceae (Compositae). Large rectangular opaque platelets with irregularly spaced holes are found only in achenes of *Helianthus annuus*. Numerous silica platelets with irregular edges and echinate sculpturing on one side are produced in the fruit of *Celtis occidentalis*. This unusual type of phytolith appears to be unique to that genus.

Various types of verrucate (bumpy) phytoliths are produced in several dicot families. Polyhedral epidermal cells filled with coarse verrucae appear to be unique to Ulmaceae. Certain types of stalked verrucate phytoliths may be specific to *Celtis*, *Morus*, *Boehmeria* or *Urtica*. Elongate verrucate phytoliths with one or both ends tapering to a point are found only in *Pilea*.

Segmented hairs are commonly produced in only two families studied, Asteraceae (Compositae) and Cucurbitaceae. Silicified hairs with four elongated segments appear to be unique to Asteraceae (Compositae). Silicified hairs of 2–3 segments with round-oval bases also appear to be unique to this family. Certain types of silicified segmented hairs are probably specific to particular Asteraceae (Compositae) genera. One of the more unusual types of silicified hairs that appears to be unique to a particular genus is produced in *Cirsium*. It is unique in that one segment has an open top and a flared base. A stem-like appendage is attached to the base.

One of the more distinctive types of dicot phytoliths is formed in rinds of *Cucurbita* fruit. These phytoliths are distinguished by deeply scalloped surfaces of contiguous concavities not found in any other taxon.

In conclusion, this study demonstrates that dicots produce many different types of diagnostic phytoliths. Some types are common in both herbaceous and woody dicots. Other types are specific to certain families and genera. Application of this classification should prove very useful in classifying fossil phytoliths, thereby allowing a more precise reconstruction of vegetative histories and paleoenvironments.

ACKNOWLEDGMENTS

I am grateful to Ralph Brooks and Ronald McGregor (McGregor Herbarium, University of Kansas) for verifying the species identification of plants collected in the field. I would like to thank William Johnson (Department of Geography, University of Kansas), Dwight Brown (Department of Geography, University of Minnesota) and the two anonymous reviewers for their critical review of this manuscript.

REFERENCES

Bozarth, S (ms) 1985a Distinctive phytoliths from various dicot species. Paper presented at 2nd Phytolith Conference, University of Minnesota, Duluth.

____1985b Distinctive phytoliths from various dicot species. *Phytolitharien Newsletter* 3(3): 7–8.

____1987a Diagnostic opal phytoliths from rinds of selected *Cucurbita* species. *American Antiquity* 52: 607–615.

____1987b Opal phytolith analysis of edible fruits and nuts native to the Central Plains. *Phytolitharien Newsletter* 4(3): 9–10.

____1988 Preliminary opal phytolith analysis of modern analogs from parklands, mixed forest, and selected conifer stands in Prince Albert National Park, Saskatchewan. *Current Research in the Pleistocene* 5: 45–46.

Geis, JW 1973 Biogenic silica in selected species of deciduous angiosperms. *Soil Science* 116: 113–130.

Jones, LH and Beavers, AH 1963 Some mineralogical and chemical properties of plant opal. *Soil Science* 96: 375–379.

Klein, RL and Geis, JW 1978 Biogenic silica in the Pinaceae. *Soil Science* 126: 145–155.

Piperno, DR 1988 *Phytolith Analysis: An Archaeological and Geological Perspective.* San Diego, Academic Press: 280 p.

Rovner, I 1971 Potential of opal phytoliths for use in paleoecological reconstruction. *Quaternary Research* 1: 343–359.

Solereder, H 1908 *Systematic Anatomy of the Dicotyledons.* Oxford, Clarendon Press.

Wilding, LP and Drees, LR 1971 Biogenic opal in Ohio soils. *Soil Science Society of America, Proceedings* 35: 1004–1010.

_____1973 Scanning electron microscopy of opaque opaline forms isolated from forest soils in Ohio. *Soil Science Society of America, Proceedings* 37: 647–650.

Wilding, LP, Brown, RE and Holowaychuk, N 1967 Accessibility and properties of occluded carbon in biogenetic opal. *Soil Science* 103: 56–61.

Wilding, LP, Smeck, NE and Drees, LR 1977 Silica in soils: Quartz, cristobalite, tridymite, and opal. *In* Dixon, JB and Weed, SB, eds, *Minerals in Soil Environments.* Madison, Wisconsin, Soil Science Society of America: 471–552.

11

PHYTOLITH TAXONOMY IN SELECTED SPECIES OF TEXAS CACTI

JOHN G. JONES and VAUGHN M. BRYANT, JR.

ABSTRACT

For this study we examined sixteen modern cactus samples. Based on descriptions of phytolith surface features and ratios of size variations, we believe it is possible to separate certain cacti at the level of genus based only on their phytoliths. In addition, we found it possible to identify individual species of cacti in the genus *Opuntia* based only on the size and morphology of their phytoliths. Another part of our study includes a test of phytolith response to environmental factors and the potential for phytolith variations within individual plants. In order to test for the potential stability of phytolith morphology in cactus fruits, we examined phytoliths from various tissues of cactus fruits and compared those data with phytolith data from the stems of the same cacti as part of our test for similarities and/or differences. Finally, we examined phytoliths from tissues in a prehistoric cactus pad 6,000 years old that was recovered from a dry rockshelter in southwest Texas.

INTRODUCTION

The first recognition that plants had crystals came in 1835 when Struve, a German botanist, observed them in living plant tissues. A few years later in 1841, a German scientist named C. G. Ehrenberg was one of the first to recognize that both pollen and phytoliths could be recovered from fossil sediment samples collected from the bottom of lakes. He became especially interested in the numerous kinds of siliceous particles he found in sediments, and called them *Phytolitharia* (Piperno 1988). Even Charles Darwin, while on his famous trip aboard the *H.M.S. Beagle*, noted that phytoliths were present as part of the dust blowing in the atmosphere near the Cape Verde Islands off the western coast of Africa.

From the 1800s through modern times the development of two disciplines, one that studies pollen and one that studies phytoliths, each evolved in a similar

Phytolith Systematics, Edited by G. Rapp, Jr. and
S.C. Mulholland, Plenum Press, New York, 1992

manner. This similarity resulted partly because pollen and phytoliths have much much in common. For example: 1) pollen and phytoliths are both produced by plants; 2) both are microscopic in size; 3) both are produced in vast numbers; 4) both can be carried to distant locations by wind currents; 5) both can be found in sediments throughout the world; and 6) both pollen and phytoliths are often used by paleoecologists, geologists, archaeologists and other scientists as clues to past events. This last point is especially important because both pollen and phytoliths can be used as: 1) index fossils reflecting changes in past environmental conditions, 2) indications of past dietary and cultural practices, 3) forensic tools in criminology and 4) indicators of different types of depositional records (*ie*, marine or terrestrial).

Many plants produce phytoliths (Esau 1965). We have focused our interest on a study of the phytoliths produced by various species of cacti. Although the majority of phytoliths are silica, others are composed of compounds such as calcium oxalate (Piperno 1988).

Rovner (1983) reports that phytoliths can occur in many parts of plants, such as the epidermis of plant leaves, in the stems of some monocotyledonous plants, in the pith region of some hardwoods, in the roots of some plants and in the needles and wood of conifers.

Phytoliths can be formed in, and shaped by, the walls of specialized plant cells called idioblast cells (Esau 1965). Dunn (1983), for example, reports that silica phytoliths are formed in those areas of a plant where certain cells lose their protoplasts and the lumen becomes filled with silica. Some phytoliths form in the cellular and intercellular spaces of plants, eg, those in the grass and sedge families, and serve the added purpose of forming a rigid skeleton-like structure that helps prevent wilting when water resources become scarce (Parry & Smithson 1958).

Calcium oxalate phytoliths tend to form in the vacuoles of specialized cells, or grow in the cytoplasm of generalized parenchymal cells (Esau 1965). Unlike many of the opaline phytoliths, which generally conform to the shapes of the plant cells in which they form, most of the calcium oxalate phytoliths found in the cacti develop into druse-shaped structures that vary in size, but only slightly in overall shape in different species. The basic druse shape seems to be governed to a great extent by the crystalline properties of calcium oxalate and any impurities associated with it. As Esau (1965) notes, calcium oxalate crystals are sometimes deposited uniformly throughout the tissues of many species of cacti. In other plants, such as the genus *Vitis*, phytoliths are found only in specialized regions, for instance, along the margins of the phloem rays (Esau 1965).

Although the number of morphological studies of phytoliths has increased in recent years, nearly all taxonomic reports have dealt with the silica bodies formed in monocotyledonous plants. Our current study suggests that phytoliths produced by many taxa in the Cactaceae offer considerable promise as useful taxonomic tools for classification and identification purposes, and as useful markers indicating the presence of cacti in sediments from previous periods.

Our interest in developing a taxonomic key for the phytoliths in taxa of the Cactaceae developed from our ongoing paleoenvironmental and archaeological research in the arid and semi-arid regions of southwest Texas and to a more limited extent, northern Mexico. Archaeologists have worked in these arid regions for many years (Johnson 1964; Dibble & Lorrain 1968; Collins 1969; Hester 1983), but only recently realized the identification value of calcium oxalate phytoliths.

Little effort has been made to collect, process or analyze phytoliths from species of cacti or from the deposits of archaeological sites. The only known phytolith study prior to our current efforts was an unpublished study of three taxa of cacti (two species of *Opuntia* and one species of *Mammillaria*) and six taxa of desert succulents (two species of *Agave*, one species of *Dasylirion* and three species of *Yucca*) conducted by Bryant (nd) in 1968.

Our studies indicate that some of the calcium oxalate and silica crystals, which develop in the tissues of plants growing in southwest Texas, can be used as diagnostic markers in archaeological deposits. For example, phytoliths are often present in midden soils, in ash deposits of fire hearths, attached to the cutting edges of stone artifacts, as part of the temper in ceramic vessels, caught in between the weave of baskets and mats and in the coprolites of prehistoric cultures.

We have concentrated our efforts on types found in the cacti for three reasons: 1) many taxa of cacti are common in the region, 2) some of the cacti phytoliths seem to be species- or at least genera-specific and 3) we find that these types of phytoliths are associated with deposits covering long chronological records in this region.

The recovery of spiny-shaped druse crystals in prehistoric human coprolites from sites in southwest Texas has already been used to document the importance of cacti as a primary food resource (Bryant 1974). Bryant used the presence of cacti phytoliths to indicate that parts of cactus plants, especially the pads, were eaten as food even though no evidence of cactus plant fibers or tissues were present in those same coprolite specimens. Bryant (1969, 1974) also noted in these early studies that it would be impossible to separate druse-type phytoliths into a key for individual species of cacti because the diversity within each genus and species appeared to be too great.

The present study is an attempt to achieve what Bryant initially believed was impossible, a key to the phytoliths of cacti. With the use of scanning electron microscopy, and to a lesser degree light microscopy, it is possible to separate taxa of the cacti according to the shape and size of the druse phytoliths. This leads to our phytolith-based taxonomic key for a limited group of cacti found in southwest Texas.

MATERIALS AND METHODS

We collected and examined calcium oxalate phytoliths from vegetative parts of 28 specimens of widely recognized cacti taxa from the southwestern United States. These plants represent 5 genera and 16 cactus species or varieties (Table 11.1). With this diversity of plant specimens, phytoliths might reflect trends over large geographical areas.

Samples were collected in different environmental regions of Texas in order to discover whether or not edaphic and/or climatic factors affected the shape and/or size of druse phytoliths in the same genera and/or species of cacti. In addition, we examined the phytoliths in five specimens of *Opuntia* fruits and one

TABLE 11.1. Cactus Specimens Used in This Study Recorded by Taxon and Collection Locality

Species	Collection Locality
Ariocarpus fissuratus (Eng.) K. Schumann	(1)
Echinocactus horizonthalonius Lem.	(1)
Echinocereus enneacanthus Eng.	(1)
E. triglochidiatus Eng.	(1)
Mammillaria heyderi Muehlenpf.	(1)
M. lasiacantha var *denudata* Eng.	(1)
Opuntia compressa var *grandiflora* (Eng.) Weniger	(5)
O. compressa var *macrorhiza* (Eng.) L. Benson	(6)
O. engelmannii var *alta* (Gr.) Weniger	(8)
O. engelmannii var *engelmannii* (SD) Weniger	(1)
O. engelmannii var *texana* (Gr.) Weniger	(1, 2, 3, 4)
O. leptocaulis DC	(1)
O. macrocentra Eng.	(1)
O. phaeacantha var *camanchica* (Eng.) Weniger	(7)
O. phaeacantha var *nigricans* (Eng.) Weniger	(1)
O. schottii Eng.	(1)

Collecting Localities in Texas
1. Langtry, Val Verde Co. (calcareous soil and desert scrub)
2. Uvalde, Uvalde Co. (calcareous soil with mesquite dominant)
3. D'Hanis, Medina Co. (calcareous soil with mesquite dominant)
4. San Antonio, Bexar Co. (calcareous soil with mesquite dominant)
5. Bastrop, Bastrop Co. (sandy loam soil in pine-juniper forest)
6. LaGrange, Fayette Co. (sandy soil in pine-juniper parklands)
7. Fort Hood, Coryell Co. (floodplain alluvial soil in grassland)
8. Boca Chica, Cameron Co. (sand dunes with beach vegetation)

piece of a 6,000-year-old *Opuntia* pad recovered from an archaeological site excavated in southwest Texas. These studies were conducted to: 1) determine if phytolith sizes and shapes found in cactus fruits were identical to those produced in the pads, and 2) determine whether or not we could extract and identify cactus phytoliths found in the fossil record of archaeological sites.

Plants were selected because of accessibility as well as applicability. When possible, both older prickly pear (*Opuntia*) pads as well as new ones representing the current year's growth were collected and examined. For comparative purposes, cacti plants representing the same genus, species or variety were collected from different geographical regions of Texas.

Most of the cactus specimens were collected in 1985 from seven locations in Texas during October and November. The rest were collected in March, 1989 (Fig 11.1). In most cases, small cactus plants consisting of less that 10 pads were collected. In 1985 we collected the entire plant, including the roots. Later, we realized that we did not need to collect an entire cactus plant because only a small piece of each pad was needed for analysis. Thus, for the specimens collected in 1989, only pieces of pads rather than entire plants were collected.

The ripe prickly pear fruits we examined were collected from selected cacti by removing an entire pad containing the fruit. Since most of the collections were made in autumn it was assumed that the uppermost pads containing ripe fruits were mature pads. Observations showed that the mature pads often were darker green than pads representing the first year of growth. Mature pads were selected because they would reflect fully developed phytoliths.

In the Paleoethnobotany Laboratory we verified taxonomic identifications by comparing each specimen with vouchered herbarium specimens and descriptive botanical keys.

Preliminary studies showed that phytoliths extracted from different layers of the same *Opuntia* pads often contained at least three readily identifiable forms of calcium oxalate druse crystals. The specimens were dissected and separated into three parts: the pith, cortex and epidermal regions. In many cacti, there are interior cell layers that are not easily separated from the single epidermis layer. These layers are often referred to as the "hypodermis." Thus we have used the term epidermis/hypodermis to refer to a group of cell layers that includes the epidermis and the cell layers immediately below the epidermis. These combine to form the outer portion of the stem between the cortex and the plant's outer surface.

Each of these three stem tissues was processed separately for phytoliths. The pith portion of the *Opuntia* pad was defined at the region located in the middle of the pad where most of the vascular network of fibers are found. This portion of the *Opuntia* pad is generally spongy and whitish in color. The cortex region was defined as the part of the pad found directly outside the layer of vascular network and extending to, but not including, the epidermis/hypodermis of the plant.

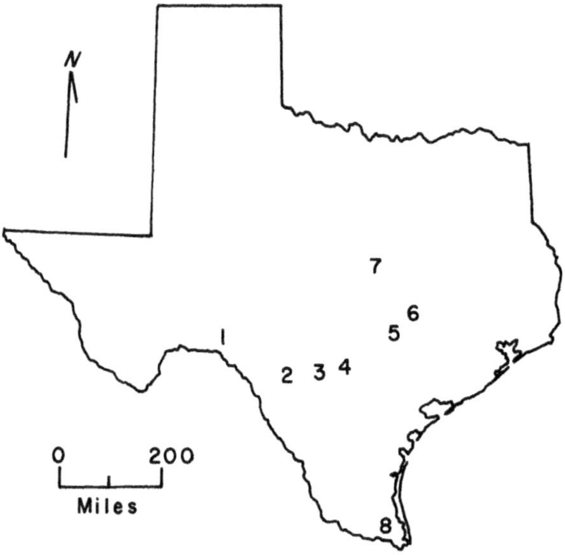

Fig 11.1. Collecting localities in Texas. 1. Langtry, Val Verde Co. (heavy limestone soil and desert scrub); 2. Uvalde, Uvalde Co. (limestone soil with mesquite dominant); 3. D'Hanis, Medina Co. (limestone soil with mesquite dominant); 4. San Antonio, Bexar Co. (limestone soil with mesquite dominant); 5. Bastrop, Bastrop Co. (sandy soil in pine-juniper forest); 6. La Grange, Fayette Co. (sandy-loam soil in pine-juniper parkland); 7. Fort Hood, Coryell Co. (floodplain alluvial soil in grassland); 8. Boca Chica, Cameron Co. (sand dunes near the coast with beach vegetation).

The portion of each *Opuntia* plant selected for dissection and processing was located between 1–3 cm interior to the outer edge of the pad. We generally collected a 2-cm square portion of the pad that was free of spines. After removing the small square pad, the various tissue layers of the pad were separated. The *Opuntia* fruits were divided into three layers. When each fruit is cut in half along its longest axis, three well-defined tissues are apparent: 1) an interior pith or pulp section which is often dark red in color and consists of spongy pulp and seeds; 2) a lighter cortex tissue immediately outside the pulp layer; and 3) a thin outermost layer of epidermal/hypodermal tissue.

All genera were treated similarly. As with the species of *Opuntia*, we divided specimens into pith, cortex and epidermis/hypodermis. We also searched for phytoliths in the roots of a cactus species that had a fleshy taproot. Phytoliths were recovered from all samples except those cut from regions of epidermis/ hypodermis in the genus *Echinocereus*.

Each method used for the extraction of phytoliths from plant tissues has had some drawbacks. One common extraction technique is dry-ashing, where the plant tissue is incinerated at high temperatures (Rovner 1983). Our experiments showed that this technique caused an alteration in the shape of silica and calcium oxalate crystals in cacti and destroyed the druse crystals. Other phytolith extraction techniques in common use involve: 1) sulfuric acid and sodium sulfate solutions, 2) chromic acid, 3) a solution of chromic and sulfuric acid and 4) a solution of nitric and perchloric acid (Rovner 1983).

These four extraction techniques have two things in common. They are all dangerous and all are designed for the extraction of silica-based phytoliths. We are not certain of the extent of degradation each of these extraction techniques produces on calcium-based phytoliths. Therefore, we did not use any of them for our phytolith extractions.

Instead, we used an extraction technique consisting of a mixture of hydrogen peroxide and potassium dichromate (Jones 1988). Plant tissue samples were collected while they were still fresh, and each was cleaned with distilled water and placed in a 400-mL glass beaker. The beakers were then placed in a fume hood and a solution of 30% hydrogen peroxide solution for about two minutes; approximately 0.25 grams of potassium dichromate crystals was then added. The sample was stirred until the crystals dissolved and a thorough mixture was obtained. Once mixed, rapid oxidation of organic material occurred.

After the oxidation reaction stopped, a few additional drops of hydrogen peroxide were added. This renewed the reaction until all remaining organic plant material was digested. The entire oxidation process rarely took more than twenty minutes.

Following the digestion process, the phytoliths were collected. As a final step, the phytoliths were rinsed several times in distilled water before being stored in 95% ethanol. We caution that many of these chemical agents are toxic.

The phytoliths discussed in this report were examined and measured using the Nikon compound microscope. Each specimen was mounted in a medium of glycerol. From each sample, 100 phytoliths were measured. Some specimens were also examined using a Jeol 25 scanning electron microscope.

RESULTS

In the results listed below we have noted only whether each phytomorph was present or absent in each sample.

Non-*Opuntia* Genera

Echinocactus. *E. horizonthalonius* Lem. was collected near Mile Canyon located just outside the town of Langtry in Val Verde County, Texas. The rocky calcareous and limestone soil of that area has a high pH, and water is scarce throughout most of the year.

The phytoliths from the pith and cortex regions of this species occur as 1) small square-ended shaped crystals, 2) blunt or weakly-pointed druse shaped crystals (Fig 11.2A), and 3) weakly-pointed druse with large crystal development (Fig 11.2B). An initial study of the epidermis/hypodermis region of *Echinocactus* indicated no phytoliths. Closer examination of the outermost layer of the epidermis/hypodermis showed the presence of very small phytoliths. These phytoliths from the epidermal/hypodermal region were irregularly shaped, rounded masses. When viewed under a light microscope they appeared to be similar in size and structure. These phytoliths may be composed of silica.

The size of the phytoliths found in *Echinocactus horizonthalonius* averaged: 1) 96.5 microns in the pith region, 2) 108 microns in the cortex area and 3) 13.65 microns in the epidermis/hypodermis. Size extremes ranged from 10 microns in the epidermis/hypodermis to 180 microns in the cortex region.

Echinocereus. The two species studied were: 1) *Echinocereus enneacanthus* Eng. and 2) *Echinocereus triglochidiatus* Eng. Both species were collected near Mile Canyon located just outside the town of Langtry in Val Verde County, Texas.

The two species showed a similarity in phytolith development and size, but each had a great diversity of morphologies. For example, phytolith types in the pith and cortex regions included: 1) elongate polyhedral-shaped phytoliths (Fig 11.2C and D), 2) irregular masses of small-sized, square-ended phytoliths and 3) a combination of elongate polyhedral and square-ended crystals. One additional type of druse phytolith was found in tissues of *Echinocereus enneacanthus*, but not in *Echinocereus triglochidiatus*. The epidermal/hypodermal layers of both species seemed to be devoid of any type of phytoliths.

The druse-type phytoliths found in the genus *Echinocereus* were some of the largest encountered. The range of druse phytolith sizes varied from 115 to 620 microns in *Echinocereus enneacanthus*. In *Echinocereus triglochidiatus*, the range extended from 100–610 microns.

Studies of 100 phytolith counts from the pith region of *Echinocereus enneacanthus* revealed an average size for druse types of 319 microns. For the cortex region, the average phytolith size dropped to 258 microns. In *Echinocereus triglochidiatus*, the size of the druse phytoliths in the pith region averaged 370 microns and those in the cortex averaged 246 microns.

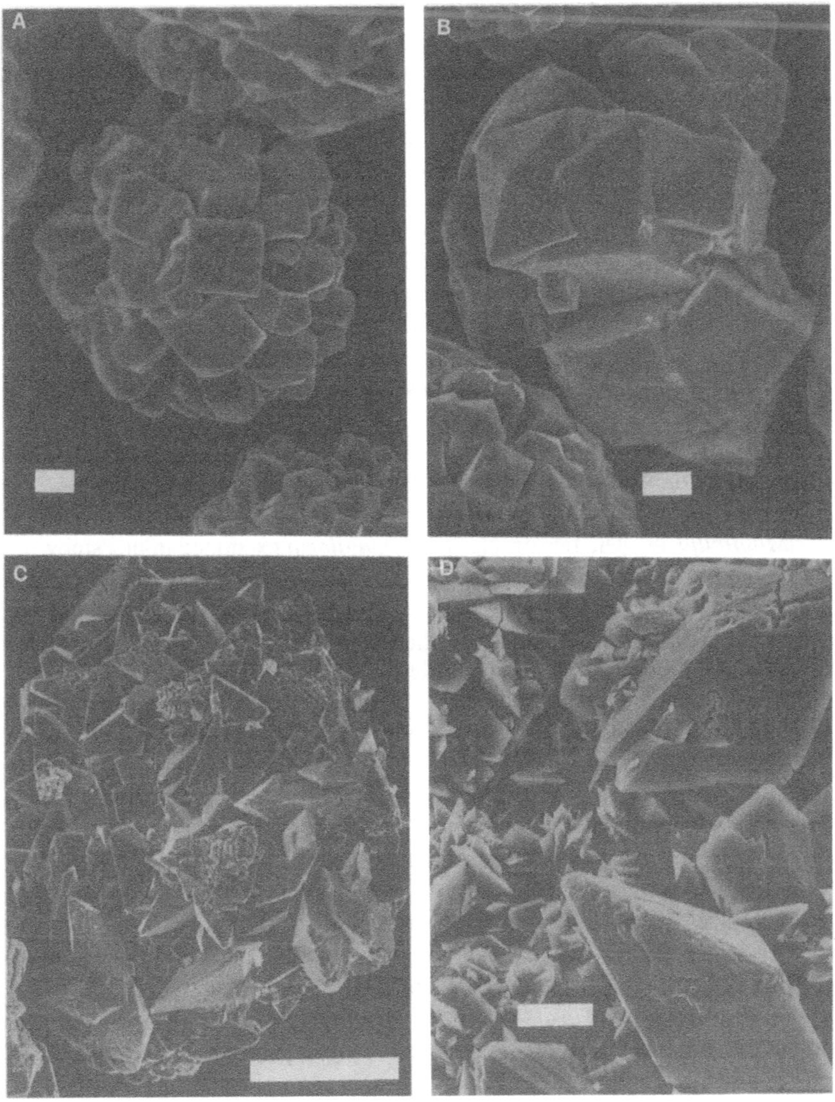

Fig 11.2. Scanning electron micrographs. A. Weakly-pointed small druse crystal from the pith and cortex of *Echinocactus horizonthalonius*. Bar is 10 microns; B. Weakly-pointed large druse crystal from the pith and cortex of *Echinocactus horizonthalonius*. Bar is 10 microns; C. Large druse phytolith from the pith and cortex of *Echinocereus enneacanthus*. Bar is 100 microns; D. Crystal development in a large druse phytolith from *Echinocereus enneacanthus*. Bar is 10 microns.

Ariocarpus. The only specimen from this genus, *Ariocarpus fissuratus* (Eng.) D. Schumann, was collected from a location near Langtry, Texas.

This small cactus is commonly called a "living rock," because it is small, grows close to the ground and blends in with the surrounding rocks and soils of the region. It posed a challenge because of its unusual growing habits. One problem was separating the tissues into pith, cortex and epidermal/hypodermal regions. This cactus is one of the few that has a large, fleshy taproot.

Phytolith types from the cortex region consisted of round to oval masses in groups of large and small square-ended crystals. Counts of 100 of these phytoliths revealed an average size of 173 microns. Phytoliths recovered from the epidermal/hypodermal tissue averaged 117 microns in diameter and were mostly rounded with a poorly developed surface morphology of small square-ended crystals. The phytoliths from the gelatinous taproot region consisted of poorly defined masses of small, square-shaped crystals and averaged 162 microns in diameter.

The total range of phytolith sizes for this one cactus species varied from phytoliths of only 45 microns to ones that were 290 microns in diameter.

Mammillaria. The two species of *Mammillaria* examined in this study were: 1) *Mammillaria heyderi* Muehlenpf., and 2) *Mammillaria lasiacantha* var *denudata* Eng. Both specimens were collected from locations near Langtry, Texas.

Examination revealed two druse-shaped phytolith shapes that were common throughout all the tissues of both species. These two shapes were: 1) masses of small square-ended crystals (Fig 11.3 A), and 2) masses of large square-ended crystals. A third phytolith type, found only in the epidermal/hypodermal regions of these two taxa, was somewhat round with an irregular surface morphology. When viewed through a light microscope, the cross section revealed a series of concentric layers of calcium oxalate resembling the layers of ice often found in the interior of a large hail stone. The layers may represent different growth periods of the crystals.

The sizes of phytoliths in *Mammillaria heyderi* ranged from 40 to 210 microns. The average size in the pith region was 143 microns. In the cortex and epidermal/hypodermal regions, the average size was 107 microns.

In *Mammillaria lasiacantha* var *denudata,* the phytoliths ranged in size from 40 to 240 microns. Average sizes were 1) 105 microns in the pith, 2) 104 microns in the cortex and 3) 122 microns in the epidermis/hypodermis. The unusually large average size found in the epidermal/hypodermal region may not be accurate. A size nearly 20 microns larger than epidermal/hypodermal phytoliths found in *Mammillaria lasiacantha* may have resulted from the accidental collection and incorporation of a few phytolith specimens from the edge of the cortex region. Future studies may clarify this.

Summary. When the phytolith types of these four genera are viewed as a unit, ten different phytolith shape types emerge (Table 11.2). Of these, only the phytolith types with small, square-ended crystals were found in the tissues of all four genera. Conversely, the large square-ended form was found only in *Ariocarpus* and *Mammillaria*. Each of the other nine basic forms occurring in these four genera appears to be unique to the genera. These preliminary results for these taxa should be viewed with caution until more extensive studies can be conducted.

Opuntia Types

Opuntia. The majority of cactus taxa studied were members of the subfamily Opuntioideae. All plants in this group are in the genus *Opuntia* even though Backeburg (1977) recommends transferring two *Opuntia* species into another genus: *Opuntia schottii* Eng., which he calls *Corynopuntia schottii* Knuth, and *Opuntia leptocaulis* DC, which he calls *Cylindropuntia leptocaulis* (Eng.) Knuth.

Many taxonomists separate the *Opuntia* group into two categories: Cylindropuntia (species with round stems) and Platyopuntia (species with flat pads). Some of the examples we studied in each category include the round stemmed *Opuntia leptocaulis* (chollas group) and those with flat pads such as *Opuntia engelmannii* var *texana* (prickly pear group). The acceptance of the taxonomic recognition of the two other genera in the subfamily Opuntioideae, *Cylindropuntia* and *Corynopuntia*, is less certain. In this study all genera are called *Opuntia*.

TABLE 11.2. Dominant Phytolith Shapes Found in Non-*Opuntia* Genera of Cacti Studied in our Survey

Phytolith Shape	Genera
1. Small, square-ended	Found in all four genera
2. Small, weakly pointed	*Echinocactus*
3. Large, weakly pointed	*Echinocactus*
4. Small globular shaped	*Echinocactus*
5. Elongate polyhedral shaped	*Echinocereus*
6. Square-ended and polyhedral	*Echinocereus*
7. Irregular shaped mass	*Echinocereus*
8. Large, and square-ended*	*Ariocarpus/Mammillaria*
9. Square-ended but round shaped	*Ariocarpus*
10. Round with "growth rings"	*Mammillaria*

* Has since been observed in both *Sclerocactus* and *Ferocactus*

We found that the phytoliths present in each of the three stem tissue types (pith, cortex and epidermis/hypodermis) in all of the species of *Opuntia* were similar. For example, the pith regions in the stems of all the *Opuntia* plants contained numerous large druse forms (Fig 11.3 B) characterized by pronounced pointed spines. Similar druse phytoliths were found in the cortex region of the stems of most *Opuntia* taxa, yet they were generally smaller in diameter than the ones located in the pith region. The cortex region in the stems of most *Opuntia* species also contained a type of druse phytolith with short but pointed spines (Fig 11.3C). Phytoliths extracted from the epidermal/hypodermal tissues in the stems were rounded and possessed minimal spine development (Fig 11.3D).

Spiny druse crystals are one of the most common calcium oxalate shapes. Similarly shaped druse forms occur in some species of the Malvaceae, Euphorbiaceae, Brassicaceae (Cruciferae) and other families as well. However, calcium oxalate phytoliths with a large size and consistent shape, such as those found in the genus *Opuntia*, are not often found in non-cactus plants.

This study revealed that phytoliths found in the stems of Cylindropuntia-type cacti, such as *Opuntia leptocaulis*, were nearly identical in morphological shape to phytoliths found in the stems of the Platyopuntia-type cacti. The only major difference was size. Measurements of 100 druse phytoliths from the stem pith region of several *Opuntia leptocaulis* plants revealed an average size range from 59.8 to 60.6 microns. Similar observations of druse phytoliths from the stem cortex region of these same plants revealed sizes that averaged from 35.3 to 35.8 microns in diameter.

Measurements showed that even though the size of the pith and cortex druse forms in each of the two main groups of *Opuntia*, Cylindropuntia and Platyopuntia, were different, the druse phytoliths located in the stem epidermal/hypodermal regions of both *Opuntia* groups were nearly identical in morphology and size. For example, in several plants of *Opuntia leptocaulis* (Cylindropuntia), the average mean diameter for 100 counts of druse phytoliths from the stem epidermal/hypodermal region ranged in size from 37.8 to 40.6 microns. This size range is nearly identical to that found in the stem epidermal/hypodermal region of the Platyopuntia group of *Opuntia*. This similarity in phytolith size may be controlled by a similar genetic or physical factor in both groups.

In an attempt to test the validity of using phytolith morphology and size as a taxonomic tool in the genus *Opuntia*, we compared the phytoliths in five different ways: 1) variations of phytolith size and morphology in different stem regions of a single pad, 2) the range of potential variations found in mature as compared to immature pads on the same plant, 3) possible phytolith production and variations that might exist in the fruits of these plants, 4) the potential range of phytolith variation that might exist within the stems of plants of a single species growing in different edaphic and ecological regions and 5) variation that might exist among different taxonomic varieties within the same genus and species.

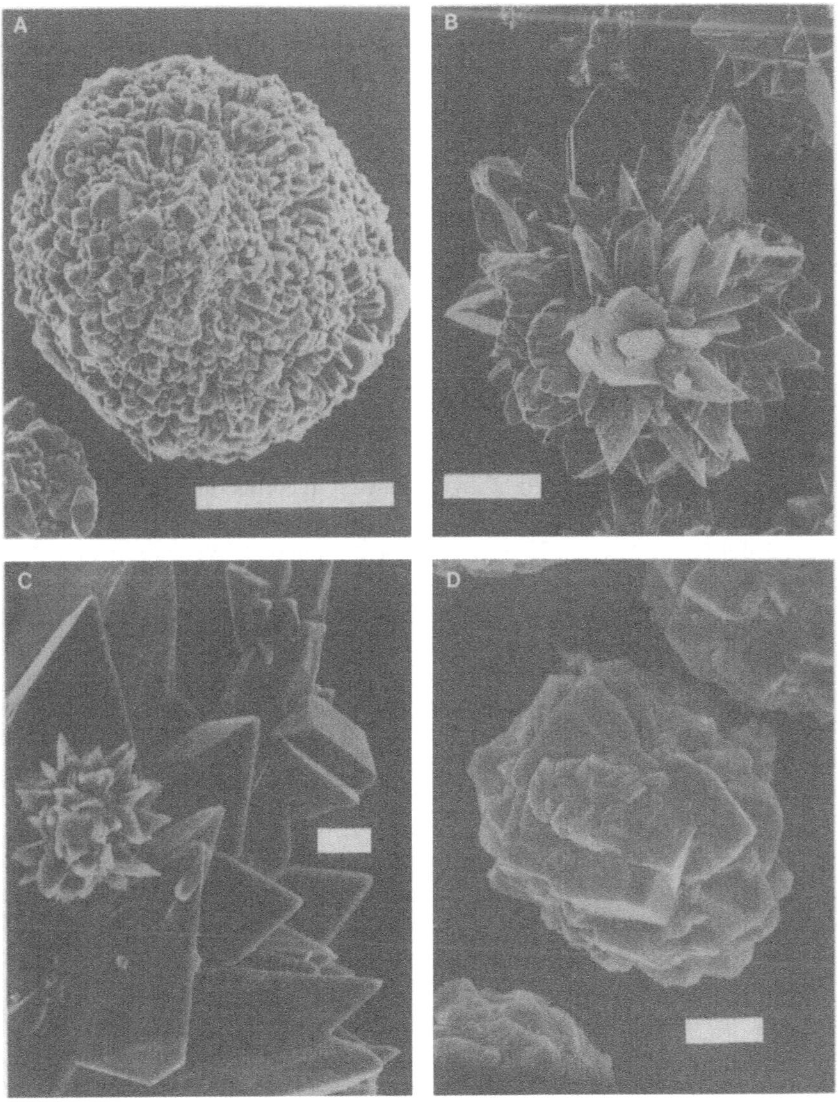

Fig 11.3. Scanning electron micrographs. A. Druse phytolith with small square-ended crystals from *Mammillaria heyderi*. Bar is 100 microns; B. Large druse phytolith from the pith of *Opuntia engelmannii* var *texana*. Bar is 100 microns; C. Small druse phytolith from the cortex of *Opuntia engelmannii* var *texana*. Bar is 10 microns; D. Epidermal druse phytolith from *Opuntia engelmannii* var *texana*. Bar is 10 microns.

Our study of possible phytolith variation within a single pad focused on the detailed examination of phytoliths recovered from one mature pad from a plant of *Opuntia engelmannii* var *texana* (Gr.) Weniger. This species and variety of *Opuntia* was chosen because it is one of the most widespread cactus types found throughout Texas and the southern regions of the United States. Field observations indicate that this taxon grows as far west as California and as far east as Florida (Weniger 1984).

The single pad studied measured 13 cm × 21.5 cm and was collected from an *Opuntia* plant growing near Langtry in southwest Texas. The pad was divided into five equal areas: 1) the lower portion of the pad that attaches to the rest of the plant, which we labeled as Part A; 2) the central portion of the pad, which was divided into three sections: the left one-third including the outer margin (Part B), the center region which we labeled Part C, and the right one-third and margin which we labeled part D; and 3) the apical region of the pad, labeled Part E. Tissue samples from the pith, cortex and epidermis/hypodermis were collected from each of the five areas and examined. Each sample was processed separately, and measurements of 100 druse phytoliths were recorded for each tissue type in each area of the plant.

As illustrated in Table 11.3 we found a few minor differences in the average size of phytoliths from one region of the pad to the next. Even so, the overall consistency of morphology and phytolith size was notable within each of the three tissue layers from all five regions of the same pad. This suggests that phytolith variation within each of the three tissue layers within a single pad of *Opuntia* is probably minimal. This study examined only one pad, and it needs to be repeated for assurance that these results represent genetically controlled factors that remain consistent throughout the pads of a single species over a large ecological range.

In the second phase of this study we examined potential phytolith size and morphological variations that exist between mature and immature pads growing on the same plant. For this study we selected examples from three different species of *Opuntia*. These included samples from *O. phaeacantha* var *nigricans*, *Opuntia engelmannii* var *texana* and *Opuntia leptocaulis*. From individual plants in each of these taxa we removed mature and immature pads. From each pad we removed portions of the three basic tissue types (pith, cortex and epidermis/hypodermis) and extracted phytoliths from each. Counts of 100 phytoliths from each sample are presented in Table 11.4, and illustrate that some variations seem to occur in phytolith size between mature and immature pads. For example, the average size of druse phytoliths in the pith region of the mature pad from *Opuntia phaeacantha* var *nigricans* is slightly larger than those in the immature pad. However, in this same plant the reverse is true for the phytoliths found in the cortex region. In that region, there appears to be a slight decrease in the size of phytoliths found in mature pads as compared to those found in the immature pads.

TABLE 11.3. Druse Phytolith Measurements within a Single Pad of *Opuntia engelmannii* var *texana* (Gr.) Weniger*

Pad area	Tissue type	Mean diameter	Range in microns
A	pith	132.98	75–240
A	cortex	61.80	30–105
A	epi/hypodermis	36.40	25–60
B	pith	128.40	70–290
B	cortex	44.60	15–85
B	epi/hypodermis	37.35	25–60
C	pith	138.25	75–290
C	cortex	56.35	35–85
C	epi/hypodermis	37.10	25–55
D	pith	142.35	90–255
D	cortex	53.65	30–85
D	epi/hypodermis	40.00	25–70
E	pith	146.60	85–285
E	cortex	56.45	25–105
E	epi/hypodermis	39.90	25–70

*Based on counts of 100 phytoliths for each tissue type in each area

TABLE 11.4. Variation in Phytolith Sizes in Mature versus Immature Pads of Individual *Opuntia* Plants*

	Species					
Tissue	*O. phaeacantha*		*O. engelmannii*		*O. leptocaulis*	
Pad type	immature	mature	immature	mature	immature	mature
Pith	156	162	186	203	60.6	59.8
Cortex	60	56.8	57.8	73.7	35.3	35.8
Epidermis/ Hypodermis	36.7	36.7	40.6	39.8	40.6	37.8

*Measurements represent averages based on 100 phytoliths per count

When we examined a mature and immature pad of *Opuntia engelmannii* var *texana*, we found a pronounced increase in the average druse phytolith size in both the pith and cortex regions of the mature pad as compared to those of the immature pad. On the other hand, one factor that may have skewed these data somewhat was the presence of a few relatively large phytoliths in the pith and cortex region of the mature pad, which increased the overall average. There does appear to be a consistent difference between phytolith size in mature versus immature pads of the species of *Opuntia*.

In various tissues of mature and immature cylindrical stems of *Opuntia leptocaulis*, we could find no trends or significant size differences. Thus, the age of the plant or the age of various parts of the plant does not necessarily control the size of the phytoliths formed in this species. Also, as noted for species in the Platyopuntia group, there seems to be very little change in the size of epidermal/hypodermal druse phytoliths from one species to the next in the Cylindropuntia group of *Opuntia*.

In summary, two results seem to be apparent: 1) individual plants in the Platyopuntia group seem to reflect more druse phytolith size variation between mature pads versus immature pads of the same individual plants than is found within the single plant examined in the Cylindropuntia group; 2) the general similarity of size from one taxa to another of epidermal/hypodermal druse type phytoliths appears to be a characteristic of both the Platyopuntia and Cylindropuntia groups of *Opuntia*. Thus, we suspect that because of this stability of size and general shape, epidermal/hypodermal druse phytoliths will not prove to be a useful taxonomic tool for differentiating the various species or even sub-genera of this genus.

A separate study was made to investigate variations in phytolith morphology and size among different plants of the same species of cactus growing under different edaphic and climatic conditions. We selected *Opuntia engelmannii* var *texana* plants because this taxon grows in almost every geographic region of Texas and is also found throughout many regions of the southern and southwestern United States.

Samples were collected from mature pads growing in four different locales, each representing a specific habitat type. We also examined tissue samples from a preserved, mature pad of this same cactus species that was recovered during the archaeological excavation of Hinds Cave, a prehistoric rockshelter located near Langtry in Val Verde County, Texas. The fossil sample was 6,000 years old, relatively large in size and compared favorably to modern specimens of *Opuntia engelmannii* var *texana*. This specimen was studied to verify that phytoliths could be easily extracted from long-dried plant specimens, and that phytolith shapes remained consistent through time. Table 11.5 shows this fossil pad contained an abundance of phytoliths with sizes within the range of other specimens of *Opuntia engelmannii* var *texana*.

Table 11.5 shows that there are marked fluctuations in the average size of pith phytoliths for each of the plants tested. We have no explanation for this apparent lack of uniformity except to note that of all the phytolith types found in the various cacti examined, the greatest size variations were among the phytoliths found in the pith region. This may reflect individual environmental factors.

We found a much greater degree of size consistency in the phytoliths of the cortex and epidermis/hypodermis regions. As seen in Table 11.5, the average size for cortical phytoliths ranged from only 57.8 to 64.8 microns in diameter and the average size of epidermis/hypodermis phytoliths ranged from 36.4 to 43.6 microns. We can offer few conclusions about the potential effect of geography, climate or soils upon the shape and/or size of phytoliths in the genus *Opuntia*. Unlike pollen studies, where extensive experiments have shown that morphological shape (and to a great degree size) of pollen types remains consistent regardless of where the plants grow, similar statements cannot yet be made for calcium oxalate plant phytoliths.

TABLE 11.5. Average Phytolith Size Based on Counts of 100 Phytoliths from *Opuntia engelmannii* var *texana* Plants Growing in Different Locales and One Sample Recovered from 6,000-year-old Deposits in an Archaeological Site

Plant Tissue	Collecting Locality				
	1	2	3	4	5
Pith	140	168	155	186	167
Cortex	61.8	64.8	62.4	57.8	56.6
Epidermis/ Hypodermis	36.4	42.4	42.1	40.6	43.6

Collection Location	Annual Rainfall	Mean Annual Temperature	Soil Composition
1. Langtry, Val Verde Co.	12"	68°F	calcareous, stony, clays and clay loam
2. Uvalde, Uvalde Co.	23"	70°F	calcareous, stony, clays and clay loam
3. D'Hanis, Medina Co.	24"	70°F	calcareous, stony, clays and clay loam
4. San Antonio, Bexar Co.	28"	70°F	dark calcareous, clay and clay loam
5. 6,000 yrs BP Hinds Cave, Val Verde Co.	—	—	site deposits

We also examined the phytoliths present in various tissues of cactus fruits from five different species or varieties of *Opuntia* because ethnographic and archaeological reports note that cactus were common food items in both prehistoric and historic times in many areas of Northern Mexico and the American Southwest. If these fruits contained phytoliths they should be present in the soils and human coprolites of many archaeological deposits.

We found phytoliths present in all the major tissue areas (pith, cortex, epidermis/hypodermis) of each cactus fruit we examined. The *Opuntia* fruits came from: 1) *Opuntia engelmannii* var *alta*, 2) *Opuntia engelmannii* var *texana*, 3) *Opuntia compressa* var *grandiflora*, 4) *Opuntia compressa* var *macrorhiza* and 5) *Opuntia leptocaulis*. Although the average sizes of phytoliths found in various tissues of the pads of these taxa varied, the sizes of the phytoliths found in the fruits of these same taxa were fairly uniform and consistent with the size gradient represented in the *Opuntia* sp pads. Again, the largest druse forms were found in the pithy part of the fruit, and the smallest were found in the epidermal/hypodermal layers.

DISCUSSION

One important application of phytolith taxonomy is the ability to collect and identify phytoliths from ancient sediments and use them as keys to the identification of plants that once grew in a specific region. The application of this technique to Southwestern archaeology in the United States would expand our ability to identify and recognize plants once used by prehistoric cultures. Since cactus plants are common throughout the arid regions of southern North America, taxonomic guides to species of cacti might prove useful for archaeologists working in this region.

This preliminary attempt revealed that for some genera of cacti, shape differences do exist and perhaps can be used to identify phytoliths from these plants present in the fossil record. However, for *Opuntia*, phytolith differences among the species and varieties vary more in size than they do in shape. This discovery is disappointing because we hoped it might be possible to use *Opuntia* phytoliths as keys to the cultural use of these plants.

Examination of phytoliths recovered from ancient human coprolites found in a number of archaeological sites excavated in southwest Texas revealed that many of them contained large numbers of druse-type phytoliths of the *Opuntia* type. This confirmed that prehistoric peoples in that region ate *Opuntia* cactus pads and fruits, but we were unable to identify which species of *Opuntia* was used because the druse phytoliths occurred over a wide size range.

One approach to size variation is to assign each species of *Opuntia* a number based on size ratios. We examined phytoliths from tissues in 19 separate *Opuntia* plants representing 10 taxa. For each tissue type in each of the 19 samples, we measured 100 phytoliths to determine the mean size. These data are presented in Table 11.6.

Our project had already established that there was very little size variation in the phytoliths found in the epidermal/hypodermal layers of all species of *Opuntia*. Therefore, we used the mean size of the phytoliths found in the epidermal/hypodermal layer of each of the 19 specimens as a "constant" against which we could compare the size range of phytoliths in the cortex and pith regions of the

TABLE 11.6. Mean Size and Derived Ratios of Phytoliths Found in the Pith, Cortical and Epidermal/Hypodermal Tissues of Ten Taxa of *Opuntia*.*

Species (P/C/E)**	N-number of Plants	Ratios	
		P/E	C/E
1. *O. macrocentra*	(N=1)		
253 / 73.5 / 35.9		7.03	2.04
2. *O. phaeacantha* var *camanchica*	(N=1)		
186 / 72.1 / 33.0		5.63	2.18
3. *O. engelmannii* var *alta*	(N=1)		
189 / 58.7 / 36.4		5.19	1.61
4. *O. phaeacantha* var *nigricans*	(N=3)		
162 / 63.3 / 39.2		4.14	1.61
5. *O. engelmannii* var *engelmannii*	(N=3)		
182 / 65.3 / 44.4		4.09	1.47
6. *O. engelmannii* var *texana*	(N=4)		
163 / 61.7 / 40.4		4.03	1.53
7. *O. compressa* var *macrorhiza*	(N=1)		
81.8 / 40.9 / 24.3		3.36	1.68
8. *O. schottii*	(N=2)		
119 / 67.7 / 38.9		3.06	1.74
9. *O. leptocaulis*	(N=2)		
60.2 / 35.6 / 39.2		1.54	.91
10. *O. compressa* var *grandiflora*	(N=1)		
35.4 / 24.0 / 27.2		1.30	.88

*For species or varieties where cortex, pith, and epidermal/hypodermal tissues from only one plant were examined, the averages derived from those 100 specimen counts were used as the mean we have listed in this table. For species or varieties where cortex, pith, and epidermal/hypodermal tissues from more than one plant were examined, the averages derived from each of the 100 specimen counts for each tissue were averaged. The resulting average was used as the mean listed in this table for that species or variety. The ratio for each *Opuntia* specimen we examined and for each of the ten different species or varieties of *Opuntia* we studied are listed in Table 11.6, and the numerical values we calculated are listed in Table 11.7.

**P-pith; C-cortex; E-epidermis/hypodermis

same plants. The mean size for each group of cortical and pith phytoliths was divided by the mean size of their epidermal/hypodermal phytoliths, establishing a numerical value for each of the taxa.

As noted in Table 11.7, the numerical values for each of the two or more varieties of the same species of *Opuntia* are quite similar. Greater differences seem to exist between different species of *Opuntia*. The exception is *Opuntia compressa*. Note that the numerical value for each of the two varieties of this species differs by a greater amount than this species differs from two of the other *Opuntia* species (*Opuntia schottii* and *Opuntia leptocaulis*). We have ruled out possible environmental factors as an explanation for this difference, because both plants of this species were collected from the same region of Texas and in the same type of habitat (sandy loam soils in the piney woods region of central Texas).

We believe that this type of numerical ratio study of the genus *Opuntia* has potential as a key to the identification of fossil phytoliths. However, we acknowledge that this study is preliminary and that a much more comprehensive examination of this technique needs to be conducted to test its usefulness for determining species separation and recognition.

We studied the phytolith morphology of five different genera of cacti. From these studies we noted several trends. First, in general, there was a similarity of crystal form within the species in each genera and more pronounced differences between the genera. Second, for many of the taxa there seemed to be a fairly consistent reduction in the size of the phytoliths as one moved outward from the pith region towards the epidermis/hypodermis.

TABLE 11.7. Ten Taxa of *Opuntia* Listed by their Numerical Value*

Genus and Species	Numerical Value
1. *Opuntia macrocentra*	3.45
2. *Opuntia engelmannii* var *alta*	3.22
3. *Opuntia engelmannii* var *engelmannii*	2.78
4. *Opuntia engelmannii* var *texana*	2.63
5. *Opuntia phaeacantha* var *camanchica*	2.58
6. *Opuntia phaeacantha* var *nigricans*	2.57
7. *Opuntia compressa* var *macrorhiza*	2.00
8. *Opuntia schottii*	1.76
9. *Opuntia leptocaulis*	1.69
10. *Opuntia compressa* var *grandiflora*	1.47

*Numerical value based on P/C ratio from Table 11.6

General observations of the phytoliths present in the cacti species we examined in this study suggest that phytolith identification at the generic level is possible for the genus *Opuntia* because, as far as we know, *Opuntia* and the related genus *Pereskia* are the only genera of cacti that produce a distinctive type of spiny druses. As in most of the other cacti genera examined in this study, the largest druse phytoliths in the genus *Opuntia* are found in the pith region, become progressively smaller in the cortex, with the smallest occurring in the epidermal/hypodermal region. We also have noted that the epidermal/hypodermal druse phytoliths in *Opuntia* appear to be more consistent in size throughout the genus as a whole than the druse phytoliths in other tissue regions. Finally, we want to stress that more taxonomic work needs to be done in classifying these enigmatic phytoliths before reliable keys can be produced.

SUMMARY

Our research on phytoliths in some genera of the Cactaceae has revealed new data. Examination of 16 taxa of cacti demonstrates that it may be possible to develop a taxonomic key based on phytoliths for many of them. We found that *Echinocactus*, *Echinocereus*, *Ariocarpus* and *Mammillaria* cactus species can be separated at the generic level based on phytolith characteristics.

Our examination of the taxa of *Opuntia* showed that phytolith types, frequencies and construction are similar throughout the genus. Our goal for this study was to search for a method of developing a taxonomic key for the phytoliths of the genus *Opuntia*. To do this we compiled data on five criteria: 1) determine the location, abundance and types of phytoliths in the stem tissues of each taxon of *Opuntia*; 2) measure, record and search for similarities and/or differences in the general morphology and crystalline structure of phytoliths found in each of the stem tissue types [pith, cortex and epidermis/hypodermis] of each taxon; 3) look for the presence or absence of types of "exotic," non-druse type phytoliths in each of the tissues of each taxon; 4) examine phytoliths produced in the stems of *Opuntia* plants of the same taxon growing in different habitats for similarities and/or differences; and 5) search for ways of comparing the phytolith data from each taxon of *Opuntia* that would allow separation of each taxon into a unique category based on its phytoliths.

Measurements of 100 phytoliths from each of the three tissue layers in the stems of each of the ten taxa of *Opuntia* showed little variation in the location, abundance or types of phytoliths present in each taxon. Likewise, we discovered that with a few exceptions, the druse type phytoliths found in the pith, cortex and epidermis/hypodermis tissues of each taxon of *Opuntia* were essentially similar in morphology. Our search for non-druse exotic type crystals in each taxon of *Opuntia* did reveal some unique shapes. We found that many of the taxa contained a few non-druse phytoliths that we have termed "exotic" because they were always rare when compared to the abundance of the druse types. We were unable to determine the origin of these exotic-type crystals and do not know enough about them to determine their potential uniqueness or value as taxonomic markers. One of the most common exotic types is the dodecahedral phytoliths we found in tissues of *Opuntia compressa* var *macrorhiza* (Fig 11.4). We have found similar types of phytoliths in taxa of the Brassicaceae (Cruciferae), Euphorbiaceae, Asteraceae (Compositae) and in species of *Yucca* and *Agave* growing in the arid habitats of southwest Texas. As yet it is unclear whether these exotic types of phytoliths are inherent to each of the taxa mentioned above, or whether these phytoliths may be formed by some type of precipitate during our chemical extraction procedure. We suspect the former, but cannot rule out the latter.

Our examination of phytoliths from stem tissues of one taxon of *Opuntia* growing in different habitats revealed a fairly high degree of consistency. This suggests that in one taxon of the genus *Opuntia*, phytolith shape and size are governed more by genetics than environment. Whether or not this consistency will prove uniform throughout all taxa of the *Opuntia* is an aspect that will need further testing.

Fig 11.4. Dodecahedral phytoliths isolated from *Opuntia compressa* var *macrorhiza*. Bar represents 10 microns.

Finally, our search for a way to compare the phytoliths of one taxon of *Opuntia* to those of other *Opuntia* taxa led us to construct numerical ratios for each of the ten taxa studied. These numerical values were based on ratios constructed by dividing the average size of pith phytoliths found in the plant stem by the average size of the epidermis/hypodermis phytoliths from the same stem. Similar ratios were also constructed for the cortex phytoliths of each taxon using the same criteria.

This research is a starting point for further studies of the phytoliths found in taxa of the Cactaceae. In this preliminary study we divided stem sections of each taxon into what we felt were three reliable types of tissues. This may not be the best way to examine the problem of phytolith size and shape variation. This study documented that within the stems of the cactus taxa, the sizes of the dominant druse phytoliths seem to occur in a gradient. The largest ones seem to form in the central pith region of the stem, whereas the smallest ones most often occur in the epidermis/hypodermis tissues.

Most of the druse phytoliths seemed to lack distinctive surface features. However, careful studies of these same phytoliths at high levels of resolution using scanning electron microscopy may reveal subtle differences in shapes that might prove unique to individual taxa. We suspect that x-ray probe analysis might reveal slightly different chemical compositions of cactus phytoliths that could be unique to individual taxa.

We believe that gross shape, spine shape and length, size and size ratios are the most useful characteristics to examine in future studies of the dominant druse phytoliths in taxa of the Cactaceae. We hope that someday sufficient comparative data will be available to permit a linking of druse phytoliths recovered from archaeological sites and non-cultural deposits from the southwestern United States and Northern Mexico to precise taxa in the Cactaceae.

ACKNOWLEDGMENTS

We wish to thank Dr. Richard G. Holloway for his technical assistance and recommendations concerning the extraction of the phytoliths from plant tissues. Graduate students Sherrian K. Edwards, Linda K. Stanley and Aleydis Van de Moortel assisted us with measuring the phytoliths. We would like to extend our special thanks to Mr. Jack Skiles of Langtry, Texas, for permitting us to make frequent cactus collecting trips to his ranch located in and around Mile Canyon near Langtry. Finally, we extend special thanks to the personnel in the Texas A&M Electron Microscope Center for permitting us to use their facilities and equipment in this study. The photographs we used for this article were developed and printed by Virginia Massey, Michael Pendleton and Aleydis Van de Moortel.

REFERENCES

Backeburg, C 1977 *Cactus Lexicon*. Dorset, Blandford Press: 828 p.

Bryant, VM, Jr (ms) 1969 Late full-glacial, and post-glacial pollen analysis of Texas sediments. PhD dissertation, Austin, University of Texas: 168 p.

Bryant, VM, Jr 1974 Prehistoric diet in Southwest Texas: The coprolite evidence. *American Antiquity* 39: 407–420.

Bryant, VM, Jr (ms) Preliminary study of phytoliths from desert plants in Val Verde County, Texas.

Collins, MB 1969 Test excavations at Amistad International Reservoir, Fall 1967. *Papers of the Texas Archaeological Salvage Project*, No 16, Austin, Texas: 103 p.

Dibble, DS and Lorrain, D 1968 Bonfire Shelter. A stratified bison kill site, Val Verde County, Texas. *Texas Memorial Museum Miscellaneous Papers* 1: 1–78.

Dunn, ME 1983 Phytolith analysis in archaeology. *Midcontinental Journal of Archaeology* 8: 287–297.

Esau, K 1965 *Plant Anatomy*. New York, John Wiley and Sons: 550 p.

Hester, TR 1983 Late Paleo-Indian occupations at Baker cave, southwestern Texas. *Bulletin of the Texas Archaeological Society* 53: 101–119.

Johnson, L 1964 The Devil's Mouth Site: A stratified campsite at Amistad Reservoir Val Verde County, Texas. *Archaeology Series*, No 6: 1–115. Department of Anthropology, Austin, The University of Texas.

Jones, JG 1988 A simplified procedure for the extraction of silica and calcium oxalate phytoliths from plant tissue. *Phytolitharien Newsletter* 5(3): 9–10.

Parry, DW and Smithson, F 1958 Silicification of bulliform cells in grasses. *Nature* 181:1549–1550.

Piperno, DR 1988 *Phytolith Analysis: An Archaeological and Geological Perspective*. San Diego, Academic Press: 280 p.

Rovner, I 1983 Plant opal phytolith analysis: Major advances in archaeobotanical research. *In* Schiffer, MB, ed, *Advances in Archaeological Method and Theory* 6: 225–266. New York: Academic Press.

Weniger, D 1984 *Cacti of Texas and Neighboring States*. Austin: University of Texas Press: 356 p.

12

SILICA DEPOSITION IN SUBTERRANEAN ORGANS

ALLAN G. SANGSTER and MARTIN J. HODSON

ABSTRACT

Silica in roots and rhizomes has not been investigated in a comprehensive or extensive manner. Excluding the grasses, the literature indicates two monocotyledons, two dicotyledons and a conifer contain silica in subterranean organs. Grasses, although not completely investigated, have been examined in more detail; major sites of deposition include the root endodermis and rhizome epidermis. Review of our knowledge of silica function, localization and deposition mechanisms indicates that many gaps still exist. Although it appears at this time that aerial parts produce phytoliths of greater taxonomic significance, subterranean organs should be systematically surveyed.

INTRODUCTION

Silica in the subterranean organs of higher plants has received considerably less attention than deposition in the aerial parts. Such fragmentary information as is available is almost entirely confined to grasses and cereals. Previous reviews of this area have emphasized anatomical diversity within the roots of the Poaceae (Gramineae) (Sangster 1978; Sangster & Parry 1981; Parry, Hodson & Sangster 1984). Since these reviews, two major developments have occurred in silica research: an explosion of interest in phytoliths from paleobotanists and archaeologists (Piperno 1988); and the application of cryotechniques in electron microscopy (Hodson & Sangster 1988). It is our intention in this article to review recent work on silica in roots and rhizomes, and to point out the large gaps in our knowledge that still exist in this area.

DISTRIBUTION OF ROOT AND RHIZOME SILICA WITHIN TAXA

The last detailed review of this topic (Parry, Hodson & Sangster 1984) concentrated entirely upon deposition within the roots of the Poaceae (Gramineae). In the present paper we intend to broaden the subject area to include the subterranean organs from all higher (vascular) plants.

Silica Deposition in Roots from Plants Other Than the Poaceae (Gramineae)

An extensive search of the literature has revealed very little information concerning silica deposition in roots of plants other than grasses and cereals. It seems either that little silica is deposited in these roots, or perhaps more likely, that little research has been carried out. In the monocotyledons (excluding the Poaceae or Gramineae), we have located just two cases. Cutler (1969: 302) reported "small silica-like particles present in many ground tissue cells" in the roots of *Staberoha cernua* from the Restionaceae (Juncales). Metcalfe (1971) gave considerable coverage to phytoliths within the aerial parts of the Cyperaceae, but gave no indication of silica deposition in the roots. It is therefore of some interest that Ollendorf, Mulholland and Rapp (1987) reported phytoliths present in the roots of *Cyperus lanceus* and *Cyperus conglomeratus* (both Cyperaceae). In neither case was the original anatomical location of these deposits stated, and this report requires substantiation.

In the dicotyledons, Metcalfe and Chalk (1983) make no mention of silica deposits in roots. Piperno (1988: 106) stated that she "would expect that in many species production is not particularly impressive." One of us (MJH) has, however, received an unconfirmed report that silica is present in the endodermis of beech roots. Also, Lanning and Eleuterius (1985) report silica (0.70%) in the roots of *Batis maritima*, a succulent plant of hypersaline marshes.

While investigating aluminum toxicity in Norway spruce (*Picea abies*), Hodson and Wilkins (1991) noted considerable amounts of silicon deposited in the root cortical walls of aluminum treated plants. This intriguing finding is the first we know of from the roots of the Gymnospermae, and deserves further investigation.

New Data on Silica Deposition in the Roots of the Poaceae (Gramineae)

Three previous review articles have covered this ground (Sangster 1978; Sangster & Parry 1981; Parry, Hodson & Sangster 1984), and here we will only consider the major conclusions, and highlight the advances since the last review. Table 12.1 is adapted from Parry, Hodson and Sangster (1984), and shows the present state of knowledge with regard to silica deposition in roots of the Poaceae (Gramineae). Although our knowledge of silica deposition in these plants is superior to that for any other group, it is still far from complete, with only twelve tribes having been investigated to any extent. The endodermis has emerged as

TABLE 12.1. Silica Distribution Patterns in the Roots of Grasses and Cereals*

Location	Type of deposit	Tribe	Genera
1a Endodermis	Silica aggregates and inner tangential wall (ITW)	Andropogoneae	eight genera (A), including *Sorghum* (C)
b Endodermis	Cell walls only	Paniceae	*Digitaria*
	(i) ITW		*Panicum virgatum*
	(ii) ITW and radial	Oryzeae	*Oryza*, rice (young)
		Bambuseae	*Sasa*, bamboo
	(iii) All four walls	Triticeae	*Triticum aestivum* (D) (3 cultivars)
			Hordeum sativum
		Aveneae	*Avena sativa*
		Phalarideae	*Phalaris canariensis* (B)
c Probably endodermis	Chemical analysis only	Chlorideae	
		Eragrosteae	
2 Intercellular spaces cortex	Extracellular	Danthonieae	*Molinia* (young)
3 All tissues epidermal sclerenchyma xylem vessels	Cell walls	Danthonieae	*Molinia* (old)
		Oryzeae	*Oryza* (old)
4 No distribution pattern		Triticeae	Wheat (one cultivar)
		Maydeae	*Zea mays* (five cultivars)
		Arundineae	*Phragmites australis*

*This table is adapted from Parry, Hodson & Sangster (1984), and only includes the references that have appeared since then: (A) Brown (1984); (B) Hodson (1986); (C) Hodson & Sangster (1989a); (D) Hodson & Sangster (1989b).

a major site for silica deposition in grass roots. So far in seven tribes, one or more of the endodermal walls have been found to be silicified. Of these, only the Andropogoneae show characteristic silica aggregates formed on the inner tangential wall, and projecting into the endodermal cell lumen. A number of investigations have shown silica deposits in tissues of the root other than the endodermis, but on present evidence these are fairly rare. Only in two tribes, the Maydeae and the Arundineae, have no deposits been found (when the investigators have been specifically looking for them).

Only four major publications have specifically mentioned silicon in roots of the Poaceae (Gramineae) since the review of Parry, Hodson and Sangster (1984). Two of these have confirmed previously known deposition patterns in the roots of sorghum and wheat (Hodson & Sangster 1989a, 1989b), and are discussed in detail below. The root was included in a survey of silica distribution in *Phalaris canariensis* (Hodson 1986). The endodermal walls were the major site of silica deposition in adventitious roots grown in solution, but there was also evidence of deposition in the epidermis and outer cortical cell layers in aerial roots. Brown (1984) investigated phytolith shapes in the grasses of the central United States. He included roots in his survey, but unfortunately did not indicate the tissues in which the phytoliths were found. Thin elongate phytoliths were noted in the roots of *Ammophila breviligulata*, *Vulpia octoflora* and *Elymus canadensis*, while thick elongates were isolated from *Leptochloa fascicularis* and *Hilaria jamesii*. Of particular interest were the short (length less than three times width) phytoliths isolated from the roots of *Oryzopsis hymenoides*, *Aristida wrightii* and *Schizachyrium scoparium*. These almost certainly represent endodermal silica deposits. In Brown's Figure 1, identifier IBIc (isolated from *S. scoparium*), the structure of the phytolith depicted is very similar to that we have seen in other members of the Andropogoneae. Brown's work has given us a tantalizing glimpse of silica deposits in the roots of many grass genera, but a more detailed survey is still required.

Silica Distribution in the Rhizomes of the Poaceae (Gramineae)

The distribution of silica deposits in the rhizomes of the Poaceae (Gramineae) has been reviewed by Sangster (1985), and little significant work has appeared since that time. Essentially only three tribes have been investigated so far, with nearly all of the work confined to the Andropogoneae. In contrast to the roots, the epidermis is a major site of silica deposition in the rhizome. The inner tangential wall of the rhizome endodermis is a common site for silica deposition in the Andropogoneae, and conical silica aggregates similar to those observed in the roots have been noted in some species. However, so few species have been investigated that it is difficult to make generalizations about silica distribution patterns in the rhizome.

POSSIBLE FUNCTIONS OF SILICA DEPOSITS IN ROOTS

Functional aspects of silica deposition in higher plants were reviewed by Sangster and Hodson (1986). Roots and rhizomes have thus far received little attention. However, several possible roles for the silica aggregates in the endodermis of *Sorghum bicolor* roots have been suggested. Ponnaiya (1960) considered that these deposits might contribute to drought resistance, a view reiterated by Doggett (1970). More recently, two publications (Maiti *et al* 1984; El Hiweris 1987) have suggested that silica deposition in the sorghum root endodermis might lead to resistance to the important parasitic angiosperm genus *Striga* (Scrophulariaceae). Significantly, silica aggregates were absent from the endodermis of some susceptible cultivars (Maiti *et al* 1984).

Recently, Hodson and Wilkins (1991) have found that silicon levels in the root cortical walls of Norway spruce (*Picea abies*) increase when the plants are treated with aluminum. These preliminary results suggest that silicon *may* be involved in complexing phytotoxic elements in roots, increasing plant tolerance to them.

LOCALIZATION AND DEPOSITION MECHANISMS

Solid-phase silica has been detected in subterranean organs, mainly roots, of members of the Monocotyledonae and one example from the Gymnospermae (conifers). In roots of the family Poaceae (Gramineae), the deposits are incorporated entirely within the cell wall, or in the case of the silica aggregates (Tribe Andropogoneae; Table 12.1), the cone-shaped tip projects beyond the wall in which the major portion of the deposit is embedded. Opal phytoliths formed by the infilling of the interior cavity (lumen) are not reported. In *Molinia caerulea* (moor grass), cortical intercellular spaces, possibly originating from cell breakdown, become infilled, resulting in deposits that are classified as extracellular. In grass roots, deposition frequently is confined to the walls of the endodermis. However, walls of the epidermis and vascular tissues may also deposit silica in older roots. Localization sites appear to be determined largely by plant factors including tissue maturation states (developmental), annual vs perennial habits, phylogeny or cultivar (genetic), and also by external factors such as root medium and the availability of soil silica (Table 12.1; Sangster & Parry 1981; Parry, Hodson & Sangster 1984; Hodson 1986; Sangster & Hodson 1986).

In the annuals, roots frequently exhibit the lowest silica content on a dry-weight basis (eg, 1.84%: oats) in comparison to higher values in above-ground parts. In grain sorghum roots, this relationship applies from about 30 days to maturity, but during early growth, root silica levels (up to 15%) exceed those of the shoot for four cultivars (Lanning & Linko 1961; Jones, Milne & Wadham 1963). The perennial grasses (genera: *Eustachys, Andropogon, Panicum*) of the Serengeti National Park, Tanzania, exhibit silica level gradients that are the reverse of those for annuals. Silica content of the roots is highest, followed by the stems and leaves. The values for all parts considerably exceeded those recorded for cultivated forages. This was assumed to be the result of perennial growth and continued grazing pressure by ungulates (McNaughton *et al* 1985; review by Hodson & Sangster 1987).

Bennett and Sangster (1982) were unable to detect silica deposits using x-ray microanalysis in roots of maize, whereas Lanning, Hopkins and Loera (1980) measured 2% d. wt. for root silica using a gravimetric technique. The two studies employed different techniques, cultivars and soils. The deposits in the latter study were not described; thus their origin cannot be ascertained (as reviewed by Piperno 1988). In cereal roots, most silica is in soluble form and is rapidly transported to the shoot (Jarvis 1987). In fresh tissues, permeated by mostly soluble silica, the effect of ashing during analysis might result in some alteration of physical states. Drum (1968) placed pieces of fresh *Betula* woody twigs in highly concentrated metasilicate solutions for 24 hours, and demonstrated the in vitro deposition of opaline silica on cell wall inner surfaces that formed replicas of various tissues.

In determining the amount of silica passing to the plant shoot, the root exerts a critical regulatory function in the overall silica cycle at the interface between the physical and biotic realms. Nonetheless, little is known about the routes of the movement and mobilization of soluble silica in roots, nor of the ionic species present at silica polymerization sites. In our previous root studies, the silicon detected by x-ray microanalysis derived mostly from solid-phase silica because the preparatory freeze-drying of specimens for electron microscopy resulted in severe ionic losses and dislocation (including silicate). Thus, the soluble silica fraction was virtually undetectable (Parry, Hodson & Sangster 1984; Hodson & Bell 1986; Hodson & Sangster 1988).

The possibility of detecting soluble silica in plant tissues arose with the emergence within the last 15 years of two low temperature electron microscopy techniques designed to minimize ion losses. These were: 1) low temperature scanning electron microscopy (LTSEM), in which frozen hydrated tissues are analyzed on a SEM cryostage (−180°C) by means of energy-dispersive x-ray (EDX) analysis; and 2) freeze substitution and transmission electron microscopy (FSTEM), in which the water contained in quench-frozen tissues is substituted by ether, followed by embedment in resin. Sections are cut dry to avoid water contact (which causes ion loss) and then subjected to EDX analysis in a TEM. In both of these techniques, the plant tissues should retain all soluble ion species in original quantities, in situ. However, in actual practice, some inaccuracies may arise as a result of the freezing process, or from EDX analysis because of surface irregularities of the specimens (Hodson & Sangster 1989b).

Translocation, Mobilization and Deposition Sites

We applied the two cryotechniques to investigate the roots of two cereals: 1) *Sorghum bicolor* L. Moench. (grain sorghum), cultivar P508 GB, and 2) *Triticum aestivum* L. cv Wheaton (dwarf spring wheat). Except for the endodermis, very low amounts of silicon were detected in protoplasts and walls of virtually all the tissues of the upper root zone (0.25 length) of both species using LTSEM. These low levels are assumed to represent soluble forms of silicon, basically monomeric silicic acid, $Si(OH)_4$. Silicon levels were too low in the protoplasts to reveal the presence of specific accumulation routes. However, the results indicated that protoplasts must receive consideration as a potential transport pathway, even if only a minor one compared to cell walls. In contrast, solid-phase silica was detected only in the walls of the uniseriate endodermis that immediately surrounds the central conducting region of the root (the stele). In sorghum, high silicon levels were present in all four walls as well as the conical deposits (silica aggregates) of the inner tangential wall (ITW). We determined that the aggregates were formed at the initiation of secondary wall formation rather than its completion as originally interpreted. Aggregates subsequently increased in size in synchrony with wall thickening. The cryotechnique thus extended and modified previous conceptions of silicon movement and localization in the sorghum root. Deposition was confined to the upper mature zone of the root, decreasing in a gradient towards the tip region where none was detected. This confirmed earlier studies (Sangster & Parry 1981; Hodson & Sangster 1989a). Wheat roots also exhibited low soluble silicon levels in most tissues and similarly high levels of deposited silicon in all endodermal walls. However, silica aggregates were not present. Similarly, a decreasing linear root gradient of silica was detected. The number of cells of the endodermal row that were silicified varied from all cells in the basal zone, to some in the central zone and none in the tip region (Hodson & Sangster 1989b).

Mobilization of soluble silicon along specific pathways was demonstrated by means of the second cryotechnique. The older root zones of seminal roots (15 cm long, 2 wk old) of wheat grown in nutrient solution containing 100 ppm SiO_2 were re-examined using FSTEM (Fig 12.1). This technique is superior because of improved specimen surface characteristics, more precise spatial resolution and greater sensitivity of the analytical system. Silicon x-ray maps indicated localization in the vacuoles and walls of the large central vessels (xylem) and in the circle of smaller peripheral metaxylem vessels. Each of the latter is joined by a two-celled "silica bridge" to the nearest endodermal cell (Fig 12.1B). A specific silicon pathway was exhibited in the wheat root in which the soluble silicon levels of cell vacuoles were higher by comparison with surrounding tissues. Small pockets of much higher silicon counts within the large central xylem vessels suggested that low molecular weight polymers or organo-complexes may have formed. Solid-phase, opaque silica was present in walls of the pathway cells across the stele of the root. Deposited silicon was localized in the endodermal walls, frequently the highest x-ray counts occurring in central areas of the ITW. Soluble silicon accumulates in endodermal cell protoplasts, most prominently in the region adjacent to the ITW, as revealed by the x-ray maps. Possibly the endodermis protoplasm is able to mobilize soluble silicon prior to wall deposition. By contrast, protoplasts of the immediately adjacent cell row (pericycle) accumulate phosphorus (Hodson & Sangster 1989b).

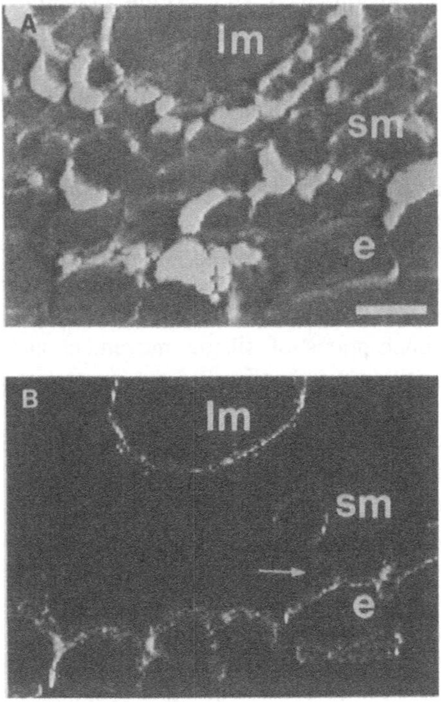

Fig 12.1. Transverse section from the basal zone of the wheat root after 14 days in nutrient solution containing 100 ppm SiO_2. Tissue prepared by freeze substitution followed by dry sectioning. A. Scanning transmission electron microscope (STEM) image of a portion of the central conducting region (stele); B. The matching silicon x-ray distribution map. Silicon is localized in the walls of a large central metaxylem vessel (lm), a small peripheral metaxylem vessel (sm), a silica bridge cell (arrow) and the endodermis (e); bar is 10 microns.

Two models are proposed for the accumulation of soluble silicic acid prior to supersaturation and the transformation to solid-phase amorphous silica gel. In the passive model, accumulation results from sap flow resulting from evaporative forces produced by shoot transpirational water loss. Polymerization is then mediated by cellular conditions, the most common factor invoked being pH. The efficacy of this model has been questioned for localized silicification in shoot structures such as trichomes (for critiques, see Kaufman *et al* 1981; Sangster & Hodson 1986; Piperno 1988). An alternative model proposes metabolic mechanisms controlling silica mobilization, for example, "by a membrane surface to which silica particles are attracted by ionic forces to form a layer of silica" as stated by Kaufman *et al* (1981: 445). The latter type of model appears more appropriate for root tissues. Non-evaporative mobilization can be postulated for the xylem-bridging cells-endodermis pathway discussed for wheat roots (Hodson & Sangster 1989b), possibly by means of silicon pumps located in the immediately adjacent cells.

In earlier studies, factors known to influence sol-gel transformations were emphasized as mechanisms for silica deposition. These included sol concentration, the presence of metal oxides (eg, aluminum), electrokinetic properties, especially pH, and cytoplasmic senescence products. These were considered to regulate opal phytolith formation within cell lumens. Deposition in cell walls was conceived as being controlled possibly by other mechanisms (Sangster 1970; Hayward & Parry 1973; Kaufman *et al* 1981; Simkiss & Wilbur 1989). Deposits in the root endodermis of sorghum are formed synchronously with cell wall thickening, which demonstrates that silicification is integrated with plant developmental processes (Sangster & Parry 1981; Hodson & Sangster 1989a). Recent studies have emphasized that the matrix properties of walls can influence the physical ordering of the deposited silica. Perry, Williams and Fry (1987) found that the carbohydrate components of the macrohair wall (*Phalaris canariensis*) that serve as a matrix for silicification changed during maturation, and that this resulted in corresponding changes in the physical ordering of the silica. Waterkeyn, Bienfait and Peeters (1982) correlated wall silicification with the presence of callose. Earlier studies reported that the deposition mechanism in cell walls of roots might be linked to the presence of specific wall components including phenols, suberin and lignin (see review by Sangster & Parry 1981). Mann (1988) emphasized that the influence of organic macromolecules is important in the regulation of growth of biominerals. The ionic environment at cell wall deposition sites has recently received attention as an important control mechanism in polymerization (Perry, Mann & Williams 1984; Hodson & Bell 1986). In the sorghum and wheat roots studied by cryotechniques, the ions most frequently detected in the protoplasm and wall deposition sites of the endodermis were phosphorus, sulphur, potassium, chlorine and calcium. The last three ions may exert some influence on deposition, but evidence is lacking (Hodson & Sangster 1989a, 1989b). In roots, mechanisms that may prevent polymerization are also significant (Simkiss & Wilbur 1989), because most soluble silicic acid is transported to the shoot.

Silicification of Rhizomes

These horizontally growing stems contain storage reserves, and by rooting at the nodes provide clonal reproductive and colonization potential. It is the major reproductive mode for grasses such as the introduced *Miscanthus sacchariflorus* (Maxim.) Hackel in southern Ontario where it does not set viable seed. Silicification of cell walls, without silica aggregates, is localized in three concentric zones: the epidermis (exterior), the cells bordering the air canals (cortex), and the central cavity (interior). The taxa *Andropogon, Sorghastrum, Schizachyrium* (tribe Andropogoneae), with no air canals or cavities, exhibit different concentric patterns of silica distribution variously involving the epidermis, the endodermis (cortex), the sclerenchyma and bundle sheaths (vascular regions). Silica aggregates are typically present on walls (ITW) of internal tissues. Developmentally, silica levels are highest in the outer walls of epidermal cells of the young apical region of *M. sacchariflorus*, decreasing sharply internally, whereas in mature internodes, silicon content is high in all three zones. Deposition is initiated early and is closely integrated with wall thickening during tissue maturation. Opal phytoliths (lumen deposits) are reported for the rhizome epidermis of the grass *Distichlis spicata*. Rhizome deposition is highly site specific, which contributes to functional efficiency of the subterranean organ. While absent from the apical region of roots, which is coated by mucilaginous material, the young growing apex of the rhizome exhibits a heavily silicified epidermal surface. Consequently, the silicified rhizome tip of *D. spicata*, for example, is able to penetrate compact shales, the silica providing compression and shearing resistance. Similarly, silica imparts rigidity around the air canals, which helps maintain critical oxygen exchange in species subject to flooding. Plants are able to mobilize silicon at specific epidermal sites in the shoot in response to tissue damage. Responses to wounding, grazing and fungal penetration are well documented. This lends support to the active or metabolic model for these aspects of deposition (Hansen *et al* 1976; Sangster 1985; Rovner 1986).

Future Perspectives

Mechanisms of transport and the accumulation or mobilization of soluble silica are poorly understood. One might speculate upon the characteristics of endodermal wall structure that render it particularly receptive to heavy silicification, whereas the particular ability of the cytoplasm to mobilize high levels of soluble silica may represent the critical factor. Directional flow of soluble silicon across the root to the endodermis requires elaboration, as do the depositional mechanisms in walls. Recent studies indicate that polymerization and the physical structure of the resulting phytoliths are influenced both by the macromolecules of the cell wall matrix and by the ionic environment. However, evidence for these interconnections is lacking. In sorghum roots, the means remain unknown by which silica aggregate localization or its formation is switched on and off in synchrony with the thickening of the endodermal ITW.

Because the amounts of soluble silicon to be measured are often minuscule, technical improvements (eg, ultra-thin windows) are required to increase sensitivity of the detectors to the lighter elements, including silicon. Finally, although phytolith production from subterranean organs should contribute considerably to sediments and soils, especially from roots, many taxa have not been tested for phytolith production (Williams 1986; Hart 1988; Piperno 1988; Hodson & Sangster 1989a, 1989b).

POTENTIAL ARCHAEOLOGICAL AND PALEOECOLOGICAL USES OF PHYTOLITHS FROM SUBTERRANEAN ORGANS

Thus far, root and rhizome phytoliths do not appear to have found a practical use in archaeological or paleoecological research (Piperno 1988). Perhaps this is to be expected as phytolith research is still in its infancy. What are the best possibilities for the future? It seems unlikely that root phytoliths will assist in studies of maize, as Bennett and Sangster (1982) could not detect silica deposits in the roots of five cultivars. The endodermal deposits of the type found in wheat (Bennett 1982; Hodson & Sangster 1989b), barley (Bennett 1982), oats (Bennett 1982) and canary grass millet (Hodson 1986) do seem to be fairly widespread in the Poaceae (Gramineae). It seems unlikely, however, that these phytoliths could be easily used to distinguish between genera or even tribes, unless on the basis of size. It is probable that phytoliths isolated from the aerial parts will be of greater taxonomic significance. Perhaps the rather striking silica aggregates found so far only in the endodermis of the roots of the Andropogoneae (Brown 1984; Hodson & Sangster 1989a) will prove to be the most useful type of root silica deposit for work in this area. Piperno (1988) has, however, suggested that these phytoliths appear similar to those in some dicotyledon woods. Work on grass rhizomes is even more sparse than work on roots, and it is evident that a thorough systematic survey of silica deposition in roots and rhizomes will be required before much further progress can be made.

ACKNOWLEDGMENTS

The STEM micrographs were produced by Douglas Holmyard, Electron Microscope Facility (Head, Dr. K.P.H. Pritzker), Mount Sinai Hospital, Toronto. The freeze substitution and cryo-SEM studies were supported by a grant from the Natural Sciences and Engineering Research Council (Canada).

REFERENCES

Bennett, DM 1982 Silicon deposition in the roots of *Hordeum sativum* Jess., *Avena sativa* L. and *Triticum aestivum* L. *Annals of Botany* 50: 239–245.

Bennett, DM and Sangster, AG 1982 Electron-probe microanalysis of silicon in the adventitious roots and terminal internode of the culm of *Zea mays*. *Canadian Journal of Botany* 60: 2024–2031.

Brown, DA 1984 Prospects and limits of a phytolith key for grasses in the central United States. *Journal of Archaeological Science* 11: 345–368.

Cutler, DF 1969 *Anatomy of the Monocotyledons. IV. Juncales.* Oxford, Clarendon Press: 357 p.

Doggett, H 1970 *Sorghum*. London, Longmans Green: 512 p.

Drum, R 1968 Silicification of *Betula* woody tissue in vitro. *Science* 161: 175–176.

El Hiweris, SO 1987 Nature of resistance to *Striga hermonithica* (Del.) Benth. Parasitism in some *Sorghum vulgare* (Pers.) cultivars. *Weed Research* 27: 305–311.

Hansen, DJ, Dayanandan, P, Kaufman, PB and Brotherson, JD 1976 Ecological adaptations of salt marsh grass, *Distichlis spicata* (Gramineae) and environmental factors affecting its growth and distribution. *American Journal of Botany* 63: 635–650.

Hart, DM 1988 The plant opal content in the vegetation and sediment of a swamp at Oxford Falls, New South Wales, Australia. *Australian Journal of Botany* 36: 159–170.

Hayward, DM and Parry, DW 1973 Electron-probe microanalysis studies of silica distribution in barley (*Hordeum sativum* L.). *Annals of Botany* 37: 579–591.

Hodson, MJ 1986 Silicon deposition in the roots, culm and leaf of *Phalaris canariensis* L. *Annals of Botany* 58: 167–177.

Hodson, MJ and Bell, A 1986 The mineral relations of the lemma of *Phalaris canariensis* L. with particular reference to its silicified macrohairs. *Israel Journal of Botany* 35: 241–253.

Hodson, MJ and Sangster, AG 1987 Recent progress in botanical research on phytoliths. *The Phytolitharien* 5(1): 5–11.

Hodson, MJ and Sangster, AG 1988 Observations on the distribution of mineral elements in the leaf of wheat (*Triticum aestivum* L.) with particular reference to silicon. *Annals of Botany* 62: 463–471.

Hodson, MJ and Sangster, AG 1989a X-ray microanalysis of the seminal root of *Sorghum bicolor* (L.) Moench. with particular reference to silicon. *Annals of Botany* 64: 659–667.

Hodson, MJ and Sangster, AG 1989b Subcellular localization of mineral deposits in the roots of wheat (*Triticum aestivum* L.). *Protoplasma* 151: 19–32.

Hodson, MJ and Wilkins, DA 1991 Localization of aluminum in the roots of Norway spruce [*Picea abies* (L.) Karst] inoculated with *Paxillus involutus* Fr. *New Phytologist* 118: 273–278.

Jarvis, SC 1987 The uptake and transport of silicon by perennial ryegrass and wheat. *Plant and Soil* 97: 429–437.

Jones, LHP, Milne, AA and Wadham, SM 1963 Studies of silica in the oat plant. II. Distribution of silica in the plant. *Plant and Soil* 18: 358–371.

Kaufman, PB, Dayanandan, P, Takeoka, Y, Bigelow, WC, Jones, JD and Iler, R 1981 Silica in shoots of higher plants. *In* Simpson, TL and Volcani, BE, eds, *Silicon and Siliceous Structures in Biological Systems*. New York, Springer-Verlag: 409–449.

Lanning, FC and Eleuterius, LN 1985 Silica and ash in tissues of some plants growing in the coastal area of Mississippi, USA. *Annals of Botany* 56: 157–172.

Lanning, FC and Linko, Y 1961 Absorption and deposition of silica by four varieties of sorghum. *Agricultural and Food Chemistry* 9: 463–465.

Lanning, FC, Hopkins, TL and Loera, JC 1980 Silica and ash content and depositional patterns in tissues of mature *Zea mays* L. plants. *Annals of Botany* 45: 549–554.

Maiti, RK, Ramaiah, KV, Bisen, SS and Chidley, VL 1984 A comparative study of the haustorial development of *Striga asiatica* (L.) Kuntze on sorghum cultivars. *Annals of Botany* 54: 447–457.

Mann, S 1988 Molecular recognition in biomineralization. *Nature* 332: 119–124.

McNaughton, SJ, Tarrants, JL, McNaughton, MM and Davis, RH 1985 Silica as a defense against herbivory and a growth promoter in African grasses. *Ecology* 66: 528–535.

Metcalfe, CR 1971 *Anatomy of the Monocotyledons V. Cyperaceae.* Oxford, Clarendon Press: 597 p.

Metcalfe, CR and Chalk, L 1983 *Anatomy of the Dicotyledons. II. Wood Structure and Conclusions of the General Introduction.* Second edition, Oxford, Clarendon Press: 197 p.

Ollendorf, AL, Mulholland, SC and Rapp, G 1987 Phytoliths from some Israeli sedges. *Israel Journal of Botany* 36: 125–132.

Parry, DW, Hodson, MJ and Sangster, AG 1984 Some recent advances in studies of silicon in higher plants. *Philosophical Transactions of the Royal Society of London Series* B 304: 537–549.

Perry, CC, Mann, S and Williams, RJP 1984 Structural and analytical studies of the silicified macrohairs from the lemma of the grass *Phalaris canariensis* L. *Proceedings of the Royal Society of London Series B* 222: 427–438.

Perry, CC, Williams, RJP and Fry, SC 1987 Cell wall biosynthesis during silicification of grass hairs. *Journal of Plant Physiology* 126: 437–448.

Piperno, D 1988 *Phytolith Analysis: An Archaeological and Geological Perspective.* San Diego, Academic Press: 280 p.

Ponnaiya, BWX 1960 Silica deposition in sorghum roots and its possible roles. *Madras Agricultural Journal* 47: 31–32.

Rovner, I (ms) 1986 Plant opal phytoliths: A probable factor in the origins of agriculture. Paper presented at the V. Gordon Childe Colloquium, Insituto de Investigaciones Anthropologicas, UNAM, Mexico City, D.F., Mexico.

Sangster, AG 1970 Intracellular silica deposition in mature and senescent leaves of *Sieglingia decumbens* (L.) Bernh. *Annals of Botany* 34: 557–570.

____1978 Silicon in the roots of higher plants. *American Journal of Botany* 65: 929–935.

____1985 Silicon distribution and anatomy of the grass rhizome, with special reference to *Miscanthus sacchariflorus* (Maxim.) Hackel. *Annals of Botany* 55: 621--634.

Sangster, AG and Hodson, MJ 1986 Silica in higher plants. *Silicon Biochemistry.* Ciba Foundation Symposium 121, Chichester, UK, Wiley: 90–111.

Sangster, AG and Parry, DW 1981 Ultrastructure of silica deposits in higher plants. *In* Simpson, TL and Volcani, BE, eds, *Silicon and Siliceous Structures in Biological Systems.* New York, Springer-Verlag: 383–407.

Simkiss, K and Wilbur, KM 1989 *Biomineralization: Cell Biology and Mineral Deposition.* San Diego, Academic Press: 337 p.

Waterkeyn, L, Bienfait, A and Peeters, A 1982 Callose et silice epidermiques: Rapports avec la transpiration cuticulaire. *Cellule* 73: 267–287.

Williams, RJP 1986 Introduction to silicon chemistry and biochemistry. *Silicon Biochemistry.* Ciba Foundation Symposium 121, Chichester, UK, Wiley: 24–39.

13

DARWIN AND DESIGN IN PHYTOLITH SYSTEMATICS: MORPHOMETRIC METHODS FOR MITIGATING REDUNDANCY

IRWIN ROVNER and JOHN C. RUSS

"And hence it becomes a task as easy as it is delightful, to trace, in the uniformity of design and infinite variety of modification in the execution, the hand of the Great Master-builder."
On the Existence of Structure in the Ashes of Plants and their Analogy to the Osseous System in Animals, Rev. J. B. Reade (1837)

ABSTRACT

Phytolith classification has generally followed pre-Darwinian paradigms, marked by idealized or simplified models, anecdotal descriptive nomenclature and lack of attention to individual variation. In this context, redundancy has appeared to present a problem in taxon identification. However, the polymorphic mode of phytolith production does not fit well the monomorphic, typological approach used for classification of other microfossils. Some recent redundancy-reducing methods include use of phytolith size, frequency and three-dimensional shape characteristics; additional morphometric characteristics derived from stereology offer additional ways to reduce redundancy. Test results indicate that these methods (which include more measures of individual variation) are more robust than methods of central tendency.

INTRODUCTION

When an early nineteenth-century naturalist, the Rev. J. B. Reade, discovered a mineral "osseous system" in plants based on silica and analogous to the calcium-based skeleton of animals, he saw it as one more example of the divine "Unity of Design" imposed on nature by the supernatural Creator. Reade was not first to discover what has since been termed "silica skeletons." Indeed, Reade thanked colleagues including Charles Lyell, the father of Uniformitarianism in geology, for calling to his attention that, among others, G. A. Struve two years

Phytolith Systematics, Edited by G. Rapp, Jr. and
S.C. Mulholland, Plenum Press, New York, 1992

earlier had reported the same phenomenon, explicitly referring to it as a plant skeleton.

In this twilight of the pre-Darwinian era such discoveries were interpreted within the paradigms of divinely inspired design and immutability of species which then dominated biology. Variation, especially individual variation soon to figure so prominently in Darwin's revolutionary paradigm, was either ignored or discounted. All species of plant or animal were unique, conforming to fixed and determinate designs or organizations without exact duplicate in whole or in part in any other species. All individuals within a species, however, conformed to the design.

Systematists sought to identify and describe commonalities that defined and differentiated each species. Thus taxonomies and classification systems, including the Linnean binomial system, derived central tendency models by reduction of variation to "mean" configurations to achieve idealized "wild types." Departures from the ideal could be ignored or, if extreme, dismissed as insignificant "sports." Each species design standard was therefore an idealized form to which no living (or fossil) individual in reality conformed perfectly. It was also by definition an overly simplified model of morphological attributes.

In spite of its triumph in general biology, Darwin's variation-based paradigm had a far lesser impact on systematic classification, including the area of microfossils. Idealized and/or simplified models still dominate taxonomic keys and systematic classifications. Reduction to central tendency is often the primary objective, as in defining a phytolith type (phytomorph) category. This is evident in most phytolith taxonomies, eg, Twiss, Suess and Smith 1969; Rovner 1971; Brown 1986; Mulholland 1986; Piperno 1988; Mulholland and Rapp 1989; Pearsall 1989.

Phytolith illustrations reinforce this condition by presenting the most characteristic view of the most typical example of a phytomorph. Drawings in particular, often little more than outlines, further simplify forms. Scales are standard size bars providing only linear measure of simple attributes such as length and width. Size variation is rarely provided or is given as maxima and minima values; not a histogram of population variance.

Shape attributes are anecdotally presented, mainly through arbitrary selection of descriptive nomenclature; eg, saddle, dumbbell, cross, fan-shaped, elongate, hat, tower, horned tower, etc. This kind of mnemonic strategy, useful to a degree, reinforces reliance on central tendency models, ignoring the reality and significance of individual variation.

Piperno, addressing the nomenclature issue, states,

> Since many phytoliths can be assigned to specific loci of origin on the basis of their morphology (sic), I prefer a system that describes and names major categories of phytoliths on the basis of their origin in living tissue (Piperno 1988: 53).

Use of anticlinal epidermis, hair base, mesophyll, sclereid, cystolith, etc., are intended to provide a nomenclature that is "more easily standardized," avoiding jargon (Piperno 1988: 53). Nonetheless, this by itself falls short of robust and effective definition of phytolith morphology. No amount of nomenclature discussion or name changing addresses the problem of individual variation in phytolith types or avoids reliance on mean form models.

If the pre-Darwinian paradigm of unique and perfect design were true, then reliance on central tendency typology would be perfectly suitable. Each plant would have a silica skeleton pattern unique to its own species and shared in precise copy by all individual species members. Unfortunately, the nature of phytolith production, contrary to the perfect design concept, creates complex problems of "multiplicity" and "redundancy" (see following section), decreasing the limits and increasing the uncertainty of phytolith taxonomic correlation.

> As was expected, every specimen tested yielded a number of discrete morphological types of opal phytoliths with widely varying frequencies. It is expected that additional types will be found in other portions of these plants and surely in other plants. In addition to multiplicity of types within a single plant, there is the problem of redundancy, types appearing in related as well as taxonomically unrelated species. Unlike pollen, whose differentiations have distinct physiological functions, opal phytoliths cannot be expected to have differentiations to the degree present in pollen. Furthermore, there is no necessary relationship between morphological distinctiveness and taxonomic significance (Rovner 1971: 349).

More unfortunately, this statement has influenced subsequent efforts at phytolith classification to the point of becoming dogma. Most unfortunately, this dogma has given critics an ostensibly valid reason to reject development of phytolith taxonomy and to impugn the validity of phytolith data interpretation. It is long past time to challenge this dogma.

THE EFFECTS OF MULTIPLICITY AND REDUNDANCY CONCEPTS

The concepts of multiplicity and redundancy in phytolith classification were created in a context of largely intuitive observations and comparisons of simplified, central tendency typologies. Type definitions and descriptions, borrowed or original, have been largely qualitative. Illustrations are simple outlines, metric information is scant, and shape attributes are embodied in arbitrary descriptive nomenclature. Few phytolith studies have progressed beyond these standards. Several valuable taxonomic studies have increased the number of families and species tested (eg, Brown 1986; Piperno 1988); and/or established new type categories (eg, Kondo 1974, 1976; Pearsall & Trimble 1983, 1984; Piperno 1984). However, they are still embedded in the central tendency, simplified morphology mode of presentation and definition. Phytolith systematics seems to remain in the context of a pre-Darwinian mode of perfect design and idealized form.

The problem of multiplicity, readily evident in the phytolith assemblages of many species, is not insurmountable. An approach was suggested some time ago.

> The associations of morphological types, however, that are shared by various species are usually inconsistent. If morphological types assemblages or 'type-sets' are investigated, it may be possible to eliminate some species from representation in a soil sample by the absence of one or more of their 'set members'. Analysis of frequency of shared types within a set may yield statistical probability for the presence of one species' type-set over another. That some level of meaningful differentiation is possible can indeed be stated with confidence (Rovner 1971: 357).

It is gratifying to see this prediction confirmed. The diagnostic value of crossbody to bilobate ratios in separating cultivated maize from wild teosinte (Piperno 1988; Pearsall 1989) is a case in point. However much one might hope for a more perfect separation in which bilobates occurred only in one taxon and crosses in the other (*ie*, no redundancy), the presence of both basic forms in both taxa (*ie*, multiplicity) can actually be used to advantage.

The redundancy problem has proved more daunting in the development of phytolith taxonomy. Redundancy is not perceived as a problem of such magnitude in other microfossil systems: eg, pollen and diatoms. Pollen and diatom testa from a single species respectively are "monomorphs" adhering to very narrow limits of variation of size and shape. In such cases, central tendency typology is usually sufficient for assignment to the level of taxonomic precision permitted by an individual fossil. As phytolith production in species is typically polymorphic, this approach, while insufficient, is the only comparable model in wide use for microfossil systematics. Thus, conventional practice forces phytolith analysis into an inappropriate mode of "monomorphistry" resulting in inevitable taxonomic confusion and simplifying distortions.

As early as 1855, Gregory compared seemingly redundant siliceous microfossils including phytoliths and constructed this argument.

> Some years ago it was supposed that most of the species . . . were no longer found to be living. But since then most of them have been found to be recent . . . Yet, all these forms were supposed, not long since, to be exclusively fossil. We cannot say that there are no species exclusively fossil, but so many that have been thought so are daily found living, that it is probable the rest may be so found too, and at all events, a very large proportion of the forms in the oldest fossil deposits are absolutely identical (sic) with the forms of the present day, as a glance at Ehrenberg's (*Mikrogeologie*) figures will prove (Gregory 1855: 71–72).

Gregory is presenting the argument of a perfect design advocate confronting evolution models that require validation by extinctions in the fossil record. Since the perfect design paradigm ignores variance in favor of reduction to simplistic central tendencies, fossil phytoliths, of subtlely distinct morphologies are inevitably classified into the same typological categories, appearing "absolutely identical" in spite of biostratigraphic separation. This is then offered as proof of the absence of extinctions, denying mutability of species and supporting perfect design in the fossil record.

Such morphometric standards are still commonplace today. Bukry, noteworthy for his reports of terrestrial plant phytoliths in deep sea core sediments, takes the following view of phytolith morphological taxonomy.

> *Zea mays*, *ie*, Indian Corn, is native to the Western Hemisphere. It produces several morphotypes of panicoid (dumbbell-shaped) and elongate phytoliths. The Pliocene phytolith assemblage from offshore North Africa at Site 397 contains the *same* panicoid morphotypes as modern *Zea mays*. Therefore, particular morphotypes that duplicate those of a native American plant can be produced by the tall grasses of Africa. *This is an example of the limited biostratigraphic value of the common forms of opal phytoliths* (Bukry 1979: 979, emphasis ours).

It is likely that Bukry's manuscript was already in press when Pearsall's (1978) report on crossbody differentiation of maize from wild grasses appeared. Nevertheless, such statements from prominent microfossil experts present a problem for phytolith systematics.

Many scholars hold similar views:

> (A systematic phytolith taxonomy for genus, *Zea*) has small chance of yielding definitive results since the phytoliths of corn cannot be reliably distinguished from those of *Arundinaria*, a distant relative. It is scarcely to be expected that they will prove to be distinguishable from those of teosinte, corn's closest relative (Anonymous Reviewer 1985).

Both Pearsall (1978) and Piperno (1984), reporting successful differentiation of maize and teosinte phytoliths, were published and available, but were not addressed by this 1985 reviewer. Such a comparison for paleobotanical analysis is invalid anyway because these grasses are quite unlikely ever to have occurred together in the same location in prehistory. Moreover, more recent work by Ollendorf, Mulholland and Rapp (1988) used phytolith taxonomy to distinguish between *Arundo donax* and its very close relative in the Arundineae, *Phragmites communis*.

> Examples of phytolith analysis in systematic botany are noticeably scarce and are not rigorously applied. The present study suggests that phytolith analysis may be useful for making identifications of certain grasses unidentified or misidentified in the field. The giant reedgrasses, *Arundo donax* and *Phragmites communis* . . . are not readily distinguished if inflorescence is lacking. The present authors propose that phytolith analysis may be useful in distinguishing among related grass species (Ollendorf, Mulholland & Rapp 1988: 213).

Separating maize and *Arundinaria* phytoliths should be relatively easy, but it is reasonably certain that a systematic study comparing size, shape variant and frequency of phytoliths of *Arundinaria*, an Old World grass taxon, to those of *Zea*, a New World taxon, has never been attempted, much less published.

However much redundancy is used by others to invalidate or inhibit development of phytolith systematics, phytolith systematists must bear part of the blame. The redundancy concept pervades the phytolith literature explicitly, for example, in Twiss, Suess and Smith 1969; Rovner 1971, 1983; Pearsall 1982; Brown 1986; Mulholland, Rapp and Ollendorf 1987; Mulholland and Rapp 1989 among others and implicitly in the rest. Robust procedures for resolving redundancy are largely lacking in that literature.

Pearsall (1989), in an excellent practical overview of phytolith analysis, specifies

> redundancy as a prominent weakness. Redundancy of phytolith production exists for a number of forms. Similar silica bodies may be produced in widely divergent plant groups. For example, silica deposited in epidermal and mesophyll leaf tissues looks similar in many taxa. At times only very broad characterization (eg, dicotyledon vs grass) can be based on phytolith assemblages (Pearsall 1989: 343).

This is an almost exact reiteration of Rovner's (1971: 357) redundancy concept regarding dicotyledon phytoliths. There is, however, one cogent semantic difference. Pearsall uses "similar" rather than "identical," implying that distinctions exist in spite of a current lack of procedures in our systematics to elucidate them. Piperno (1988) likewise addresses the issue of morphometric standards.

> As a consequence of the discrepancy in the amount of attention devoted to different plant taxa, there resulted a number of serious misconceptions about, and lacunae in, the phytolith morphological data base. Descriptions and comparisons of phytoliths from entire families were never made available to modern researchers. Even among grasses, by far and away the most intensively studied plants, *important morphological and metric attributes such as three-dimensional shape, size and relative frequency* of phytoliths in individual species remained poorly studied. Hence, the widely held notion that phytoliths could not be used to identify genera and species of grasses was derived (Piperno 1988: 51, emphasis ours).

In spite of her operational success in using shape, size and frequency attributes in maize phytolith classification, Piperno does not theorize beyond the perfect design mode of systematics. Phytolith analysis is valid, she argues, because individual phytoliths can be used as powerfully as individual palynomorphs and diatoms are.

> Some investigators . . . prematurely concluded that individual phytoliths are of little utility in the identification of plants from which they are derived, and a single species may contain several different kinds of phytoliths. However, it will be shown that individual

phytoliths often characterize particular taxa at various levels (genus, tribe, subfamily, and family), and there is a close correlation between the types of phytoliths and taxonomic affinities of plants containing them (Piperno 1988: 522–523).

Piperno's positivism notwithstanding, we believe new thinking is needed. The redundancy problem will not go away so long as phytolith systematists continue to use conventional methods to seek taxonomic "fossil directeurs"—individual wild type phytomorphs that have a 1:1 correspondence with botanical taxa. Finding one is cause for celebration, but relying solely on the quest for monomorphistry in phytolith systematics excludes the reality of Darwinian variation.

The solution to redundancy is to abandon simplistic typological procedures that may work adequately in monomorphic systems, eg, pollen and diatoms, but are questionable in phytolith taxonomy. New and different procedures specific to the reality of phytolith variation must be developed, tested and applied.

MORPHOMETRY AND VARIATION

Tentative but significant movements away from wild type systematics toward attributes of size, shape and frequency have already begun with notable success. Pearsall (1978) was first to validate this approach by demonstrating the importance of size-frequency categories of crossbodies in distinguishing cultivated maize from wild grasses, including those that also produce crossbodies. Redundancy was overcome by employing a slightly more complex and statistical quantitative approach. It employed the ability to measure the short axis (width) of a crossbody under the microscope with an eyepiece micrometer.

Advances in maize phytolith taxonomy made by Piperno (1988: 79–87) are of two kinds. Her identification of eight variant categories of three-dimensional crossbody shape is a major advance, but by itself is still a perfect design approach. Instead of one wild type category with considerable redundancy, eight wild types are used, reducing but not eliminating redundancy. Piperno eliminated redundancy by employing in addition: 1) frequency of the variant; 2) Pearsall's short axis size parameter; and 3) crossbody to bilobate ratios. Resulting data were assessed using multivariate statistics and discriminant function analysis to achieve taxonomic separation of phytolith assemblages. Considerable progress in resolving phytolith multiplicity and redundancy has occurred using common, conventional procedures.

The fundamental question concerning the problem of redundancy is whether phytoliths and phytolith populations are really redundant or whether the results proceed from inadequate morphometric standards. Use of simplified illustrations of phytoliths as in Metcalfe's (1960) seminal work on grass systematics supports a false reality of ineluctable redundancy. At such a level of (non-) measurement of phytolith populations and simplicity of illustration, "identical" phytolith types do occur in spite of obvious taxonomic boundaries. However, this is an artifact of Metcalfe's methodology, which did not include systematic study of phytolith morphological variance or frequency at the individual or population levels. It is our contention that such work is far from conclusive, has obscured rather than revealed the non-redundant reality of phytolith populations and therefore cannot be used as the standard to assess the value of phytolith analysis as a microfossil data system.

The robust capacity of size, shape and frequency parameters to resolve problems of multiplicity and redundancy in maize phytoliths has been demonstrated to date using low level morphometric procedures. Application of more powerful and sophisticated morphometric techniques promises enormous advances in phytolith systematics. Image analysis has a well-established array of methods for measuring size and shape of chaotic populations (Russ 1986; Russ & Rovner 1989a, 1989b).

Image analysis is more than simple processing of images to enhance their appearance, although some processing steps may be used to improve the ability of automatic algorithms to locate and isolate features of interest. Image analysis involves the reduction of the data in the original images (consisting of the brightness values at typically a few hundred thousand points or "pixels" in the image) to a few numbers, such as the dimensions of features present, or their quantity or simply their presence. Image analysis includes several specific fields, such as stereology, which describes the relationship between features seen in two-dimensional images of sections through structures and the three-dimensional structure that is actually present. It also includes stereoscopy which is the combination of two views of a structure or surface to determine depth in the third dimension. For the purpose of this paper, the tools of image analysis that are of greatest use are isolation and measurement of features in images, and the statistical analysis of those data.

Feature measurements may be broken down for convenience into measurements of size, position, gray-scale (or color) and shape information. Size includes the most familiar parameters, such as area or length (Fig 13.2), although we will see that more specific definitions are required to make these useful. Position measurement may include absolute location and orientation values for each object, or relationship between objects, but is of little use for specimens (eg, disarticulated phytoliths) that have been prepared for imaging in ways that erase such information. Gray-scale information includes brightness and its variation within a feature, and can be extended to include color values if they are available.

Shape descriptors are of several very different kinds. Some are formally dimensionless ratios of size parameters, such as Aspect Ratio = Length/Breadth. These are often useful for object recognition, but are not unique values. Harmonic or Fourier analysis (Fig 13.4) provides a complete and unique description of shapes, but can be difficult to interpret. Many natural objects have fractal (as opposed to Euclidean) shapes, and the fractal dimension can be used to characterize roughness (Fig 13.5). Finally, topological properties ignore size information to extract shape information from feature skeletons (Fig 13.6).

In the descriptions that follow, many examples of specific parameters in each of these classes are shown. These are all available on the particular system used in our experiments, although some are more useful than others for the purpose of phytolith identification. Although they can be determined by the computer in fractions of a second, their manual measurement is time-consuming, effectively precluding the testing of their usefulness for identification of a real population of objects. Measurement of area (used for at least 150 years) involves tracing an object, cutting out the shape, and weighing it. Weighing a known area of paper provides a reasonably fast tool, but not one that could be used for thousands of objects. A string placed carefully around an object can be lifted off to measure perimeter (or pulled tight and then removed to measure taut-string perimeter), but this technique is likewise tedious and prone to error.

The Prism system (Dapple System, 355 W. Olive Ave., Sunnyvale, CA 94086) operates on a Mac II computer and provides a variety of feature-specific measurements for each feature. These are divided into categories of size, shape, topology, position and gray-scale information. Checking the item in the dialog box selects the parameter for measurement, either for ranking, or to be displayed in the information window for each feature as it is selected with the cursor, or to write data for all features present in the binary image to a disk file for statistical analysis. The final group of derived parameters described below may be calculated after data have been collected and saved. All parameters are measured in the current scale units (converted from pixel dimensions) except as noted.

Measures of Size

Area: the total number of contiguous pixels (eight-neighbor connected) in the feature, converted to the current scale units.
Filled Area: the total number of pixels within the feature, including any internal holes (because the feature is defined as eight-neighbor connected, the holes are four-connected).
Convex Area: the area within a 32-sided bounded polygon formed by rotating axes in 16 steps (11.25 degrees) and locating the extreme points, which form the vertices of the polygon.
Perimeter: the total boundary distance of the feature using a chain-code definition where each link in the chain is either 1 or $\sqrt{2}$ pixels long (for orthogonal and diagonal directions), converted to the current scale units.
External Perimeter: the total boundary distance of the feature using a chain-code definition, but not including any boundaries around internal holes.

Convex Perimeter: the perimeter of the 32-sided polygon described above for convex area.
Mean Hole Diameter: the mean equivalent circular diameter (as defined below) for any internal holes in the feature.
Length: the maximum chord length between points on the boundary of the feature. This is determined by rotating the axes in 16 steps of 11.25 degrees and finding the extreme points. The worst-case error is the arc cosine of 5.625 = 0.995 (one pixel in 200).
Breadth: the minimum caliper dimension obtained in the same way as Length, described above.
Fiber Length: estimated length along the midline of a curved feature, calculated as (Perimeter/2) − 2 (Area/Perimeter). This is generally greater than the Length described above.
Width: the estimated mean width of a curved feature (minimum caliper dimension) calculated as $(4 \cdot \text{Area})/(\pi \cdot \text{Length})$. This is generally less than the Breadth described above.
Equivalent Diameter: a mean linear dimension of the feature calculated as the diameter of the circle with the same area as the feature.
Volume (prolate): the volume of prolate (football-shaped) ellipsoid of revolution with major and minor axes equal to the length and width defined above.
Volume (oblate): the volume of an oblate (discus-shaped) ellipsoid of revolution with major and minor axes equal to the length and width defined above.
Surface (prolate): the surface area of the prolate ellipsoid described above, divided by the formfactor (described below) to adjust for roughness and deviation from ideal shape.
Surface (oblate): the surface area of the oblate ellipsoid described above, divided by the formfactor (described below) to adjust for roughness and deviation from ideal shape.

Measures of Shape

These are dimensionless parameters except as noted.

Hole Count: the number of internal holes within the feature.
Formfactor: $4\pi\text{Area}/\text{Perimeter}^2$, which is 1.0 for a circle and less for features with irregular boundaries.
External Formfactor: the formfactor as described above, but using the filled area and external perimeter.
Roundness: $(4 \cdot \text{Area})/(\pi \cdot \text{Length})^2$, which is 1.0 for a circle and less for features that are elongated.
Convexity: the ratio of the convex perimeter to the feature perimeter, which is less than 1.0 for features that are not convex.

Solidity: the ratio of the feature area to the convex area, which is less than 1.0 for features that are not convex.
Extent: the ratio of the feature area to the area of a minimum bounding parallelogram defined by length and breadth as described above.
Compactness: the ratio of the equivalent diameter to the length.
Hole Fraction: the fraction of the filled area of the feature occupied by holes.
Mean Hole Formfactor: the mean formfactor of any internal holes within the feature, calculated as described above from their area and perimeter.
Aspect Ratio: of length to breadth.
Elongation: the ratio of fiber length to width.
Curl: the ratio of length to fiber length.
Fractal Dimension: a measure of the self-similar roughness of the feature boundary determined as the slope of a plot of the log of or the total perimeter vs the log of pixel size as the feature image is progressively coarsened.

Measures of Topology

These are all determined by reducing the feature to its skeleton; this is done automatically, and the original features restored afterwards. The data are grouped with other feature-specific measurements for display or writing to a disk file for analysis. Euler's topological relationships apply to these values, not all of which are independent.

Number of End Points: the number of terminal vertices in the skeleton, which correspond to the number of "points" in the feature.
Number of Nodes: the number of internal branch points or vertices where links meet.
Number of Loops: the number of closed loops in the network, equal to the "genus" of the feature.
Number of Links: the number of connections between internal nodes.
Mean branch length: the average length of links that end at terminal vertices or end points, determined by counting the pixels within each branch using weights of 0.948 for orthogonal and 1.340 for diagonal touching: this gives more accurate overall measurements of length than the simple 1.0 and $\sqrt{2}$ values.
Mean link length: the average length of links whose two ends are both at internal nodes as opposed to terminal end points.

Measures of Position

These are not used for identification of features in this application, because the feature location in the image is a chance result of preparation techniques.

X-Center of Gravity: the X-coordinate of the point at which the feature would balance if cut from cardboard, from the left side of the image. This point need not lie within the feature area if it is concave or contains holes.
Y-Center of Gravity: the Y-coordinate of the point at which the feature would balance if cut from cardboard, from the bottom of the image.
Orientation Angle: the orientation of the Length vector described above (a multiple of 11.25 degrees).
X-Feret: the projected width of the feature parallel to the X (horizontal) axis, in pixels.
Y-Feret: the projected height of the feature parallel to the Y (vertical) axis, in pixels.
X-Location: the X-coordinate of the center of the bounding rectangle defined by the Feret's dimensions, from the left side of the image.
Y-Location: the Y-coordinate of the center of the bounding rectangle defined by the Feret's dimensions, from the bottom of the image.
Moment Angle: the angle in degrees of the vector from the Location coordinates to the Center of Gravity coordinates.
Adjusted Count: a measure of the probability that features with this size and shape will intersect the bounds of an image of the current size if they are randomly placed. Calculated as (Wx Wy)/(Wx-Fx) (Wy-Fy) where W and F are the image width and Feret's diameters, respectively, the value (greater than 1) is used to provide edge correction in feature distributions.
Contains Objects: a True/False flag for features containing holes, indicating whether other features lie within those holes.

Measures of Gray-Scale

These are all measured on the designated gray-scale image using the binary image as a reference mask. In a color system, these directly extrapolate to either RGB (red, green, blue) or HIS (hue, intensity, saturation) color information. In most cases, the RGB data format is used for acquisition, but the HIS format is more meaningful for feature measurement as well as for image processing.

Brightness: the pixel brightness (on the scale 0 = white, 255 = black) averaged for the pixels within the feature.
Contrast: the average brightness difference for all pairs of touching pixels within the feature.
Texture: the slope of a plot of brightness difference between pairs of pixels within the feature as a function of their distance apart in pixels (correlates with surface roughness in many cases).
Density: converts the mean brightness value to a density value. By default, this

is just a logarithmic conversion, but if there is a user-defined calibration curve present, it will be used instead.

Derived Parameters

The standard program provides the following transform models; the user can add additional ones if required.

Spherical Volume: the volume of a sphere whose projected area is the same as the area of the feature.
Spherical Surface: the surface of that sphere, divided by the feature formfactor to adjust for surface irregularities.
Indentation Depth: the mean depth of indentations around the periphery of the feature, calculated as 2 (C-A)/ (P-T) where C is the Convex Area, A is the Area, P is the Perimeter, and T is the "taut string" or convex perimeter.
3-D Aspect Ratio: the estimated three-dimensional aspect ratio of an elongated solid viewed in projection in a random orientation, calculated as $1+(4/\pi) \cdot ((\text{Length/Width}-1))$.
Radius of Curvature: the mean radius of curvature of the feature midline, calculated iteratively as the solution to L/(2 R) = Sin (F/2 R) for R (the Radius of curvature) in terms of L (the Length) and F (the Fiber Length) of the feature.
Arc of Curvature: the angular extent of a curved feature calculated as 2 Arc Sin (L/2R) where R is the Radius of curvature and L is the Length.
Nearest Neighbor Distance: the distance from the feature's center of gravity to that of the nearest other feature in the image or data file.
Nearest Neighbor Direction: the angle in degrees of the vector from the feature's center of gravity to that of the nearest other feature in the image or data file.
Integrated Brightness: the sum of the individual pixel brightnesses (0 = white, 255 = black) for all pixels in the feature.

The following illustrative figures show some of these measurement parameters. Area and Perimeter are common to our everyday experience, but may or may not include internal holes or indentations along the periphery as shown in Figure 13.1. The convex boundary is also known as a "taut-string" boundary. The most common linear measures of size are shown in Figure 13.2 (Length, Breadth, Width and Fiber Length).

Shape is a more difficult concept to comprehend than is size, because few of the words have consistent meaning in our everyday experience. Instead of descriptive adjectives, humans often employ nouns to communicate shape information (an "L-shaped" building, a "horse-like" animal, etc.). Furthermore, shape parameters are often incomplete: they characterize only one aspect of shape, and in most

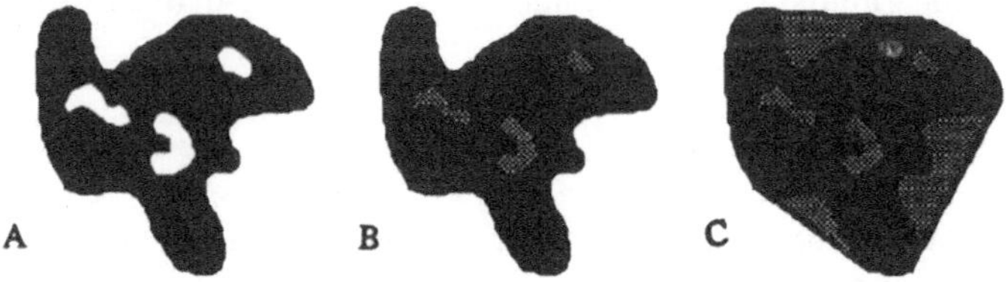

Fig 13.1. Illustration of a feature with its filled and convex area and perimeter. A. Feature area; B. Filled feature area; C. Convex feature perimeter and area.

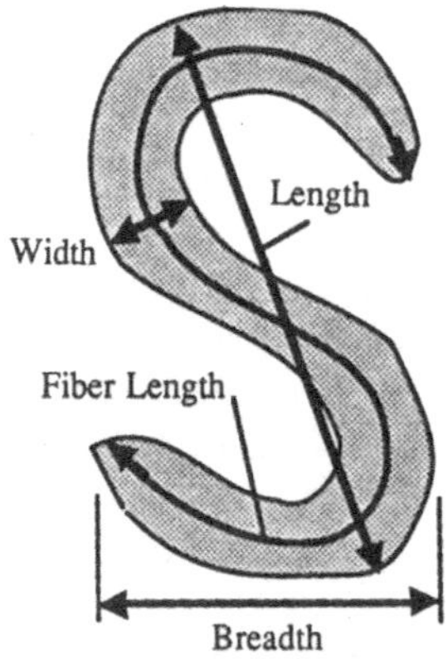

Fig 13.2. A feature showing the Length, Breadth, Width and Fiber Length.

cases it is easy to construct a great many features that differ widely in one shape parameter but are essentially identical in others. Figure 13.3 shows this for the case of Formfactor and Roundness, two of the more common examples of dimensionless ratios formed by combining size parameters.

Dimensionless ratios of size parameters are not new to biological systematics and have been used extensively in anthropometry and forensic anthropology. The most common example is probably cephalic index: 100 times head (or skull) length divided by breadth. This is the reciprocal of aspect ratio as defined above, and is used to provide a dimensionless index of "head shape."

It is possible to encode all of the shape information in a few parameters by using Harmonic or Fourier analysis, in which the feature shape is "unrolled" as shown in Figure 13.4 and expressed as a series of sine and cosine terms. As few as 30–50 of these terms can be used to reconstruct the original feature to a pre-

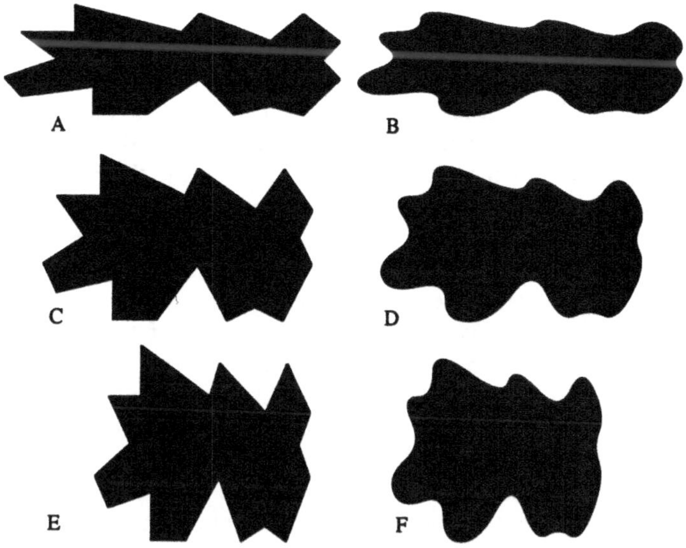

Formfactor		*Roundness*	
A 0.311	B 0.447	A 0.260	B 0.278
C 0.332	D 0.467	C 0.497	D 0.502
E 0.318	F 0.477	E 0.623	F 0.649

Fig 13.3. A series of related features that can be grouped differently by similar values of Formfactor and Roundness (Appeared in *American Antiquity* 54: 787. Reprinted by permission).

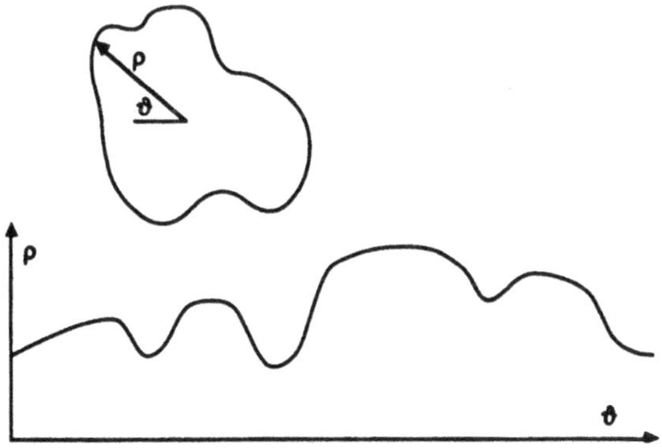

Fig 13.4. Example of unrolling a feature profile as a plot of $\rho(\vartheta)$ to determine the Fourier shape descriptors. $\rho(\vartheta) = a_0 + a_1 \cos(\vartheta) + b_1 \sin(\vartheta) + a_2 \cos(2\vartheta) + b_2 \sin(2\vartheta) + \ldots$

cision equivalent to that of the original image measurement. The magnitude of the parameters is not easily interpreted by a human, but it has been shown in a number of specific instances that some of the numbers can be used in a statistical sense for the identification of chaotic shapes. This includes the identification of sand grains deposited as sediment from different rivers, the shape of foraminifera as a function of ocean temperature, etc. One drawback to the method is the difficulty of selecting the particular coefficients that are important in each instance.

Many natural objects have been shown to exhibit a fractal shape in which the surface or boundary roughness is "self-similar," that is, has the same degree of roughness in a statistical sense regardless of the image magnification. In these cases, the fractal dimension, determined from a plot of the perimeter of the feature vs. the size of the measuring tool used to determine it, provides a very compact descriptor of that roughness. Figure 13.5 shows several examples.

Related to shape parameters, but obtained by reducing the features so that only the topological (as opposed to metric) information remains, are parameters determined from the skeletons of the features. Some of these, such as the number of end points or numbers of loops (Figure 13.6), are extremely powerful shape descriptors that are not easily estimated from the classical shape parameters that use ratios of size values.

REDUNDANCY REVISITED

Russ and Rovner (1989b) report demonstration blind test results (using the Image-Plus expert vision computer system, Dapple Systems, Sunnyvale, CA 94086) on five populations of maize and teosinte phytoliths. The original intent was to replicate and verify Pearsall's size-frequency crossbody discriminator and Piperno's crossbody to bilobate ratio parameter by statistical grouping of size, shape and frequency into teosinte and maize categories, respectively. After SEM photos were acquired by the computer system, fourteen size and shape parameters were selected and recorded for all short cells. Data from two populations, arbitrarily chosen, were statistically evaluated to determine which parameters best distinguished bilobates from crossbodies. A 4-parameter discriminant function was created that mathematically described and distinguished bilobate and crossbodies at 98% or higher confidence intervals.

Statistical comparison of population variance and homogeneity for all five cases was instituted. A test of Pearsall's parameter, crossbody short-axis size, comparable to the computer's breadth parameter, was intended but quickly rendered unnecessary. Analysis of variance of population length was selected for no better reason than it was ahead of breadth on the software menu. Both cross-body and bilobate populations were tested because the computer measures them simultaneously.

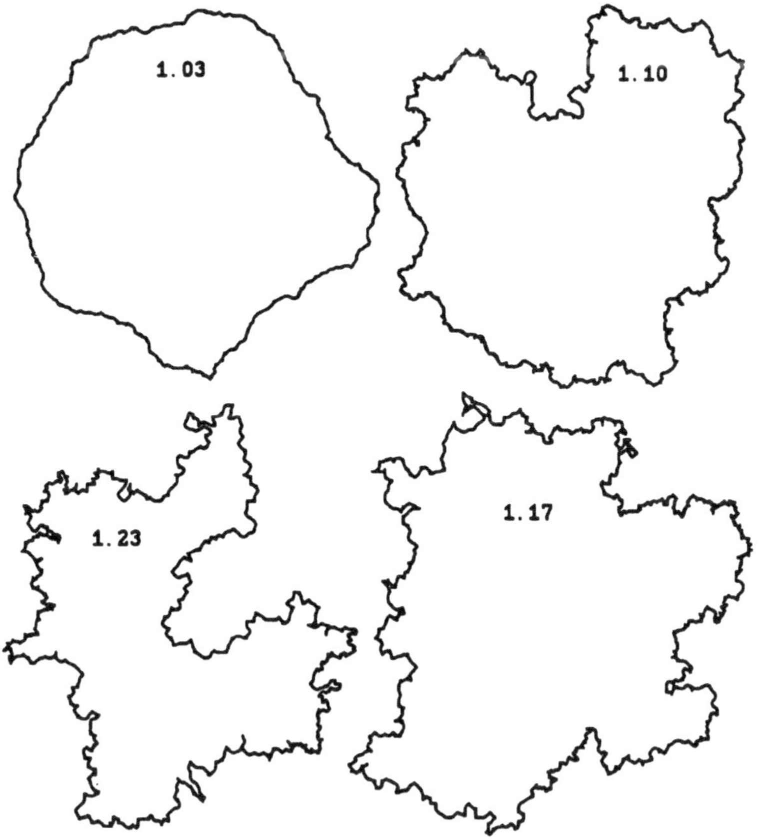

Fig 13.5. Feature outlines with varying fractal dimensions characterizing their roughness.

The results were unexpected and startling. Bilobate length variance outperformed crossbody variance in grouping/separating maize and teosinte phytolith populations. It required less than one hour of effort to define and separate the two short cell categories morphometrically to discover that it is a mistake to separate the short cell categories taxonomically. With one exception (ANOVA F = 87.32%), either of the two teosinte populations could be separated from any of the three maize populations with better than 95% confidence based on a statistical evaluation of the simple univariate length parameter on all short cells.

Several hypotheses were immediately generated for further testing. First, Pearsall's high frequency of large size crossbodies in maize may be a coincident factor in that *all* maize short cell populations reflect larger size compared to those in teosinte. Second, in spite of the apparent statistical superiority of bilobates to distinguish maize from teosinte, observers (unaided by image analysis) rejected this category because it is far more chaotic morphologically in favor of the somewhat inferior, but morphologically simpler crossbody group. Third, the demonstration test was accomplished by barely scratching the surface of computer-assisted image measurement capabilities. The problems of multiplicity and redundancy, apparent and overwhelming to the unaided human eye, can be resolved effectively with the assistance of computer image analysis.

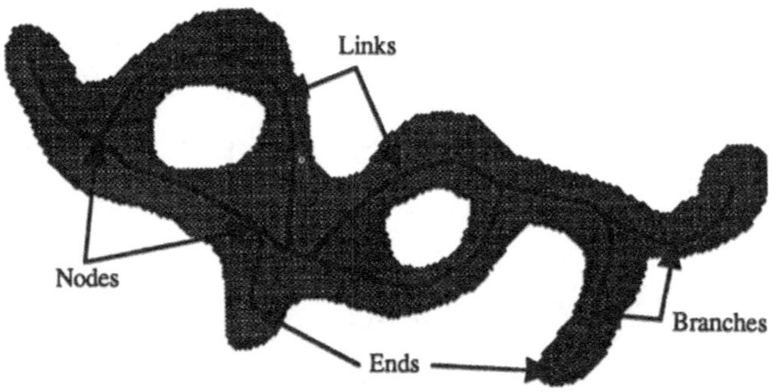

Fig 13.6. A feature with its skeleton, containing 4 Ends and Branches, 6 Nodes, 2 Loops and 7 Links.

The length parameter is not by itself likely to be effective in the broader context of other grasses, some of which are known to have remarkably large bilobates exceeding 80 microns in length (J. Noguera, personal communication 1985). However, this feature can surely be accommodated using larger sample size, discriminant function analysis of other phytomorph categories and the reservoir of univariate and multivariate feature measurement parameters yet to be tested.

Machine-assisted classification was tested against the unassisted human mind during the course of the demonstration test. The phytolith expert (Rovner) was given, one at a time, the same training series of SEM photos the image expert (Russ) used in the computer, and was asked to count only the bilobates present. Russ kept a running tabulation that Rovner did not see. On the second pass, Rovner was asked to count only crossbodies in each photo. The computer-assisted classification had separated the training sample of 72 phytoliths into 38 bilobates and 34 crossbodies. Rovner separated the sample into 38 bilobates and

36 crossbodies, classifying a total of 74 phytoliths from a sample of 72! A deviation of 39–33 or 37–35, etc., could easily be rationalized as random error, human intuition and experience versus rigid statistical machine definition, etc. However, in at least two repeated cases Rovner, observing intuitively that a phytolith fell into an "indecision" zone, arbitrarily decided to "include" every time. Thus, at least two particles included in the bilobate group were also included in the crossbody group on the second pass.

Needless to say, the test was "rigged" by Russ to invite this kind of error, but results are instructive. This kind of mistake cannot be made with computer assistance. Whenever the computer encounters a rogue phytolith in the statistical indecision zone, the computer stops and "asks" for help from the expert, who informs the machine which category, if any, to place the rogue into. The machine then recalculates the statistical algorithm defining the category, narrowing the indecision zone so that the classes are better defined and fewer errors and uncertainties will occur in the future.

This method constitutes a major benefit to consistency and precision in systematics. The machine cannot classify a particle if critical parameters remain imbedded in the intuitive construct of the observer's mind. When the machine stops and asks for assistance in placing a particle that the human can classify intuitively, it is demanding that the intuitive logic missing from its instruction be made explicit and transparent. By contrast, when classification logic is transferred from human to human, obscured parameters in the instructor's intuition may remain hidden if the "student" substitutes his/her own intuition to fill logical gaps.

The regression of an optimal context line to provide high-confidence separation of these types was human-directed in this case. A more recent study (Russ 1989) employed an automatic classification and learning program. This study was undertaken with a small population (20–30 of each) of chaotic natural types of nuts (walnuts, Brazil nuts, acorns, almonds, hickory nuts, etc.). Context lines were derived by stepwise multiple regression, and distribution histograms were constructed along them. The system then continued to learn by refining these distributions, using a method ("contextual learning") that shares some similarities with neural networks and other artificial intelligence techniques. After a total of more than 1000 nuts had been examined, fewer than 2% of the nuts were misclassified in spite of completely random orientation in their presentation combined with their natural variation in size, shape and appearance.

The importance of unequivocal classification logic in phytolith systematics was highlighted when it became clear recently that the Russ and Rovner (1989b) original intent to replicate Pearsall's crossbody parameter was flawed from the outset. In retrospect, we now know that obscured intuitive parameters distinguishing crossbodies and bilobates entered the training exercise so that Pearsall's definition was not followed. A Rovner-selected population of crossbodies had little likelihood of matching a set selected by Pearsall's criteria.

> I defined the cross-shape as a panicoid short cell with indentations on at least three of four sides which was no more than two eyepiece micrometer units longer than wide. Converted to absolute size, crosses are square to rectangular in outline and no more than 9.16 microns longer than wide. This artificial classification, formally creating the cross category, allowed me to separate crosses and dumbbells (bilobates) and to create two clearly defined types. This is the key to the success of the identification technique (Pearsall 1989: 329).

In geometric terms, Pearsall's defining factor is limited elongation. Pearsall's cross shape must have an aspect ratio (length/width) less than or equal to 1 + (9.16/width). Since aspect ratio is not sensitive to changes at the margins, Pearsall's criteria ignores such factors as depth or embayment of the 3 or 4 indentations that help define the cross-shape.

For Rovner, the unassisted definition of a crossbody incorporated limited elongation, 3 or 4 indentations *and* embayments that were more or less equal, usually shallow. This last attribute is not explicit in Pearsall's definition, yet is present (intuitively?) in her illustrations of crossbodies in Figure 5.14 and 5.16 (Pearsall 1989: 329, 335) and most other illustrations in the literature. Rovner's crossbody is closer to the definition used by Mulholland.

> . . . I took cross to mean a body, in planar view, with four roughly equal lobes separated by four equal indentations. This is a more restricted definition than Pearsall's (Mulholland 1986: 43).

Rovner's "unassisted" bilobates, on the other hand, have embayments that are unequal—the lateral embayments are typically deep, forming a narrow waist, whereas any terminal embayments are shallow. In addition, a cross often has a dark nugget or "dot" in its center (see Piperno 1988, Plates 23, 25, 27, 28; Mulholland 1986, Fig 4); whereas a bilobate may have one or more dark nuggets at a more terminal placement, often above and below the narrow waist. It was, therefore, startling to see two obvious (?) narrow-waisted classic "Rovner bilobates" in Pearsall's (1989) Figure 5.15a captioned as "Variant I cross from *Zea mays*." Undoubtedly, both phytoliths fit the crossbody category in Pearsall's elongation formula, which ignores marginal embayment. Rovner's and Mulholland's anecdotal classifications and the Rovner-Russ computer algorithm, which uses three perimeter-sensitive shape factors, would likely classify them as bilobates.

Russ and Rovner developed a classification algorithm: (−3.95 × Formfactor) (−1.41 × Solidity) (−1.22 × Aspect Ratio) (+ 4.26 × Convexity), yielding values near 1 for bilobate, and near 2 for crossbody. Both ANOVA and Wilcoxon (nonparametric) tests indicated that the algorithm separated the phytomorph classes in all populations at the 99.9% confidence level. The value of aspect ratio was expected given the importance of elongation in Pearsall's algorithm. However, formfactor, convexity and solidity all use perimeter in some way in the calculation. Thus three variables in the Russ-Rovner algorithm are sensitive to changes at the margin, such as depth of embayment, ignored in Pearsall's definition.

If the 72 short cells in the training set were subjected to Pearsall's classification algorithm, it is unlikely that the computer-separated groups would be replicated exactly. Even so, this in no way alters the fundamental agreement in the results obtained by the two different algorithms. In fact, computer-assisted image analysis of maize short cell phytoliths has made it quite clear that every time Pearsall measured as a crossbody any bilobate (which would be rejected by Rovner or Mulholland), the statistical separation of maize and teosinte would be enhanced. Maize phytoliths and teosinte phytoliths are *not* alike—populations can be separated and identified with confidence.

The importance of embayment was not recognized until computerized feature analysis exposed the intuitive discrepancy; there may be others. For example, Mulholland and Rapp (1989: 37, Table 1) provide the following distinctions between bilobates and crossbodies in their "lobate" category: a. Cross— "Minimum diameters of base approximately equal"; Bilobate—"Minimum diameters of base unequal." Even without formal testing, it is clear that Mulholland-Rapp separated groups will suffer as much significant deviation as between a Pearsall, Rovner and/or Russ-Rovner attempt.

Mulholland and Rapp have dedicated considerable effort to testing phytolith taxonomies for consistency and reproducibility.

> We wanted to test how many phytoliths would give a consistent and reproducible count. We tried different methods of counting up to a 1000 and in one case up to 1500 in groups of 100. I would count the first 100, record them; count the second 100, record them separately; and so on. First, I tested Chi square on all the 100's against each other. I tried combining all the 200's against each other and then the 300's . . . The results seemed to say the more you counted, the worse your statistical variability got. I am not sure what that means, but I have a feeling that the classification scheme I was using was based on variables that were not significant. I may have been picking up a lot more variability than exists. We will be able to determine this, only when we have a more refined classification system (Mulholland comments in Rovner 1986: 125).

Without the assurance of consistency and replication provided by the precision of computer-assisted measurements, phytolith systematics will fail to persuade the research community that phytolith analysis carries the same compelling validity as does palynology. Results of phytolith analysis will remain suspect and systematic development will be inhibited by critical peer review. The most powerful tools currently at our disposal are computer-assisted feature measurement and statistical analysis.

A second phase demonstration test of maize short cell phytolith populations was recently conducted on images provided by the Archaeometry Laboratory, University of Minnesota-Duluth. Image analysis was performed on two 100 count phytolith populations as a blind test of populations taken from different parts of the same maize leaf. Mulholland, Rapp and Ollendorf (1987) reported significant differences in phytolith type frequencies from different locations on a maize leaf. This time the test was "rigged" by the phytolith expert. The biometric goal of this test was to determine if truncated leaf phytolith populations could confuse size, shape or frequency parameters useful in separating maize from teosinte. If the distinguishing parameters from the first test succeeded in separating the two populations in this second test, then image analysis fails. The results were correctly negative, vindicating the technique. As indicated in Table 13.1, the length parameter that appeared unexpectedly so powerful in distinguishing *inter*-species (maize vs. teosinte) populations did *not* distinguish *intra*-species phytolith populations from different parts of the same leaf. This suggests that distribution of short cells may vary throughout a leaf, but significant aspects of short cell morphology remain stable.

The same result occurred in testing the shape parameter, formfactor, useful in separating short cell categories. Formfactor also appears to be stable regardless of leaf location. Curiously, not all shape factors are stable in this way. The shape factors of roundness and aspect ratio did distinguish the two locational populations at very high levels of statistical confidence. It may eventually prove significant that roundness was rejected as a parameter used to separate crossbodies and bilobates, yet it does distinguish short cell populations locationally. It is not clear which specific cause and effect variables are at play here, but this may well be clarified with further morphometric analysis of larger samples from a number of maize taxa.

The results of image analysis tests conducted thus far on *Zea* phytolith short cell populations are enormously encouraging. Statistically tested parameters that appear to distinguish morphological categories (crossbodies vs bilobates) do not

TABLE 13.1. Comparison of Short Cell Phytolith Populations from Different Locations on a Single Maize Leaf

	Length	Formfactor	Roundness	Aspect Ratio
Mean	20.421	0.5483	0.4683	1.780
S.D.	4.823	0.0844	0.1012	0.531
Mean B:	19.819	0.5406	0.4124	2.015
S.D.	5.002	0.1035	0.1084	0.552
ANOVA F	0.750	0.336	14.210	9.359
Prob. Diff.	39.17%	43.03%	99.95%(!)	99.71%(!)
Wilcoxon U	4714	4714	3458	3560
Wilcoxon Z	0.6988	0.6328	3.768	3.518
Significant?	NO @ 5%	NO @ 5%	YES @ 1%	YES @ 1%

distinguish intra-species locational populations or inter-species populations. Those that distinguish locational populations are not significant in separation of type or taxa; and factors that separate taxa are not affected by location or type parameters. Surely this is the ideal basis upon which to develop effective, consistent and reproducible phytolith classification, not only for *Zea*, but for Poaceae (Gramineae) as a whole and many other families in the plant kingdom.

REFERENCES

Brown, DW 1986 Taxonomy of a midcontinent grasslands phytolith key. *In* Rovner, I, ed, Plant opal phytolith analysis in archaeology and paleoecology: Proceedings of the 1984 Phytolith Research Workshop, North Carolina State University, Raleigh, North Carolina, *Occasional Papers of The Phytolitharien* No 1: 67–86.

Bryant, VM Jr 1989 Book review. *American Scientist* 77: 294.

Bukry, D 1979 Comments on opal phytoliths and stratigraphy of Neogene silicoflagellates and coccoliths at Deep Sea Drilling Project Site 397 off northwest Africa. *In* Shamback, JD, ed, *Initial Reports of the Deep Sea Drilling Project.* Washington, DC, US Government Printing Office 49: 977–1010.

Gregory, W 1855 On the presence of Diatomaceae, Phytolitharia and sponge spicules in soils which support vegetation. *Proceedings of the Botanical Society, Edinburgh:* 69–72.

Kondo, R 1974 Opal phytoliths—the relations between the morphological features of opal phytoliths and the taxonomic groups of gramineous plants. *Pedorojisuto* (Pedologist) 18: 2–10.

____1976 On the opal phytoliths of tree origins. *Pedorojisuto* (Pedologist) 20: 176–190.

Metcalfe, CR 1960 *Anatomy of the Monocotyledons I. Gramineae.* Oxford, Clarendon Press: 731 p.

Mulholland, SC 1986 Identification of plants in a sediment. *In* Rovner, I, ed, Plant opal phytolith analysis in archaeology and paleoecology: Proceedings of the 1984 Phytolith Research Workshop North Carolina State University, Raleigh, North Carolina, *Occasional Papers of The Phytolitharien* No 1: 41–51.

Mulholland, SC and Rapp, GR Jr 1989 Characterization of grass phytoliths for archaeological analysis. *Materials Research Society Bulletin* 14(3): 36–39.

Mulholland, SC, Rapp, GR Jr and Ollendorf AL 1987 Variation in phytoliths from corn leaves. *Canadian Journal of Botany* 66: 2001–2008.

Ollendorf, AL, Mulholland, SC and Rapp, G Jr 1988 Phytolith analysis as a means of plant identification: *Arundo donax* and *Phragmites communis. Annals of Botany* 61: 209–214.

Pearsall, DM 1978 Phytolith analysis of archeological soils: Evidence for maize cultivation in Formative Ecuador. *Science* 199: 177–178.

____1982 Phytolith analysis: Applications of a new paleo-ethnobotanical technique in archaeology. *American Anthropologist* 84: 862–870.

____1989 *Paleoethnobotany: A Handbook of Procedures.* San Diego, Academic Press, Inc: 470 p.

Pearsall, DM and Trimble, MT 1983 Phytolith analysis of soil samples: Report 18. *In* Clark, JT and Kirch, PV, eds, Archaeological investigations of the Mudlane-Waimea-Kawaihae Road Corridor. Island of Hawai'i: An interdisciplinary study of an environmental transect, *Bernice P. Bishop Museum Departmental Report Series* 83-1: 472–497.

_____1984 Identifying past agricultural activity through soil phytolith analysis: A case study from the Hawaiian Islands. *Journal of Archaeological Science* 11: 119–133.

Piperno, DR 1984 A comparison and differentiation of phytoliths from maize (*Zea mays* L.) and wild grasses: Use of morphological criteria. *American Antiquity* 49: 361–383.

_____1988 *Phytolith Analysis: An Archaeological and Geological Perspective.* San Diego, Academic Press, Inc: 280 p.

Reade, JB 1837 On the existence of structure in the ashes of plants and their analogy to the osseous system in animals. *The London, Edinburgh and Dublin Philosophical Magazine and Journal of Science* 11: 13–17.

Rovner, I 1971 Potential of opal phytoliths for use in paleoecological reconstruction. *Quaternary Research* 1: 345–359.

_____1983 Plant opal phytolith analysis: Major advances in archaeobotanical research. *In* Schiffer, M, ed, *Advances in Archaeological Method and Theory* 6: 225–266. New York, Academic Press.

_____ed 1986 Plant opal phytolith analysis in archaeology and paleoecology: Proceedings of the 1984 Phytolith Research workshop, North Carolina State University, Raleigh, North Carolina, *Occasional Papers of The Phytolitharien*, No 1: 147 p.

Russ, JC 1986 *Practical Stereology.* New York, Plenum Press: 185 p.

_____1988 *Computer Assisted Microscopy: The Measurement and Analysis of Images.* Raleigh, North Carolina, North Carolina State University Engineering Extension Service: 453 p.

_____1989 Computerized object recognition using contextual learning. *Journal of Computer Assisted Microscopy* 1: 105.

Russ, JC and Rovner, I 1989a Expert vision systems in archaeometry. *Materials Research Society Bulletin* 14(3): 40–44.

_____1989b Stereological identification of opal phytolith populations from wild and cultivated *Zea mays. American Antiquity* 54: 784–792.

Struve, GA 1835 (ms) De silica in plantis nunnullis. PhD dissertation, Berlin.

Twiss, PC, Suess, E and Smith, RM 1969 Morphological classification of grass phytoliths. *Soil Science Society of America, Proceedings* 33: 109–115.

14

ANNOTATED BIBLIOGRAPHY OF PHYTOLITH SYSTEMATICS

SUSAN C. MULHOLLAND, ELIZABETH J. LAWLOR and IRWIN ROVNER

INTRODUCTION

Literature dealing with phytolith systematics is scattered in a wide variety of journals reflecting the diverse disciplines that contribute to or use phytolith analysis. Morphological information has been reported in geology, soil science and archaeology contexts as well as botany. In addition, much research is published in languages other than English; indeed, major centers of phytolith activity span the globe from Germany to Japan. The diverse nature of phytolith research and the wide range of applications have produced a literature that is often difficult for active participants to encompass. Newcomers to the field often "reinvent the wheel" in the search for morphological information. Clearly, a central source of phytolith systematic data would be useful.

The bibliography addresses this need by presenting annotations of citations dealing with phytolith systematics. The authors selected citations from their individual bibliographies; one (Rovner) has been compiling phytolith references for well over two decades. The resulting lists were merged for this chapter. Although not comprehensive (particularly in regard to non-English language publications), the bibliography represents a first compilation of works of this nature. The annotations provide a brief summary of the contents in regard to systematic information. Evaluations are given only when descriptions are particularly clear or useful. Lists of figures, plates, photographs and tables are provided; some abbreviations for procedures are listed below.

The major portion of the bibliographic research was done by Rovner, not only in selection of citations pertaining to systematics but also because the bibliographies of the other authors are based on his bibliography. Annotations were done by Mulholland and Lawlor. A portion of this material is based upon work supported under a National Science Foundation Graduate Fellowship to Lawlor. Any opinions, findings, conclusions or recommendations expressed in this publication are those of the author(s) and do not necessarily reflect the views of the National Science Foundation.

Phytolith Systematics, Edited by G. Rapp, Jr. and
S.C. Mulholland, Plenum Press, New York, 1992

Key to Abbreviations

EDX - energy-dispersive x-ray spectrometry
EM - electron microscopy
EPM - electron probe microanalysis
ITW - inner tangential wall
OTW - outer tangential wall
SEM - scanning electron microscopy
SIAM - soft x-ray image analysis
TEM - transmission electron microscopy
WDX - wavelength-dispersive x-ray spectrometry
XRD - X-ray diffraction

Elements are often abbreviated to chemical symbol (*ie*, Si for silica, Ca for calcium).

Standard botanical abbreviations are used as listed below:

cv - cultivar
sp - species (single)
spp - species (multiple)
var - variety

CITATIONS AND ANNOTATIONS

Akai, S 1939 (On the ash figures of leaves of the rice plants transplanted from the different kinds of nursery beds and their susceptibilities to the blast disease.) *Annals of the Phytopathological Society of Japan* 9: 223–235. (Japanese, English summary). Comparison of seedlings from humid and arid soils shows more phytoliths (especially bulliform cells) from the humid soil. Silicification of bulliform cells is inversely correlated to blast disease. 15 tables.

Amos, GL 1952 Silica in timbers. *Australia Commonwealth Scientific and Industrial Research Organization Bulletin, Melbourne* 267: 1–55. Lists silica content and distribution in tissues for 400+ taxa in 32 families. Non-accumulating genera and families lacking siliceous timbers are also listed. 4 plates (32 photos), 6 figures, 2 appendices.

Amos, GL and Dadswell, HE 1948 Siliceous inclusions in wood in relation to marine borer resistance. *Australia Commonwealth Scientific and Industrial Research Organization, Journal of the Council for Scientific and Industrial Research* 21: 190–196. Differences in marine borer resistance of some tropical woods are correlated to total Si content and silica inclusions. Calcium oxalate and silica crystals have been reported in various species. 2 plates (8 photos), 1 table.

Andrejko, MJ and Cohen, AD 1984 Scanning electron microscopy of silicophytoliths from the Okefenokee swamp-marsh complex. *In* Cohen, AD, Casagrande, DJ, Andrejko, MJ and Best, GR, *The Okefenokee Swamp: Its Natural History, Geology, and Geochemistry.* Los Alamos,

New Mexico, Wetland Surveys: 468–491. Phytoliths from 6 sedge and grass spp are compared in detail. 8 figures (38 photos), 3 tables.

Andrejko, MJ, Cohen, AD and Raymond, R, Jr 1983 Origin of mineral matter in peat. *In* Raymond, R, Jr and Andrejko, MJ, eds, *Mineral Matter in Peat: Its Occurrence, Form, and Distribution.* Los Alamos National Laboratory, Publication LA-9907-OBES: 3–24. Mineral matter in peat is derived from several sources. Authigenic minerals include silica and calcium oxalate deposits formed in plants. No photos, drawings or descriptions of phytolith shape or distribution. 6 figures, 2 tables.

Artschwager, E 1925 Anatomy of the vegetative organs of sugar cane. *Journal of Agricultural Research* 30: 197–221. Internal anatomy of *Saccharum officinarum* (1 var) is detailed. Brief mention of epidermal silica cells (see plate 5). 25 plates, 8 figures.

_____1930 A comparative study of the stem epidermis of certain sugarcane varieties. *Journal of Agricultural Research* 41: 853–865. Epidermal patterns of stem material differ between sugarcane varieties and can be used for identification. Various epidermal cells, including silica cells, are described (14 var). 9 figures, 1 table, 1 key.

_____1940 Morphology of vegetative organs of sugarcane. *Journal of Agricultural Research* 60: 503–549. A "descriptive outline" of cultivated sugarcane morphology is presented to facilitate identification of varieties. Numerous morphological characters are described, including distribution patterns of silica cells. 10 plates, 14 figures, 3 tables.

_____1942 A comparative analysis of the vegetative characteristics of some variants of *Saccharum spontaneum. US Department of Agriculture Technical Bulletin* 811: 1–55. Morphological comparison of *Saccharum spontaneum* (23 var) indicates a wide range of characters exhibited by this complex species. Vegetative characters, including silica cell patterns, identify varieties and distinguish this group from the noble canes. 26 figures, 4 tables, 1 key.

Baas, P, van Oosterhoud, E and Scholtes, CJL 1982 Leaf anatomy and classification of the Olacaceae, *Octoknema* and *Erythropalum. Allertonia* 3: 155–210. Leaf anatomy (described for all genera of Olacaceae) forms the basis for a new classification. Silicified mesophyll and crystalliferous epidermal cells are of taxonomic significance at generic level; silica bodies and crystal components are useful below genus level. 39 figures (37 photos), 3 tables, 1 key, 26 generic descriptions.

Bair, RA 1966 Leaf silicon in sugarcane, field corn and St. Augustinegrass grown on some Florida soils. *Proceedings of the Soil and Crop Science Society of Florida* 26: 64–70. Low cane production correlates with Si deficiency in soils. Dry weight of Si in lbs/tons is reported for 4 spp (including corn and cane) grown on various soils. Note: sawgrass is a sedge. 7 tables.

Baker, G 1959a Opal phytoliths in some Victorian soils and "red rain" residues. *Australian Journal of Botany* 7: 64–87. Opaline bodies in sediments are identified as phytoliths derived from plants rather than sponge spicules. 255 particles are illustrated from soils and a "red rain" residue. 5 figures, 3 tables.

_____1959b A contrast in the opal phytolith assemblages of two Victorian soils. *Australian Journal of Botany* 7: 88–96. Differences in phytolith soil assemblages are correlated with different vegetation. 158 particles are illustrated from the soils. 4 figures, 2 tables.

_____1959c Fossil opal-phytoliths and phytolith nomenclature. *Australian Journal of Science* 21: 305–306. Brief discussion of the word "phytolith." Present usage is for "microscopic bodies of mineral matter secreted by a plant." 5 categories are briefly described.

_____1960a Hook-shaped opal phytoliths in the epidermal cells of oats. *Australian Journal of Botany* 8: 69–74. Oats and other grasses in Australia develop hook-shaped phytoliths composed of a solid core and delicate sheaths on the distal portion. The layers reflect differences in timing of deposition. 26 shape types are illustrated. 2 figures.

_____1960b Phytoliths in some Australian dusts. *Royal Society of Victoria, Melbourne, Proceedings* 72: 21–40. Phytoliths are common components of dust. 7 general shape groups are described and illustrated. 11 figures, 1 table.

_____1961a Opal phytoliths and adventitious mineral particles in wheat dust. *Mineragraphic Investigations Technical Paper* 4: 3–12. Wheat dust contains sharp small phytoliths that could be factors in transportation workers' health problems. 33 shape types are illustrated from wheat dust. 1 figure (8 photos).

_____1961b Opal phytoliths from sugar cane, San Fernando, Philippine Islands. *Queensland Museum, Brisbane, Memoirs of the Queensland Museum* 14: 1–12. Processing releases sharp airborne particles that can be implicated as health risks. 20 shape types are illustrated from sugarcane. 1 figure (12 photos), 3 tables.

_____1968 Micro-forms of hay-silica glass and of volcanic glass. *Mineralogical Magazine* 36: 1012–1023. "Microtecktites" from ocean sediments could be fused plant phytoliths or volcanic glass form. Glass bodies from a fired haystack include fused and distorted forms. 2 figures (85 photos).

Baker, G, Jones, LHP and Wardrop, ID 1961 Opal phytoliths and mineral particles in the rumen of the sheep. *Australian Journal of Agricultural Research* 12: 462–472. Phytoliths occur in plant feed and survive transport through the rumen to fecal material. 26 shape types are illustrated from 2 grass and 1 legume plant feed. 1 figure, 3 tables.

Balabko, PN, Prikhodka, VE and Ammosova, YM 1980 (Silica bioliths in plants and certain forest soils.) *Biologicheski Nauki (Moscow)* 12: 92–96. (Russian, unpublished translation by Glesser, C). Soils contain a great variety of silica bioliths (phytoliths and diatoms). 10 figures, 1 table.

Balan Menon, PK 1965 Guide to distribution of silica in Malayan woods. *Malayan Forester* 28: 284–288. Silica deposition in tissues is listed for 177 Malayan woods from 22 families. 1 table.

Bamber, RK and Lanyon, JW 1960 Silica deposition in several woods of New South Wales. *Tropical Woods* 113: 48–53. 9 spp sampled at increasing distances from the cambium were wet-ashed. Si was deposited not in new ray parenchyma, but in the cell lumen of inner sapwood as small particles and eventually as mature aggregates of possible taxonomic significance. 8 figures (8 photos), 1 table.

Bardin, SW and Bish, DL 1983 The occurrence of calcium oxalate minerals within aquatic macrophytes from Okefenokee Swamp. *In* Raymond, R, Jr and Andrejko, MJ, eds, *Mineral Matter in Peat: Its Occurrence, Form and Distribution.* Los Alamos National Laboratory, Publication LA-9907-OBES: 53–62. Whewellite, a crystalline form of calcium oxalate, was detected by XRD in 4 genera. Only *Nymphaea* and *Gordonia* exhibited definite crystals. 5 figures (5 photos).

Bargagli-Petrucci, SG 1902 Concrezioni silicee intracellulari nel legno secondario di alcune Dicotiledoni. (Intracellular silica concretions from secondary wood of some dicotyledons.) *Malpighia* 17: 23–27. (Italian). Discusses silica concretions in 8 families of dicots. No illustrations.

Barkworth, ME 1981 Foliar epidermes and taxonomy of North American Stipeae (Gramineae). *Systematic Botany* 6: 136–152. 10 each of 49 spp were studied as spodograms and wet-ashed. Focus is on silica bodies, which are described and measured and were mostly dumbbell to nodular but rectangular in some spp. 18 figures, 2 tables.

Barkworth, ME and Everett, J 1987 Evolution in the Stipeae: Identification and relationships of its monophyletic taxa. *In* Soderstrom, TR, Hilu, DW, Campbell, CS and Barkworth, ME, eds, *Grass Systematics and Evolution.* Washington, DC, Smithsonian Institution Press: 251–264. Description of groups within the Stipeae includes silica body morphology. 4 figures (15 photos).

Bartoli, F and Guillet, B 1977 Étude comparée des diagrammes phytolithiques et polliniques d'un podzol des Vosges gréseuses. (Comparative study of phytolith and pollen diagrams from a podzol on vosges sandstone.) *Comptes-Rendus Hebdomadaires des Seances de l'Académie des Sciences Paris, Series D, Sciences Naturelles* 284: 353–355. (French, English abstract). Movement of microfossils though sediment is controlled by species factors. Local vegetation changes are indicated by phytolith information. 1 figure (6 photos), 2 tables.

Benbasat, E 1976 (Studies on opaline silica deposition in the vegetative organs of some Poaceae grasses in Bulgaria.) *Fitologiya (Bulgarska Akademiya Na Naukite)* 5: 75–90. (Bulgarian, English abstract, unpublished translation by Howell, R.) Spodograms of 50 species (27 genera) indicate Si is deposited in walls of all epidermal cells. Phytoliths, however, form only in the specialized silica cells. 6 plates (36 photos), 1 table.

Bennett, DM (ms) 1980 The occurrence and accumulation of silicon in temperate cereals. PhD dissertation, University College of North Wales: 74 p. Much information on deposition of Si in various tissues and cell types of wheat, oats and barley. Si deposition in barley leaf tips is outlined; rye epicarp hairs were studied in addition to the other 3 crops. Species and cultivar differences in Si deposition were noted. 35 plates, 6 figures.

_____1982a An ultrastructural study on the development of silicified tissues in the leaf tip of barley (*Hordeum sativum* Jess). *Annals of Botany* 50: 229–237. Leaf tip silicification patterns in barley are similar in epidermal cells, sclerenchyma and intercellular spaces. Chemistry and mechanisms of silicification are discussed. 3 figures (18 photos).

_____1982b Silicon deposition in the roots of *Hordeum sativum* Jess, *Avena sativa* L. and *Triticum aestivum* L. *Annals of Botany* 50: 239–246. Si in seminal roots of barley, oats and wheat is confined to the endodermis, although specific locations vary between species. A positive function of strengthening is postulated. 2 figures (22 photos).

Bennett, DM and Parry, DW 1980 Electron-probe microanalysis studies of silicon in the elongating basal internodes of *Avena sativa* (L.), *Hordeum sativum* (Jess.) and *Triticum aestivum* (L.). *Annals of Botany* 45: 541–547. Si in the endodermal layer of stems of oats, barley and wheat follows 2 types of accumulation. Endodermal silicification appears to be a strengthening factor. 2 plates (33 photos).

_____1981 Electron-probe microanalysis studies of silicon in the epicarp hairs of the caryopses of *Hordeum sativum* Jess., *Avena sativa* L., *Secale cereale* L. and *Triticum aestivum* L. *Annals of Botany* 48: 645–654. Elemental distribution and morphology of epicarp hairs is given for wheat, oats, barley and rye. Possible function and significance for cancer are discussed. 5 figures (20 photos).

Bennett, DM and Sangster, AG 1981 The distribution of silicon in the adventitious roots of the bamboo, *Sasa palmata*. *Canadian Journal of Botany* 59: 1680–1684. 2 samples per root were freeze-dried and studied by SEM and EPM. Si was confined to the endodermal cell walls. 7 figures (7 photos), 1 table.

_____1982 Electron-probe microanalysis of silicon in the adventitious roots and terminal internode of the culm of *Zea mays*. *Canadian Journal of Botany* 60: 2024–2031. 5 mature maize cultivars were studied by SEM, EDX and EPM. Roots had no solid Si deposits, unlike some other grasses. The highest Si levels in terminal internodes were in peripheral tissues, suggesting possible selection for insect resistance. Si shapes are not addressed. 23 figures (17 photos).

Bertoldi de Pomar H 1970 Fitolitos y zoolitos: Su significado geológico en sedimentos continentales. (Phytoliths and zooliths: Their geologic significance in continental sediments.) *Boletin de la Asociacion Geologica de Córdoba* 1: 21–31. (Spanish). Phytolith production in grasses, sedges, palms and the Podostemaceae is reviewed. 2 plates (87 drawings).

_____1971 Ensayo de clasificación morfológica de los silicofitolitos. (Morphological classification of silicophytoliths.) *Ameghiniana* 8: 317–328. (Spanish, English abstract). Silica phytolith morphological classification is proposed. Groups (based on size) and morphotribes (based on shape) are defined. 1 table.

_____1972 Opalo organogeno en sedimentos superficiales de la llanura Santafesina. (Biogenous opaline particles found in superficial sediments from the Santa Fe Province Plains.) *Ameghiniana* 9: 265–279. (Spanish, English abstract). Description of bioliths includes that of phytoliths from sedges, grasses, palms and the Podostemaceae. Line drawings are given. 2 plates (93 figures).

_____1975 Los silicofitolitos: Sinopsis de su conocimiento. (Silicophytoliths: Synopsis of knowledge.) *Darwiniana* 19: 173–206. (Spanish, English summary). Summary of silica phytolith information includes distribution, morphology, deposition and the silica cycle. 7 plates (56 photos), 4 figures, 5 tables.

_____1980 Analisis comparativo de silicobiolitos de diversos sedimentos continentales Argentinos. (Comparative analysis of silicobioliths from diverse continental Argentine sediments.) *Revista, Asociación Geológica Argentina* 35: 547–557. (Spanish, English abstract). Examination of six sediments indicates some bioliths are widespread (euritopic) and some are unique to an environment (stenotopic). 7 plates (43 groups), 1 figure.

Besson, A 1946 Richesse en cendres et teneur en silice des bois tropicaux. (Richness in ashes and purport of silica of tropical forests.) *L'Agronomie Tropicale* 1: 44–56. (French). Ash and Si content of over 100 species of tropical wood are reported (percentages). 3 figures (2 photos), 1 table.

Blackman, E 1968 The pattern and sequence of opaline silica deposition in rye (*Secale cereale* L.). *Annals of Botany* 32: 207–218. Si deposition in leaf and internode material is highest at leaf bases and culm internode apices. Silicification increases over time and shape types change, reflecting increasing epidermal differentiation. 5 figures, 5 tables.

_____1969 Observations on the development of silica cells of the leaf sheath of wheat (*Triticum aestivum*). *Canadian Journal of Botany* 47: 827–838. Description of silica cell formation from cell differentiation to mature phytoliths. Both silica cells and rods are examined. 6 plates (50 photos), 15 figures.

_____1971 Opaline silica bodies in the range grasses of southern Alberta. *Canadian Journal of Botany* 49: 769–781. 26 spp (12 genera) were studied by light microscopy. Representative Si bodies for each genus are described in detail, drawn to scale, and discussed in relation to grass taxonomy. 30 figures (18 photos), 2 tables.

Blackman, E and Parry, DW 1968 Opaline silica deposition in rye (*Secale cereale* L.). *Annals of Botany* 32: 199–206. Distribution of silica bodies in leaf culm and inflorescence material was examined in cleared sections. Deposition in other cell types and cell walls is discussed briefly. 22 photos.

Blanchard, E and Chaussin, J 1933 Le Blé plante a silice. (Silica in wheat plants.) *Comptes-Rendus Hebdomadaires des Seances de l'Academie des Sciences, Paris, Series D, Sciences Naturelles* 197: 188–190. (French). Total and soluble Si in wheat is given. 1 table.

Bombin, M (ms) 1984 On phytoliths, late Quaternary ecology of Beringia, and information evolutionary theory. PhD dissertation, University of Alberta: 163 p. 300 grass and 100 sedge spp were studied with light microscopy and SEM. Morphological patterns, names, locations are not listed. Classifies Si by anatomical origin (see Piperno 1983) and uses own terminology for shapes (eg, "selliform" for "saddle"). Suggests tracing grass evolution through phytolith geometry. 23 figures (57 photos), 1 table, 1 appendix.

Bonnett, OT 1972 Silicified cells of grasses: A major source of plant opal in Illinois soils. *Illinois Agricultural Experiment Station, Bulletin* No 742. Detailed report of grass silicification focusing on description of cell types and degree of Si. 3 subfamilies (Panicoids, Chloridoids, Festucoids) were identified. 13 figures (66 photos).

Borissow, G 1924 Über die eigenartigen Kieselkörper in der Wurzelendodermis bei Andropogon-Arten. (On the peculiar silica cells in the root endodermis of *Andropogon* species.) *Berichte der Deutschen Botanischen Gesellschaft* 42: 366–380. (German). Description of silica bodies in endodermal tissue of *Andropogon* roots. 19 figures.

_____1925 Rasdorskys Körperchen beim Ravenna-Gras. (Rasdorsky's particles in Ravenna grass.) *Berichte der Deutschen Botanischen Gesellschaft* 43: 178–184. (German). Description of particles in *Erianthus ravennae*. 9 figures.

Bowdery, D (ms) 1984 Phytoliths: A multitude of shapes. BA essay, Australian National University: 102 p. Survey of phytolith analysis as a technique for recovery of botanical information. Various laboratory techniques were tested; phytolith classification and orientation from several sources were presented. Distribution of phytolith shapes in Australasian taxa is summarized from the literature. Distinctive phytoliths from 3 spp are illustrated. 10 figures (9 photos), 9 tables, 4 appendices.

_____1985 Phytolith studies in Australian prehistory. *Phytolitharien Newsletter* 3(3): 4–5. Photo of a phytolith from *Eucalyptus bridgesiana*. No description or comparisons. 1 photo.

_____1989 Phytolith analysis: Introduction and applications. *In* Beck, W, Clark, A and Head, L, eds, *Plants in Australian Archaeology, Tempus* 1: 161–196, University of Queensland, Anthropology Museum. A review article of phytolith analysis with emphasis on techniques and applications. Illustrates phytoliths from 11 spp. 12 photos, 8 tables.

Bozarth, SR (ms) 1985a An analysis of opal phytoliths from rinds of selected Cucurbitaceae species. MA thesis, University of Kansas: 59 p. The results summarized in Bozarth (1987) are given in detail here. Especially useful for its review of deciduous tree leaf and gymnosperm phytoliths, detailed laboratory methods and details about the reference collection. 5 figures (17 photos), 8 tables.

_____1985b The potential for identifying maize in archaeological sites with silicified ringed tracheids produced in cobs of selected maize varieties. *Phytolitharien Newsletter* 3(2): 3. These potentially diagnostic phytoliths tended to be larger in maize cobs than in other reference plants, and one was recovered from an archaeological soil sample.

_____1986 Morphologically distinctive *Phaseolus, Cucurbita*, and *Helianthus annuus* phytoliths. *In* Rovner, I, ed, Plant opal phytolith analysis in archaeology and paleoecology: Proceedings of the 1984 Phytolith Research Workshop, North Carolina State University, Raleigh, North Carolina, *Occasional Papers of The Phytolitharien* 1: 56–66. Spheres with scalloped surfaces were found in all domesticated and wild squash rinds studied and not in 60 other reference plants (not described here). "Phaseolus" phytoliths were hook-shaped hairs but may also occur in other families, according to Piperno's discussion in same volume. 3 figures (12 photos), 7 tables.

_____1987a Diagnostic opal phytoliths from rinds of selected *Cucurbita* species. *American Antiquity* 52: 607–615. *C. maxima* and *C. pepo* varieties from central Great Plains, US, had numerous spherical or hemispherical, deeply scalloped phytoliths. *C. foetidissima* had some, but dipper gourds (*Lagenaria siceraria*) and 60 other reference plants did not. Recovered some from an archaeological context. 2 figures (8 photos), 2 tables.

_____1987b Opal phytolith analysis of edible fruits and nuts native to the Central Plains. *Phytolitharien Newsletter* 4(3): 9–10. Reports distinctive shapes in hackberry. Others not distinctive (3 spp). 14 spp lack phytoliths.

_____1990 Diagnostic opal phytoliths from pods of selected varieties of common beans (*Phaseolus vulgaris*). *American Antiquity* 55: 98–104. Beans and pods were compared to 132 spp from the central plains, US. 30 hook-shaped hairs were significantly wider 6 microns from the tip in *P. vulgaris* than in 6 other dicots, including wild *P. polystachios*. Recovered some from archaeological contexts. 1 figure (4 photos), 2 tables.

Brandis, D 1907 Remarks on the structure of bamboo leaves. *Transactions of the Linnean Society, Series 2, Botany*, London 7: 69-92. Discussion of bamboo leaf anatomy includes silica deposits (pp 81–84). 4 plates (45 photos).

Brazle, FK, Harbers, LH and Owensby, CE 1979 Structural inhibitors of big and little bluestem digestion observed by scanning electron microscopy. *Journal of Animal Science* 48: 1457–1463. Leaf surfaces contain rows of phytoliths, trichomes and stomata. Si and cutin appear to inhibit digestion by rumen microorganisms. 14 photos.

Brown, DA 1984 Prospects and limits of a phytolith key for grasses in the central United States. *Journal of Archaeological Science* 11: 345–368. Multiple samples of different parts of 112 grasses and 12 non-grasses were dry-ashed. Phytolith shapes are defined, drawn to scale, and tabulated by species, genus and tribe. C_3 and C_4 plants may be distinguishable. Moisture and phytolith size appear to be related. 3 figures, 6 tables.

_____1986 Taxonomy of a midcontinent grasslands phytolith key. *In* Rovner, I, ed, Plant opal phytolith analysis in archaeology and paleoecology: Proceedings of the 1984 Phytolith Research Workshop, North Carolina State University, Raleigh, North Carolina, *Occasional Papers of The Phytolitharien* 1: 89–102. Defines and illustrates phytolith shapes and provides a key to 112 spp, taking different parts into account. Provides identifications to tribe or subfamily levels. 1 figure, 6 tables.

Brown, FBH 1920 The siliceous skeleton of tracheids and fibers. *Torrey Botanical Club Bulletin* 47: 407–424. 13 spp (of 500 observed) are discussed and represented by illustrations of *Tecoma* sp (a South American wood). Methods included Schulze solution wet-ashing and dry-ashing. Longitudinal rods, branching more in some species than others, all contained Si although composition varied in families, genera and species. Desilicification caused almost complete loss of xylem rigidity. Taxonomy and terminology may be outdated. 5 figures, 1 table.

Brown, WV 1958 Leaf anatomy in grass systematics. *Botanical Gazette* 119: 170–178. Transverse leaf sections of 101 spp were studied. 6 major groups are identified based on the presence of inner bundle sheaths, structure of outer bundle sheaths and chlorenchyma arrangement between bundles. Provides clear if probably dated review of grass evolution. 19 figures (18 photos).

Brydon, JE, Dore, WG and Clark, JS 1963 Silicified plant asterosclereids preserved in soil. *Soil Science Society of America, Proceedings* 27: 476–477. "Branched" particles of opal from soil correlate to Douglas fir asterosclereids (British Columbia, Canada). 2 figures.

Caceres, MR 1950 Los caracteres anatómicos foliares de *Munroa mendocina* y *Blepharidachne benthamiana*. (Leaf anatomical characteristics of *Munroa mendocina* and *Blepharidachne benthamiana*.) *Revista Argentina de Agronomia* 17: 233–240. (Spanish). Leaf anatomy indicates a panicoid type, leading towards reclassification as a chloridoid. Phytoliths are illustrated in situ. 5 figures.

_____1951 La anatomía foliar de "*Scleropogon brevifolius*" y sus relaciones taxonómicas. (Leaf anatomy of *Scleropogon brevifolius* and its taxonomic relations.) *Revista Argentina de Agronomia* 18: 129–135. (Spanish). The leaf anatomy is a panicoid type, indicating a change in taxonomic placement to chloridoids. Phytoliths are described. 3 figures.

_____1956 Los caracteres histofoliares de "*Schismus barbatus*" y sus relaciones taxonomicas. (Histofoliar characteristics of *Schismus barbatus* and its taxonomic relations.) *Revista Argentina de Agronomia* 23: 109–114. (Spanish). Leaf anatomy of *Schismus barbatus* is described as phragmitoid (intermediate between festucoid, panicoid and eupanicoid). Silica-cells are illustrated in situ. 3 figures.

_____1958 La anatomia foliar de las "Pappophoreae" de Mendoza y su valor taxonomico. (Leaf anatomy of the Pappophoreae of Mendoza and their taxonomic value.) *Revista Argentina de*

Agronomia 25: 1–11. (Spanish). Leaf anatomy of 8 species is described, indicating a panicoid type (chloridoid subtype). Phytoliths are illustrated in situ. 5 figures.

Campos, AC de and Labouriau, LG 1969 Corpos silicosos de gramíneas dos Cerrados II. (Silica bodies of grasses of the Cerrados II.) *Pesquisa Agropecuraria Brasileira* 4: 143–151. (Portuguese, English abstract). From Labouriau 1983: Leaf, stem and inflorescence material from herbarium specimens was combination wet- and dry-ashed. Camera lucida drawings of silica bodies from 10 grass spp. 6 figures, 1 table.

Cavalcante, PB 1968 Contribuição ao estudo dos corpos silicosos das gramineas Amazonicas. I. Panicoideae (Melinideae, Andropogoneae, e Tripsaceae). [Contribution to the study of silica bodies from Amazonian grasses. I. Panicoideae (Melinideae, Andropogoneae and Tripsaceae)]. *Boletim do Museu Paraense Emilio Goeldi, Botanica,* No 30. (Portuguese, English summary). From Labouriau 1983: Leaf, stem and inflorescence material from herbarium specimens was dry-ashed. Camera lucida drawings of silica bodies from 25 grass spp. 26 plates.

Chaffey, NJ 1983 Epidermal structure in the ligule of rice (*Oryza sativa* L.). *Annals of Botany* 52: 13–21. Description of abaxial epidermis of ligule includes the distribution of Si. Concentrations in silica cells and hair tips are correlated to low levels of epicuticular wax but other factors seem indicated also. 1 figure (6 photos), 1 table.

Chapuis, JL 1980 Méthodes d'étude du régime alimentaire du Lapin de Garenne, *Oryctolagus cuniculus* (L.) par l'analyse micrographique des fèces. (The use of microtechniques for the identification of the dietary items in the feces of European wild rabbit, *Oryctolagus cuniculus*.) *Terre Vie* 34: 159–198. (French, English summary). Epidermal fragments, including phytoliths, in feces provide dietary information. Identification of plants by phytoliths alone is more time consuming. 2 plates (12 photos), 6 figures, 8 tables.

Chattaway, MM 1953 The occurrence of heartwood crystals in certain timbers. *Australian Journal of Botany* 1: 27–38. Masses of crystals were observed in the heartwood of 52 Australian timber species of 19 families. Association with fungal hyphae is shown in some cases. Crystal appearance is dissimilar to calcium oxalate; those in one species are identified as ellagic acid. 7 figures (21 photos), 2 tables.

_____1955 Crystals in woody tissues. Part I. *Tropical Woods* 102: 55–74. Location of 5 crystal types in over 1000 genera of tropical woody species is listed. 5 figures, 5 tables.

_____1956 Crystals in woody tissues. Part II. *Tropical Woods* 104: 100–124. Distribution of rhomboidal, square or diamond-shaped crystals in 89 families. Location within tissue types and some important modifications are described. 5 figures, 1 table.

Chatters, RM (ms) 1941 A study of the ash of some American commercial woods with special reference to the existence of siliceous skeletons in the residues of cell walls. PhD dissertation, University of Michigan: 72 p. Examination of 17 conifers and 4 hardwoods indicates that siliceous sheets or tubes reflect original cell shape. However, an organized skeletal system is not present. 15 figures, 2 tables, 1 appendix.

_____1963 Siliceous skeletons of wood fibers. *Forest Products Journal* 13: 368–372. 18 dry-ashed North American spp (14 gymnosperms) yielded potentially diagnostic phytoliths but nothing to suggest that woody plants have a Si skeletal system. Includes details of both successful and unsuccessful methods and a review of plant Si research from 1813 through 1938. 7 photos, 1 table.

Cherouvrier, A, Gueguen, A and Lefeuvre, JC 1975 Essai de détermination du régime alimentaire d'animaux herbivores à l'aide des phytolithes siliceux des Graminées et des Cypéracées. Description, après étude en microscopie électronique à balayage, des principaux types de phytolithes rencontrés. (Determination of diet of herbivorous animals based on the analysis of siliceous phytoliths from Gramineae and Cyperaceae. Description of silica bodies based on scanning electron microscope study.) *Comptes Rendus Hebdomadaires des Seances de l'Academie des Sciences, Paris, Series D, Sciences Naturelles* 281: 839–842. (French). Dietary regimes can be inferred from grass and sedge phytoliths in feces. 2 plates (26 photos).

Chowdbury, KA and Ghosh, SS 1954 Plant remains from Hastinapura. *Ancient India* 10: 120–137. Charred material and fragments from mud brick are identified as rice. Epidermal anatomy (including silica bodies) is illustrated for sugarcane. 5 plates (48 photos), 3 figures, 3 tables.

Clark, CA and Gould, FW 1975 Some epidermal characteristics of paleas of *Dichanthelium, Panicum,* and *Echinochloa. American Journal of Botany* 62: 743–748. Description of apical portion of paleas in 22 spp of 3 grass genera, focuses on shape and distribution of silicified papillae. Differences between genera were noted, as were some unusual features in 3 spp. 27 photos.

Clifford, HT and Watson, L 1977 *Identifying Grasses: Data, Methods and Illustrations.* St. Lucia, Queensland, University of Queensland Press: 146 p. Discusses merits of various types of keys. Data is presented on 297 attributes for 223 Australasian grass genera. Silica body presence and shape is included. 33 plates (198 photos).

Collins, SM (ms) 1979 Phytoliths as indicators of plant use at ancient Troy. MS thesis, University of Minnesota: 181 p. Comparison of 29 plant samples to 40 sediments indicates possible use of wheat and kalamia at ancient Troy. Phytoliths are described from 7 grasses, a reed and a sedge. The classification has been extensively modified by later work. 18 figures, 4 appendices.

Collins, SM, Rapp, G, Jr, Gifford, JA, Rondina, D and Thomson, M 1986 Phytolith analyses of samples from Voyageurs National Park, Minnesota. *In* Lynott, MJ, Richner, JJ and Thompson, M, eds, Archeological Investigations at Voyageurs National Park: 1979 and 1980, *Midwest Archeological Center, Occasional Studies in Anthropology No* 16: 363–398. Analysis of plants and sediments focused on grasses; computer analysis of sediments was made in terms of plant phytolith assemblages. The classification scheme has been extensively revised by Mulholland (1987). 8 figures (16 photos), 5 tables.

Coppenet, M, Ducet, G, Gurillot, J and Kahane, E 1947 Sur la silice de la paille et du grain de blé. (On silica of the straw and grain of wheat.) *Annales Agronomiques* NS 17: 564–567. (French). Percentage of Si and other mineral constituents is given for wheat straw and grain. No photos or descriptions.

Cummings, L Scott (ms) 1989 Coprolites from medieval Christian Nubia: An interpretation of diet and nutritional stress. PhD dissertation, University of Colorado: 204 p. Analysis of plant fossils and parasites indicated population nutritional stresses. Grass, sedge, palm, legume and okra phytoliths are discussed (pp 78–83) but not illustrated. 36 figures, 20 tables.

Dahlgren, RMT and Clifford, HT 1982 *The Monocotyledons. A Comparative Study.* Academic Press, New York: 87–90. Description of about 100 characters, including distribution, in the monocots is a review designed to assess taxonomic value and relative reliability. See especially section on silica bodies (pp 87–90, 2 figures) and calcium oxalate raphides (pp 90–92, 2 figures) for brief discussion of occurrence in families and shapes. An evaluation of superorder affinities is also presented.

Darbyshire, SJ and Aiken, SG 1986 *Zizania aquatica* var. *brevis* (Poaceae): A 1983 distribution survey and a scanning electron microscope study of epidermal features. *Naturaliste Canadien (Rev. Ecol. Syst.)* 113: 355–360. Description of epidermal features, including silica bodies, of the leaf and anthecia is presented. Differences are noted by comparison to var *aquatica.* 10 figures (9 photos).

Davies, I 1959 The use of epidermal characteristics for the identification of grasses in the leafy stage. *British Grassland Society Journal* 14: 7–16. Epidermal characteristics of 40 grasses assessed for taxonomic significance. Some characteristics, including silica bodies, are affected by growth stage. 1 plate (9 photos), 3 figures, 4 tables.

Dayanandan, P and Kaufman, PB 1973a A scanning electron microscope study of isolated guard cells. *Proceedings of the Electron Microscopy Society of America, 31st Annual Meeting*: 454–455. Description of shape and surface texture of guard cells from 9 genera (*Psilotum, Equisetum, Cycas, Ceratozamia, Pinus, Ephedra, Welwitscha, Euphorbia, Allium*). No mineralization data but shape morphology is useful. 4 photos.

_____1973b Stomata in *Equisetum*. *Canadian Journal of Botany* 51: 1555–1564. 6 spp and 2 hybrids were studied by SEM and light microscopy. Si knobs on stomata may distinguish between subgenera, and were similar despite other differences between field-grown and greenhouse plants. Detailed descriptions and clear illustrations. 30 figures (29 photos).

_____1976 Trichomes of *Cannabis sativa* L. (Cannabinaceae). *American Journal of Botany* 63: 578–591. Glandular and non-glandular hairs were compared to those on *Humulus lupulus* (hops) and *Lantana camera*. Fresh, dried, and burned samples were studied with SEM and EDX. Clear, detailed descriptions. 62 figures (50 photos).

Dayanandan, P, Kaufman, PB and Franklin, CI 1983 Detection of silica in plants. *American Journal of Botany* 70: 1079–1084. Methyl red, crystal violet lactone and silver ammine chromate stain silica in plant tissue. Examples from 11 *Oryza* spp. 15 photos.

Dayanandan, P, Hebard, FV, Baldwin, VD and Kaufman, PB 1977 Structure of gravity-sensitive sheath and internodal pulvini in grass shoots. *American Journal of Botany* 64: 1189–1199. 28 spp were studied with SEM and x-ray, and thin sections with light microscopy. Among other structural characteristics, pulvini remain unsilicified though Si accumulates above and below them. 43 figures (43 photos), 1 table.

Decker, HF 1964 An anatomic-systematic study of the classical tribe Festuceae (Gramineae). *American Journal of Botany* 51: 453–463. Examination of 135 genera indicates that Festuceae tribe is artificial. Characteristics (including silica body shapes) are used to identify several major phylogenetic groups. 14 figures, 1 table.

Deflandre, G 1963 Les phytolithaires (Ehrenberg): Nature et signification micropaléontologique, pédologique, et géologique. (Phytoliths [Ehrenberg]: Nature and micropaleontological, pedological, and geological significance.) *Protoplasma* 57: 234–259. (French). Review of Ehrenberg's work on phytoliths focuses on morphology and classification. A parataxonomic approach is advocated; 4 paragenera are defined. 3 plates (54 photos), 69 figures.

Dengler, NG and Lin, EY-C 1980 Electron microprobe analysis of the distribution of silicon in the leaves of *Selaginella emmeliana*. *Canadian Journal of Botany* 58: 2459–2466. Si is present throughout the cell walls in sclereids and epidermal cells. Some stomata also have high proportions of Si. 12 figures (11 photos), 2 tables.

Dinsdale, D, Gordon, AH and George, S 1979 Silica in the mesophyll cell walls of Italian Rye grass (*Lolium multiflorum* Lam. cv. RvP.). *Annals of Botany* 44: 73–77. Freeze-dried samples were studied with TEM and EDX. Si content was similar with all methods. Shapes are hardly discussed. 1 plate (6 photos), 2 figures.

Djamin, A and Pathak, MD 1967 Role of silica in resistance to Asiatic Rice Borer, *Chilo suppressalis* (Walker), in rice varieties. *Journal of Economic Entomology* 60: 347–351. A negative correlation between Si content and resistance to Asiatic rice borer was noted on field tests of 12 rice var. Silica bodies are thought to interfere with the boring activities. 5 figures (3 photos), 1 table.

Dore, WG 1960 Silica deposits in leaves of Canadian grasses. *Canadian Society of Agronomy, Proceedings* (of the 6th Annual Meeting, Guelph, Ontario) 6: 96–99. Brief presentation of silica body location (over veins only vs intervein and vein areas) and shape (rectangular, simple rounded or lozenge, saddle-shaped, dumbbell-shaped, butterfly-shaped) in about 40 genera.

Dumitrica, P 1973 Phytolitharia. *In* Kaneps, AG, ed, *Initial Reports of the Deep Sea Drilling Project* 13(2):940–943. Washington, DC, US Government Printing Office. Description of phytoliths recovered from ocean cores in the Mediterranean is by parataxa, *ie*, artificial taxonomic units as used by Ehrenberg (1854). Illustrations and descriptions are not accompanied by attempts to identify the plant origin. 1 plate (51 figures).

Dunn, ME 1983 Phytolith analysis in archaeology. *Midcontinental Journal of Archaeology* 8: 287–297. Tests of Pearsall's (1978) maize-indicative phytoliths show large crosses in *Tripsacum dactyloides* and *Sorghum halepense*. No data are presented on percentages of the crosses or size ranges. 1 figure, 2 tables.

Duval-Jouve, MJ 1872 Sur une forme de cellules épidermiques qui parait propre aux Cypéracées. (On a form of epidermal cells that appear characteristic of Cyperaceae.) *Comptes-Rendus Hebdomadaires des Seances de l'Academie des Sciences, Paris, Series D, Sciences Naturelles* 75: 371–372. (French). Conical epidermal cells appear unique to sedges (as opposed to Typhaceae, Juncaceae and Gramineae). No illustrations.

Edwards, WN 1935 The systematic value of cuticular characters in recent and fossil angiosperms. *Biological Review* 10: 442–459. Summary of epidermal (cuticular) research focuses on spodograms in recent and fossil monocots and dicots. Phytoliths (including calcium types) are mentioned. No illustrations but good summary of earlier literature.

Ehrenberg, CG 1846 Zusätze zu den Mittheilungen über die vulkanischen Phytolitharien der Insel Ascension. (Promise of the report of the volcanic phytolitharien of Ascension Island.) *Monatsbericht der Koiglich Preussischen Akademie der Wissenschaften zu Berlin*: 191–202. (German). Comparison of phytoliths from volcanic contexts to 30 Egyptian grasses and sedges. Lists plant origin of the various "Phytolitharia" (parataxonomic).

_____1854 *Mikrogeologie. (Microgeology)*. (German). Leipzig, Leopold Voss, 2 vols. Description and illustration of microfossils extracted from sediments worldwide. Phytolitharia are classified by genus and species designations. Numerous drawings.

Eleuterius, LN and Lanning, FC 1987 Silica in relation to leaf decomposition of *Juncus roemerianus*. *Journal of Coastal Research* 3: 531–534. Relative Si content of leaves increases as decomposition advances. Leaves 8 years old are just siliceous casts. No descriptive data are provided. 1 figure.

Ellis, RP 1979 A procedure for standardizing comparative leaf anatomy in the Poaceae. II. The epidermis as seen in surface view. *Bothalia* 12: 641–671. Essential reference. Gives nondichotomous "keys," each with about 40 characters drawn to scale and assigned numbers, for 9 structures including silica bodies; discusses silicification of the other 8 structures. Does not tie characters to tribes or genera. Summarizes anatomical terminology and taxonomic significance of structural variability. More than 250 figures.

Engel, W 1953 Untersuchungen über die Kieselsäure-Verbindungen im Roggenhalm. (Investigation on the silica acid bound in rye stalks.) *Planta* 41: 358–390. (German). Percentage data of silica under various experimental conditions. 11 photos, 8 tables.

Esau, K 1977 *Anatomy of Seed Plants*. Second edition, New York, John Wiley & Sons: 550 p. Textbook cited widely for standard terminology; see especially pp 147–180. Discusses calcium oxalate and Si bodies very briefly.

Figueiredo, RCL and Handro, W 1971 Corpos silicosos de gramíneas dos Cerrados V. (Silica bodies of grasses from Cerrados V.) *III Simposio sobre o Cerrado*: 215–230. (Portuguese, English summary). From Labouriau 1983: Leaf, stem and inflorescence material from herbarium specimens was combination wet- and dry-ashed. Camera lucida drawings of silica bodies from 13 grass spp. 13 figures.

Flores, EM, Espinoza, AM and Kozuka, Y 1977 Observaciones sobre la epidermis foliar de *Zea mays* L. al microscopio electrónico de rastreo. (Observations on the leaf epidermis of *Zea mays*

L. by scanning electron microscope.) *Revista de Biologia Tropical* 25: 123–135. (Portuguese, English abstract). Epidermal characteristics (hairs, stomata, silica-cells) are described for maize leaf. 18 figures.

Fontana, P 1954 Studies on the deposition of silica on the leaves of the grass *Panicum maximum*. *Revista Brasiliera de Biologica* 14: 35–40. Silica extracted from leaves was measured for optical density, suggesting an enzymatic nature of silica deposition. 1 photo of the spicules (trichomes). 6 figures, 1 table.

Frey-Wyssling, A 1930a Vergleich zwischen der Ausscheidung von Kieselsäure und Kalziumsalzen in der pflanze. (Comparison between the secretions of silica and calcium in plants.) *Berichte der Deutschen Botanischen Gesellschaft* 48: 184–191. (German). Relates mechanism of deposition to particle chemistry and cell/tissue type. No figures.

_____1930b Über die Ausscheidung der Kieselsäure in der Pflanze. (On the secretion of silicic acid in plants.) *Berichte der Deutschen Botanischen Gesellschaft* 27: 179–183. (German). Review of silica in plants relates deposits to transpiration. Tropical and subtropical flora and tissues near transpiration are favored for deposition. No illustrations.

Frison, E 1948 On some typical forms of silica found in plants. *Microscope* 7: 38–45. Short survey of silica in plants mentions tabashir, *Equisetum*, diatoms, grasses, palms, ivory-nuts, stegmata and silica in wood. Resistance to shipworm boring is related to silica content of some woods. 3 photos.

Frohnmeyer, M 1914 Die Entstehung und Ausbildung der Kieselzellen bei den Gramineen. (The origin and development of silica-cells in the Gramineae.) *Bibliotheca Botanica* 21: 1-41. (German). Description of development of silica cells and silica bodies in grasses. Three types are defined. 2 plates (33 figures).

Fujiwara, H 1976 Investigation on the remains of crops in ancient times by plant opal analysis. *Kokogaku Sasshi (Archaeological Journal, Japan)* 62(2): 148–156. Description of leaf anatomy. Note photos of phytoliths from plants. 3 plates (13 photos), 5 figures.

Garber, LW (ms) 1966 Influence of volcanic ash on the genesis and classification of two spodosols in Idaho. MS thesis, University of Idaho: 155 p. The origin of two spodosols was related to volcanic ash content. Two types of phytoliths were recognized; one was correlated to Douglas fir (pp 72–77). Presence of phytoliths in Douglas fir was related to ash content of the soil. 16 figures, 14 tables.

Gascho, GJ 1977 Silicon status of Florida sugarcane. *Soil and Crop Science Society of Florida, Proceedings* 36: 188–191. Si levels in sugarcane and field soils are related to sugar tonnages and freckling of the plants. High Si in soil and tissues prevents freckling. No photos/figures or descriptions of phytoliths or tissue/cell location data. 4 tables.

Geis, JW (ms) 1972 Biogenic silica in selected plant materials. PhD dissertation, State University of New York, Syracuse University: 102 p. 3 grasses, 36 deciduous angiosperms and leaf litter representing tall-grass prairie and forest of central Illinois, US, were wet-ashed. Clear, detailed descriptions. 15 plates (125 photos), 5 tables, 7 appendices.

_____1973 Biogenic silica in selected species of deciduous angiosperms. *Soil Science* 116: 113–130. Partial summary of dissertation above. Si content varied within families and genera. Many shapes resembled grass phytoliths, but some epidermal cell differences may be diagnostic. 6 plates (51 photos, much clearer than dissertation), 1 table.

_____1978 Biogenic opal in three species of Gramineae. *Annals of Botany* 42: 1119–1129. Examination of all parts of *Andropogon gerardi*, *Sorghastrum nutans* and *Panicum virgatum* indicates phytoliths occur in many tissues and cell types. A type unique to roots is formed by silicification of the ITW of the endodermis. 3 plates (58 photos), 1 table.

_____1986 Characteristics of biogenic opaline silica in angiosperm and coniferous trees. *Phytolitharien Newsletter* 4(2): 3–11. Samples show fragments predominating; over 93% of phytoliths from forest leaf litter were finer than 20 um. Potentially diagnostic phytoliths present in foliage (and only foliage) in Ulmaceae, Moraceae and Aceraceae. Fabaceae and Lauraceae had less Si. Posthumous publication of 1984 presentation to American Association for the Advancement of Science. 3 tables.

Genua, JM and Hillson, CJ 1985 The occurrence, type and location of calcium oxalate crystals in the leaves of fourteen species of Araceae. *Annals of Botany* 56: 351–361. Type and tissue location of crystals is reported for 14 spp. Druses, raphides and prismatics were identified as calcium oxalate; crystal sand and variations are other calcium compounds. 4 figures (9 photos), 3 tables.

Gibson, AC and Horak, KE 1978 Systematic anatomy and phylogeny of Mexican columnar cacti. *Annals of the Missouri Botanical Garden* 65: 999–1057. Examination of 64 spp (Mexico, West Indies, South and Central America) yielded data on morphology, biochemistry and anatomy. Both silica bodies and calcium oxalate crystals are reported. 34 figures (32 photos), 2 tables.

Gibson, RJH 1893 On the siliceous deposit in the cortex of certain species of *Selaginella*, Spr. *Annals of Botany* 7: 355–367. 16 of 52 spp wet-ashed had Si bodies, described in terms of the plant parts. Taxonomy and terminology may be out of date. 10 figures.

Gottwald, H 1980 "Louro Preto" - found to be the first silica-bearing cordia (*Cordia glabrata*, Boraginaceae). *International Association of Wood Anatomists, Bulletin* NS 1: 55–59. 'Louro preto' is identified to *Cordia glabrata*. Si inclusions and oxalate crystals are present in the wood. 6 photos.

Gould, FW 1968 *Grass Systematics.* New York, McGraw-Hill. Systematic treatment of the grasses in the US includes silica cell shape information (based on Metcalfe 1960) in description of subfamilies and tribes (pp 97–118). Information is also presented in leaf epidermis section (pp 36–38, 1 figure).

Gould, FW and Shaw, RB 1983 *Grass Systematics.* College Station, Texas, Texas A & M University Press: 397 p. Comprehensive treatment of grass taxonomy includes data on silica bodies following Metcalfe (1960). Note brief description (pp 36–38, 2 figures) and occurrence in subfamilies and tribes (pp 107, 111–130).

Govindarajalu, E 1968 The systematic anatomy of South Indian Cyperaceae: *Cyperus* L subgen, *Kyllinga* (Rottb.) Suringar. *Journal of the Linnean Society of London. Botany* 62: 41–58. Anatomical description and comparison of 7 taxa indicate two broad groups. Leaf and culm descriptions include silica bodies but not in detail. 5 figures.

_____1975 Systematic anatomy of South Indian Cyperaceae: *Eleocharis* R. Br., *Rhynchospora* Vahl and *Scleria* Bergius. *Adansonia* 14: 581–632. Description and comparison of *Eleocharis* (8 spp), *Rhynchospora* (5 spp) and *Scleria* (11 spp) indicate separation into 3, 2 and 2 groups, respectively. Anatomical information includes distribution and description of silica bodies. 10 plates, 3 keys.

Govindarajalu, E and Thanyakumar, S 1977 The vegetative anatomy of *Freycinetia sumatrana* Hemsl. and *F. rigidifolia* Hemsl. along with the comparative study of shoot apex organization in *F. rigidifolia* and *Pandanus tectorius* Soland. *Adansonia* 17: 59–76. Description and comparison of 2 spp of *Freycinetia* (Pandanaceae) from the Andamans show distinct differences. Occurrence of silica bodies and crystals is one anatomical characteristic noted. 5 plates.

Grob, A 1896 Beitrage zur Anatomie der Epidermis der Gramineenblätter. (Contribution on anatomy of the epidermis of Gramineae leaves.) *Bibliotheca Botanica* 36: 1–122. (German). Description of grass leaf anatomy, including silica cells. 10 plates.

Gueguen, A, Cherouvrier, A and Lefeuvre, JC 1975 Essai de détermination du régime alimentaire d'animaux herbivores à l'aide des phytolithes siliceux des Graminées et des Cypéracées.

Application à l'étude du regime alimentaire des Orthoptères Acridiens. (Determination of the diet of herbivorous animals based on the analysis of the siliceous phytoliths of Gramineae and Cyperaceae. Application to diet of the Acridid Orthoptera.) *Comptes Rendus Hebdomadaires des Seances de l'Academie des Sciences, Paris, Series D, Sciences Naturelles* 281: 929–932. (French). Grass silica phytoliths from feces yield information on dietary regimes. 2 plates (25 photos), 2 tables.

Guignard, G 1988 Les phytolithes chez le genre *Dactylis* tétraploïde. (Phytoliths in the tetraploid genus *Dactylis*.) *Canadian Journal of Botany* 66: 1047–1053. (French, English abstract). *Dactylis* phytoliths are described and compared between wild and cultivated plants. Morphological affinities are confirmed. 24 figures (photos), 2 tables.

Haberlandt, G 1914 *Physiological Plant Anatomy*. London, MacMillan and Co, Ltd. This book on anatomy includes some data on Si and calcium salts in plants. Note silicified hairs of *Urtica* (pp 129–131, 2 figures), cystoliths (pp 536–537, 1 figure), silica deposits (pp 537–539, 1 figure) and deposits in wood (p 685).

Handreck, KA and Jones, LHP 1968 Studies of silica in the oat plant. IV. Silica content of plant parts in relation to stage of growth, supply of silica, and transpiration. *Plant and Soil* 29: 449–459. Si content of *Avena* is correlated to growth stage and Si supply. Similar distribution patterns may be related to transpiration. 2 figures, 5 tables.

Hansen, DJ, Dayanandan, P, Kaufman, PB and Brotherson, JD 1976 Ecological adaptations of salt marsh grass, *Distichlis spicata* (Gramineae), and environmental factors affecting its growth and distribution. *American Journal of Botany* 63: 635–650. Investigation of environmental conditions and adaptations of salt marsh grass. Leaf anatomy, including silica bodies, is described; silicified trichomes receive special attention as a possible cooling mechanism. 37 figures (27 photos).

Harbers, LH, Raiten, DJ and Paulsen, GM 1981 The role of plant epidermal silica as a structural inhibitor of rumen microbial digestion in steers. *Nutrition Reports International* 24: 1057–1066. Two grasses (*Festuca arundinacea* and *Andropogon gerardi*) grown with and without Si were subjected to rumen digestion at 30 days and 60 days growth. Si at 30 days prevented cuticular rupture by microbes, but by 60 days other structures prevented rupture in both preparations. 6 figures (26 photos).

Harbers, LH and Thouvenelle, ML 1980 Digestion of corn and sorghum silage observed by scanning electron microscopy. *Journal of Animal Science* 50: 514–526. Degradation of corn and sorghum after digestion in steers was assessed by SEM. Si in epidermis of leaves and stems appears to inhibit digestion. 14 figures (37 photos).

Hart, DM 1988 The plant opal content in the vegetation and sediment of a swamp at Oxford Falls, New South Wales, Australia. *Australian Journal of Botany* 36: 159–170. Phytoliths are described for 10 spp: *Xanthorrhoea resinosa, Gahnia sieberana, Entolasia stricta, Eucalyptus gummifera, Leptospermum flavescens, Baksia aspleniifolia, Acacia schinoides, Casuarina distyla, Pteridium esculentum,* and *Gleichenia dicarpa*. Spheres, rods and sheets are most common; no diagnostics were identified. 3 figures (18 photos), 2 tables.

Hauke, RL 1978a Microreplicas as a technique for rapid evaluation of surface silica micromorphology in *Equisetum*. *American Fern Journal* 68: 37–40. Si deposition patterns on surfaces of *Equisetum* are taxonomically useful. Microreplicas on plastic cover slips are recommended as easier and quicker than SEM analysis. 6 photos.

_____1978b A taxonomic monograph of *Equisetum* subgenus *Equisetum*. *Nova Hedwigia* 30: 385–455. Thorough review of the subgenus based on 8 spp, which are described. Includes morphology, classification, distribution, evolution and other material. Surface Si micromorphology is included. 65 figures, 3 tables.

_____1979 *Equisetum ramosissimum* in North America. *American Fern Journal* 69: 1–5. Identification of *E. ramosissimum* vs *E.* x *ferrissii* was made from SEM analysis of surface Si micromorphology. 10 photos (4 of silica).

Hayward, DM and Parry, DW 1973 Electron-probe microanalysis studies of silica distribution in barley (*Hordeum sativum* L.). *Annals of Botany* 37: 579–591. Si deposition in culm, leaf and inflorescence varies with age of tissue and cell type. Metabolic control rather than passive deposition from transpiration is suggested by deposition within the leaves. 3 plates (36 photos), 4 figures.

_____1975 Scanning electron microscopy of silica deposition in the leaves of barley (*Hordeum sativum* L.). *Annals of Botany* 39: 1003–1009. Examination of in situ and disaggregated bodies shows rods, "hats", prickle bodies and other cell types. Two processes are suggested, one involving the secondary wall and one the cell lumina. 2 plates (24 photos).

_____1980 Scanning electron microscopy of silica deposits in the culms, floral bracts, and awns of barley (*Hordeum sativum* Jess.). *Annals of Botany* 46: 541–548. Si deposits are described for internodes, bracts and awns of barley. The types of cells are described, including shape. 2 plates (26 photos).

Helbaek, H 1959 How farming began in the Old World. *Archaeology* 12: 183–189. Reviews hypotheses about domestication and spread of Old World crops. Silica skeletons are one type of archaeological evidence for plants and cultivation. 11 photos.

_____1960 Cereals and weed grasses in phase A. *In* Braidwood, RJ and Braidwood, LS, eds, *Excavations in the Plain of Antioch I*, University of Chicago Oriental Institute Publications Volume 61, University of Chicago Press, Appendix II: 540–543. Imprints and silica skeletons of emmer, barley, oat grass and rye grass were recovered from sherds. 11 photos (3 of silica skeletons).

_____1961 Studying the diet of ancient man. *Archaeology* 14: 95–101. Brief review of various botanical remains found in archaeological contexts. Silica skeletons of four crops are illustrated. 14 figures.

Hillis, WE and de Silva, D 1979 Inorganic extraneous constituents of wood. *Holzforschung* 33: 47–53. 11 woody spp were examined by SEM and x-ray analysis. Si, Ca and Al salt deposits were identified in ray parenchyma, vessels, and fibrous elements. 8 photos, 3 tables.

Hirata, T, Saiki, H and Harada, H 1972 (Observation of crystals and silica inclusions in parenchyma cells of certain tropical woods by scanning electron microscope.) *Kyoto Daigaku. Nogakubu. (Bulletin of the Kyoto University Forests)* 44: 194–199. (Japanese, English summary, unpublished translation by Williams, V and Mulholland, S.) Si and calcium oxalate crystals are described from 18 spp (9 families) of tropical woods. Different morphology and structure distinguish the two materials. 1 table.

Hodson, MJ 1986 Silicon deposition in the roots, culm and leaf of *Phalaris canariensis* L. *Annals of Botany* 58: 167–177. Specimens were studied by SEM, EPM and light microscopy. Si deposited in all parts is clearly described and amply illustrated. 5 figures (37 photos).

Hodson, MJ and Bell, A 1986 The mineral relations of the lemma of *Phalaris canariensis* L., with particular reference to its silicified macrohairs. *Israel Journal of Botany* 35: 241–253. Mineral concentrations and locations in the lemma were determined during plant life. Macrohairs received special attention; Si entered through cell walls early. 2 figures, 3 tables.

Hodson, MJ and Parry, DW 1982 The ultrastructure and analytical microscopy of silicon deposition in the aleurone layer of the caryopsis of *Setaria italica* (L.) Beauv. *Annals of Botany* 50: 221–228. Si content of the grain of foxtail millet occurs in the pericarp and aleurone layer. High esophageal cancer rates may be related to these deposits. 2 figures (5 photos).

Hodson, MJ and Sangster, AG 1988 Silica deposition in the inflorescence bracts of wheat (*Triticum aestivum*). I. Scanning electron microscopy and light microscopy. *Canadian Journal of*

Botany 66: 829–838. Si deposition in various cell types is reported for wheat glume, palea and lemma. Glumes and lemma are similar but the palea differs in Si distribution. 48 figures.

____1989a Silica deposition in the inflorescence bracts of wheat (*Triticum aestivum*). II. X-ray microanalysis and backscattered electron imaging. *Canadian Journal of Botany* 67: 281–287. Description of Si deposition (location and morphology) in wheat glume, palea and lemma includes prickles, papillae, macrohairs and short cells. 18 photos.

____1989b Subcellular localization of mineral deposits in the roots of wheat (*Triticum aestivum* L). *Protoplasma* 151: 19–32. X-ray microanalysis indicated four minerals formed biomineralized structures. Si occurred in walls and lumina of endodermal, metaxylem and contiguous cells. Opaque intracellular deposits of phosphorus, calcium and manganese were also detected. 45 figures (15 photos), 3 tables.

Hodson, MJ, Sangster, AG and Parry, DW 1982 Silicon deposition in the inflorescence bristles and macrohairs of *Setaria italica* (L.) Beauv. *Annals of Botany* 50: 843–850. Inflorescence bristles contain silicified prickle hairs. The structures may be connected to high rates of esophageal cancer in China. 2 figures (13 photos), 1 table.

____1984 An ultrastructural study on the development of silicified tissues in the lemma of *Phalaris canariensis* L. *Proceedings of the Royal Society of London, Series* B 222: 413–425. Si deposition occurs after panicle emergence in trichomes and epidermal long cells. An active mechanism seems to be present. 6 plates (20 photos), 2 tables.

____1985 An ultrastructural study on the developmental phases and silicification of the glumes of *Phalaris canariensis* L. *Annals of Botany* 55: 649–665. Samples were harvested at 7 ages, studied with TEM and EDX, and dry-ashed for light microscopy. Si deposition was mainly in abaxial epidermis, proceeding from cell walls inward, and beginning in papillae and prickle hairs before panicle emergence, followed by silica cells and long cells. 7 figures (29 photos).

Hoffman, FM and Hillson, CJ 1979 Effects of silicon on the life cycle of *Equisetum hyemale* L. *Botanical Gazette* 140: 127–132. Si is demonstrated to be an essential element for growth of the sporophyte, although the gametophyte doesn't seem to be as sensitive. Larger silicified papillae correlate to higher Si concentrations. 8 figures.

Holm, T 1899 On the anatomy of some North American species of *Scleria*. *American Journal of Science* 7: 5–12. Describes anatomy of root, rhizome, stem and leaf for 11 spp. Siliceous incrustations on bulliform cell walls are discussed. 6 figures, 1 table.

Hryniewiecki, B and Kurtz, W 1936 La répartition des cônes siliceux dans les cellules des Cypéracées et leur corrélation. (The distribution of silica cones in the cells of Cyperaceae and their correlation.) *Bulletin International de l'Academie Polonaise des Sciences et des Lettres Series* B: 33–54. (French). Location of silicified cones in leaf and stem is presented for 79 sedge spp. 2 plates (20 figures), 1 table.

Hurghisiu, I 1974 Continutul în silice din *Phragmites australis* (Cav) Trin et Stend, *Typha angustifolia* L şi *Scirpus lacustris* L, crescute în condiţii ecologice variate. (Silica content of *Phragmites australis*, *Typha angustifolia*, *Scirpus lacustris*, grown in various ecological conditions.) *Celuloza Si Hirtie* 23: 10–11. (Romanian). Discusses silica content of several plant parts in June and September. 1 figure, 1 table.

Jefferies, TA 1916 The vegetative anatomy of *Molinia caerulea*, the purple heath grass. *The New Phytologist* 15: 49–71. Describes anatomy of root, rhizome, stem and leaf. Brief mention and illustration of silica bodies in leaf. 9 figures.

Johnston, A, Bezeau, LM and Smoliak, S 1967 Variation in silica content of range grasses. *Canadian Journal of Plant Science* 47: 65–71. Transplanted grasses, grass mixtures and a transect across southern Alberta, Canada, were sampled. Si content varied between years, but not consistently related to annual or seasonal rainfall. Urolithiasis implications are discussed, but not phytolith shapes. 3 figures, 4 tables.

Jones, LHP and Handreck, KA 1967 Silica in soils, plants, and animals. *Advances in Agronomy* 19: 107–149. Review of Si occurrence from soils through plants to animals includes discussion of uptake, transport, deposition and function. Mostly background on Si cycle but includes some photos of silicified cells. 1 figure (4 photos), 7 tables.

Jones, LHP, Milne, AA and Sanders, JV 1966 Tabashir: An opal of plant origin. *Science* 151: 464–466. Description of tabashir, a relatively large deposit of Si deposited in interior of bamboo stems. Specific gravity, index of refraction, x-ray diffraction, chemical composition, optical examination and electron diffraction indicate tabashir differs from other silica phytoliths. 1 figure (6 photos).

Jones, LHP, Milne, AA and Wadham, SM 1963 Studies of silica in the oat plant, II. Distribution of the silica in the plant. *Plant and Soil* 18: 358–371. Amount and type of phytoliths in oat tissues was studied. Wet ashing preserves details better than dry ashing. 6 plates (24 photos), 3 figures.

Kam, YK and Stone, BC 1970 Morphological studies in *Pandanaceae*. IV. Stomate structure in some Mascarene and Madagascar *Pandanus* and its meaning for infrageneric taxonomy. *Adansonia* 10: 219–246. Stomatal type is examined for 20 spp of *Pandanus*. Anatomical data is evaluated in terms of the 10 taxonomic sections. Note: Silica is only in table 2. Not descriptive. 8 plates (21 photos), 3 tables, 50 figures.

Kanno, I and Arimura, S 1958 Plant opal in Japanese soils. *Soil Science and Plant Nutrition (Soil and Plant Food)* 4: 62–67. Phytoliths recovered from soils were compared to those from 3 grasses: *Sasa* sp, *Imperata* sp, and *Miscanthus sinensis*. Distribution and characteristics of phytoliths are briefly discussed. 7 figures, 5 tables.

Kaufman, PB, Bigelow, WC, Petering, LB and Drogosz, FB 1969 Silica in developing epidermis cells of *Avena* internodes: Electron microprobe analysis. *Science* 166: 1015–1017. Concentrations of chemical compounds affect growth and cell development, including silica-cork pairs. 35 figures, 1 table.

Kaufman, PB, Bigelow, WC, Schmid, R and Ghosheh, NS 1971 Electron microprobe analysis of silica in epidermal cells of *Equisetum*. *American Journal of Botany* 58: 309–316. 2 spp had different phytolith patterns from each other and from grasses, but they may not be taxonomically significant within *Equisetum*. 20 figures (19 photos), 1 table.

Kaufman, PB and Cassell, SJ 1963 Striking features in the development of internodal epidermis in the oat plant (*Avena sativa*). *The Michigan Botanist* 2: 115–121. Description of epidermal cell types in oat internodes, including brief treatment of developmental sequences. 8 figures (6 photos).

Kaufman, PB, Dayanandan, P, Franklin, CI and Takeoka, Y 1985 Structure and function of silica bodies in the epidermal system of grass shoots. *Annals of Botany* 55: 487–507. More than 37 spp studied with TEM, SEM, x-ray and light microscopy. C_3/C_4 differences in Si are discussed. Si did not facilitate light transmission in experiment. 7 figures (33 photos), 3 tables.

Kaufman, PB, Dayanandan, P, Goldoftas, M, Lau, E, Srinivasan, J, Clark, JM, Hollingsworth, P, Mardinly, J and Bigelow, WC 1978 Analysis of primary deposition sites for silica (SiO_2*nH_2O) in panicoid and festucoid grasses and in scouring rushes (*Equisetum*) by scanning electron microscopy and energy-dispersive microanalysis. *Microbeam Analysis Society Proceedings 13th Conference.* Ann Arbor, Michigan, pp 82A–82C. Deposition sites in 5 grasses and a horsetail are specific to tissue and cell types. Patterns of silicification are related to function of Si.

Kaufman, PB, LaCroix, JD, Dayanandan, P, Allard, LF, Rosen, JJ and Bigelow, WC 1973 Silicification of developing internodes in the perennial scouring rush (*Equisetum hyemale* var. *affine*). *Developmental Biology* 31: 124–135. Peels were studied with SEM and EPM. Si in internodes occurred first in stomata, then papillae, then long cells. Shapes are not compared to other plants. 25 figures (25 photos), 1 table.

Kaufman, PB, LaCroix, JD, Rosen, JJ, Allard, LF and Bigelow, WC 1972 Scanning electron microscopy and electron microprobe analysis of silicification patterns in inflorescence bracts of *Avena sativa*. *American Journal of Botany* 59: 1018–1025. Si deposition in cells of oats inflorescence mostly occurs in silica cells. Trichomes and long cells also contain some Si. 21 figures, 1 table.

Kaufman, PB, Petering, LB and Smith, JG 1970 Ultrastructural development of silica-cork cell pairs in *Avena* internodal epidermis. *Botanical Gazette* 131: 173–185. Development of silica-cork cell pairs is described for oats. 17 figures, 1 table.

Kaufman, PB, Petering, LB and Soni, SL 1970 Ultrastructural studies on cellular differentiation in internodal epidermis of *Avena sativa*. *Phytomorphology* 20: 281–309. Stomata, silica-cork pairs and trichomes differentiate by cell division and/or cell expansion. All develop from short cells but by different paths. 24 figures (30 photos).

Kaufman, PB, Soni, SL, Bigelow, WC and Ghosheh, NS 1973 Scanning electron microscopic analysis of the leaf and internodal epidermis of the rice plant (*Oryza sativa*). *Phytomorphology* 23: 52–58. Description of the surface fine-structure includes information on silica cells as well as other cell types. 5 figures (14 photos).

Kaufman, PB, Soni, SL, LaCroix, JD, Rosen, JJ and Bigelow, WC 1972 Electron-probe microanalysis of silicon in the epidermis of rice (*Oryza sativa* L.) internodes. *Planta* 104: 10–17. Si accumulates in the entire epidermis (except stomata) throughout much of the internode; accumulation in the intercalary meristem favors silica cells over long cells. 8 photos, 1 table.

Kaufman, PB, Takeoka, Y, Carlson, TJ, Bigelow, WC, Jones, JD, Moore, PH and Ghosheh, NS 1979 Studies on silica deposition in sugarcane (*Saccharum* spp.) using scanning electron microscopy, energy-dispersive x-ray analysis, neutron activation analysis, and light microscopy. *Phytomorphology* 29: 185–193. Differences in Si content in internodes of sugarcane cultivars reflect number of epidermal silica cells. A possible function to increase transmittal of light to interior tissues is proposed. 21 photos, 1 table.

Kellogg, EA and Campbell, CS 1987 Phylogenetic analyses of the Gramineae. *In* Soderstrom, TR, Hilu, DW, Campbell, CS and Barkworth, ME, eds, *Grass Systematics and Evolution*. Washington, DC, Smithsonian Institution Press: 310–322. Discussion of grass phylogeny is based on 33 characters. Si body direction (horizontal vs vertical) is included for 6 subfamilies and over 50 genera. 3 figures, 1 table.

Kido, M and Yanatori, S 1960 (Physiological and ecological studies of rice plant in well-drained and ill-drained fields. 6. Changes of some constituents in root of rice plant.) *Nihon Sakumotsu Gakkai Kiji (Proceedings of the Crop Science Society of Japan) (Japanese Journal of Crop Science)* 29: 40–42. (Japanese, English summary). Si deposition in root epidermis and central cylinder is correlated to concentrations of phosphorus and nitrogen. 9 figures (8 photos).

____1963 (Physiological and ecological studies of the rice plant in well-drained and ill-drained fields. 8. Distribution of silicified tissues of rice culm and silicate content in each internode as affected by different cultural conditions.) *Nihon Sakumotsu Gakkai Kiji (Proceedings of the Crop Science Society of Japan) (Japanese Journal of Crop Science)* 31: 237–240. (Japanese, English summary). Amount and distribution of Si in culm internodes varies with growth stage and field conditions (good drainage, chemical applications). 13 figures (8 photos), 1 table.

____1964 (Studies on silicified tissues of leaf in rice plant by spodogram - hydrogen chloride treatment method.) *Nihon Sakumotsu Gakkai Kiji. (Proceedings of the Crop Science Society of Japan) (Japanese Journal of Crop Science)* 33: 115–118. (Japanese, English summary.)

Silicification sequence of cells in leaf blade is complete by emergence. Dumbbells and motor cells (bulliform) are shown. 18 photos.

Klein, RL and Geis, JW 1978 Biogenic silica in the Pinaceae. *Soil Science* 126(3): 145–156. 3 - 45 samples of 14 spp from the northeastern US were studied with SEM, EDX, WDX and light microscopy. The low Si content varied within genera and between locations but not between crown positions on a tree. Si shapes, perhaps diagnostic of family and genera, are described in detail and compared with angiosperm trees and grass. 5 plates (51 photos), 3 tables.

Kobayashi, M 1986 (Motor cell-silica bodies of *Sasa* and allied genera from Hachijojima, Izu Islands, Japan, with a special reference to the origin of *Sasa kurilensis* in the Izu Islands.) *Journal of Phytogeography and Taxonomy* 34: 31–35. (Japanese, English summary.) Phytoliths from 3 species are identifiable. Silicified bulliform (motor) cells are described. 2 figures (6 photos).

Koeppen, RC 1980 Silica bodies in wood of arborescent Leguminosae. *International Association of Wood Anatomists Bulletin* 1: 180–184. Survey of 382 genera of woody Leguminosae indicates irregular spherical silica bodies form in 8 genera of subfamily Caesalpinoideae. 4 photos, 1 table, 1 key.

Komarek, EV, Komarek, BB and Carlysle, TC 1973 The ecology of smoke particulates and charcoal residues from forest and grassland fires: A preliminary atlas. *Tall Timbers Research Station*, Miscellaneous Publication No 3. Particulates from fires are examined to determine their role in atmospheric cleansing. Plant residues from controlled burns are illustrated by SEM. Phytoliths (opaliths) are present in tissue fragments and isolated particles. 136 figures.

Kondo, R 1974 (Opal phytoliths - the relationship between the morphological features of opal phytoliths and the taxonomic groups of gramineous plants.) *Pedorojisuto* (*Pedologist*) 18: 2–10. (Japanese). A general review of phytolith morphology based on previous literature. 3 figures, 1 table.

_____1976 (On the opal phytoliths of tree origins.) *Pedorojisuto (Pedologist)* 20: 176–190. (Japanese, English summary). Si content and morphology varies in phytoliths from tree leaves. Phytoliths from sediments were identified as tree in origin. 4 plates (22 photos), 1 table.

_____1977 Opal phytoliths, inorganic, biogenic particles in plants and soils. *Japan Agricultural Research Quarterly* 11: 198–203. Briefly compares phytoliths from grasses and trees. Short survey of application to sediments. 3 plates (18 photos), 1 table.

Kondo, R and Iwasa, Y 1981 Biogenic opals of humic yellow latosol and yellow latosol in the Amazon region. *Research Bulletin of Obihiro University* 12: 231–239. Phytoliths from 2 soil types were classified by shape to plant type (palm, conifer or deciduous trees, gramineous). Differences between the soil phytolith assemblages are correlated to original vegetation. 4 figures (43 photos), 1 table.

Kondo, R and Kasahara, Y 1935 Vergleichende Untersuchungen über Aschenbilder der Spelzen von *Chaetochloa, Panicum, Echinochloa, Sacciopsis* und *Syntherisma*. (Comparative investigation on ash-pictures of glumes of *Chaetochloa, Panicum, Echinochloa, Sacciopsis* and *Syntherisma.*) *Berichte des Ohara Instituts für Landwirtschaftliche Forschungen* 4: 491–513. (German). Ashed spodograms of hullspelze (inflorescence material) is compared for 5 genera. Cell shape and distribution (including phytoliths) are described. 3 plates (10 photos), 34 figures, 5 tables.

Kondo, R and Peason, T 1981 (Opal phytoliths in tree leaves. 2. Opal phytoliths in dicotyledonous angiosperm tree leaves.) *Research Bulletin of Obihiro University*, series 1, 12: 217–230. (Japanese, English summary). Morphology and deposition sites for 160 spp (46 families) of dicot angiosperms show patterns. Broad-leaf trees are distinguished from grasses, palms and conifers. 1 plate, 4 figures (68 types), 1 table.

Kondo, R and Sase, T 1986 (Opal phytoliths, their nature and application.) *Daiyonki Kenkyu (Quaternary Research)* 25: 31–63. (Japanese). Note the 5 plates (99 photos) of phytolith forms from plants.

Kondo, R and Sumida, T 1978 (The study of opal phytoliths of tree leaves. I. Opal phytoliths in gymnosperm and monocotyledonous angiosperm tree leaves.) *Nihon Deojeo Hiryeogaku Zasshi (Journal of the Science of Soil and Manure)* 49: 138–144. (Japanese). Lists phytolith types in 37 spp. Shape types are illustrated. 92 photos.

Kunoh, H and Akai, S 1977 Scanning electron microscopy and x-ray microanalysis of dumbbell-shaped bodies in rice lamina epidermis (*Oryza sativa*). *Torrey Botanical Club Bulletin* 104: 309–313. Presents the three-dimensional description of dumbbells. The major constituent is Si. 10 figures.

Kurmann, MH (ms) 1981 An opal phytolith and palynomorph study of extant and fossil soils in Kansas. MS thesis, Kansas State University: 79 p. The study is presented in two parts: phytoliths and palynomorphs. Microfossils were extracted from soils located under tall grass, shortgrass and woodland vegetation, as well as a paleosol. Differences in the phytolith assemblages are correlated to the vegetation. 49 figures (48 photos), 2 tables.

_____1985 An opal phytolith and palynomorph study of extant and fossil soils in Kansas (USA). *Palaeogeography, Palaeoclimatology, Palaeoecology* 49: 217–235. Phytoliths and palynomorphs in sediments from shortgrass prairie, tall grass prairie and a woodland were analyzed and compared to a paleosol. Phytolith assemblages differed between the three modern plant communities. 5 plates (53 photos), 2 tables.

Kutuzova, RS 1968 Silica transformation during the mineralization of plant residues. *Soviet Soil Science* 7: 970–978. Decomposition of plant tissues was studied to document the release of phytoliths. Spruce, birch, aspen and fern leaves treated to simulate decomposition released both solid phytoliths and amorphous Si. 3 figures (32 photos), 2 tables.

Labouriau, LG 1983 Phytolith work in Brazil, a minireview. *Phytolitharien Newsletter* 2(2): 6–11. Reviews 8 articles on phytolith systematics and others on taphonomy. 7 have camera lucida drawings of silica bodies from 133 grass spp and report massive silicification of large continuous portions of leaf epidermis, especially in cerrado grasses.

Labouriau, LG, Mosquim, PR and Morhy, L 1973 Desposição de silica nas folhas de *Casearia grandiflora* St. Hil. (Deposits of silica in leaves of *Casearia grandiflora* St. Hil.). *Anais Academia Brasileira de Ciencias* 45: 545–563. (Portuguese, English summary). From Labouriau 1983: Combination wet- and dry-ashed material (leaf, stem inflorescence) from herbarium specimens was compared to dry-ashed samples; no differences were seen. This cerrado tree (Flacourtiaceae) had very localized Si deposits in leaf epidermis, often shaped as half domes, that increased in number with leaf age. Other non-grasses with potentially distinctive phytoliths were in Palmae, Myrtaceae, Cyperaceae, Bromeliaceae and Dilleniaceae. Camera lucida drawings of silica bodies.

Lanning, FC 1960 Nature and distribution of silica in strawberry plants. *American Society for Horticultural Science, Proceedings* 76: 349–358. Separate parts (including achenes) of 8 cultivars and 3 wild spp were dry-ashed and studied with x-rays and light microscopy of spodograms. All parts had Si, with more in older plants. Si varied among varieties and soil types. Shapes are not discussed. 2 figures (5 photos), 7 tables.

_____1961a Silica and calcium in black raspberries. *American Society for Horticultural Science, Proceedings* 77: 367–371. Separate aerial parts of new and fruiting canes all had Si, with more in leaves, bark and in older plants. Shapes are not discussed. 3 figures (3 photos), 1 table.

_____1961b Calcite in *Lesquerella ovalifolia* trichomes. *Science* 133: 380. Trichomes are shown by XRD to contain calcite and a small amount of Si. Ash and calcium content of leaves and trichomes are determined. 1 photo.

_____1963 Silicon in rice. *Journal of Agriculture and Food Chemistry* 11: 435–437. Si content of plant parts and a spodogram of the leaf sheath are determined. Mineralogical analysis indicates all Si is opal. 5 figures (1 photo).

_____1964 Ash, silica, and calcium in *Lesquerella ovalifolia*. *Kansas Academy of Science, Transactions* 67: 481–485. Dry-ashed samples, like other Cruciferae, had high concentrations of calcite in trichomes on the leaves and stems, but not in seed pods. Si was less variable, occurring mostly in leaves, then roots, then stems. 4 collection locations did not vary in Si. 2 figures (2 photos), 3 tables.

_____1966a Silica and calcium deposition in the tissues of certain plants. *Advancing Frontiers of Plant Science* 13: 55–66. Si and calcium content of parts from 11 spp (grasses, sedges, other monocots and dicots) differed between samples. Distribution of particles is shown in spodograms. 6 figures (7 photos), 3 tables.

_____1966b Relation of silicon in barley to disease, cold, and pest resistance. *Journal of Agricultural and Food Chemistry* 14: 636–638. Different parts of 5 var were analyzed in the fall and spring. Si content and depositional patterns were determined. No direct relation between total Si content and various detrimental factors was detected. 1 figure (5 photos), 1 table.

_____1966c Relation of silicon in wheat to disease and pest resistance. *Journal of Agricultural and Food Chemistry* 14: 350–352. Si content and distribution in parts of 11 var were determined in fall and spring. No direct relation between total Si content and disease or pest resistance was found. 1 figure (6 photos), 1 table.

_____1972a Ash and silica in *Juncus*. *Torrey Botanical Club Bulletin* 99: 196–198. 2 spp were studied as spodograms and dry-ashed for x-ray diffraction. Rows of Si spheres and fibers were noted. 1 figure (2 photos).

_____1972b Ash, SiO_2, CaO, Fe_2O_3 and Al_2O_3 in *Helianthus*. *HortScience* 7: 184–186. Various parts of 2 dry-ashed spp were studied with XRD and light microscopy of spodograms. Ca and Si were highest in hispid hairs, then leaves, then achenes. Hair Si is depicted in blurred photos but not described or measured. 3 figures (7 photos), 1 table.

_____1977 Ash, silicon, calcium, and copper in "Concord" grape leaves. *HortScience* 12: 121–122. Dry-ashed samples and spodograms were studied with SEM and light microscopy. Ca occurred in veins and small oval particles in leaves. Si occurred in leaf tips, but shapes are not described. Si and Ca content increased with age. 2 figures (2 photos), 1 table.

Lanning, FC and Eleuterius, LN 1978 Silica and ash in the salt marsh rush, *Juncus roemerianus*. *Gulf Research Reports* 6: 169–172. Leaf Si content of 3 populations averages the same but rhizomes differ (related to soil Si content). Si content increases with age; maximum content is less than in *Juncus interior*. 2 figures (1 photo), 2 tables.

_____1981 Silica and ash in several marsh plants. *Gulf Research Reports* 7: 47–52. Ash and silica content of 3 grasses, a sedge and a dicot varied in amount and mineral composition. Silica cell distribution patterns are given. 2 figures (8 photos), 2 tables.

_____1983 Silica and ash in tissues of some coastal plants. *Annals of Botany* 51: 835–850. 44 southeastern US plants were studied with SEM, EDX and light microscopy. Si content varied with species, not with habitat. *Euchlaena*, *Tripsacum* and *Manisuris* had possible maize confusers. Descriptions are not detailed. 6 figures (36 photos), 5 tables.

_____1985 Silica and ash in tissues of some plants growing in the coastal area of Mississippi, USA. *Annals of Botany* 56: 157–172. Ash and silica content of 27 spp (grasses, sedges, other monocots and dicots) varied in amount and mineralogy. Deposition patterns indicate preference for certain epidermal areas--hairs, ridges, silica bodies. Inflorescences are especially heavily silicified and may be related to cancer. 7 figures (33 photos), 3 tables.

_____1987 Silica and ash in native plants of the central and southeastern regions of the United States. *Annals of Botany* 60: 361–375. 11 grasses and 1 composite were separated into different parts studied by SEM, EDX and light microscopy. Valuable discussion includes Si contents, depositional patterns and implications for health and archaeology. Among other conclusions, C_3 grasses produced more oval phytoliths than C_4. Text and caption give different identifications for Figure 1B. 5 figures (42 photos), 3 tables.

Lanning, FC and Garabedıan, T 1963 Distribution of ash, calcium, iron, and silica in the tissues of young *Sorghum* plants. *Kansas Academy of Science, Transactions* 66: 443–448. Ash and silica content of sorghum plant parts was determined in 8 hybrids. Si deposition is shown in spodograms. 1 figure (3 photos), 2 tables.

Lanning, FC, Hopkins, TL and Loera, JC 1980 Silica and ash content and depositional patterns in tissues of mature *Zea mays* L. plants. *Annals of Botany* 45: 549–554. Separate parts from 18 var were dry-ashed, then wet-ashed and studied with SEM; spodograms were studied with light microscopy. Si content was highest in leaf blades and sheaths and lowest in kernels, pith and ear husk. Content but not pattern varied with geographic location. Morphology is not discussed in detail. 3 plates (10 photos), 2 tables.

Lanning, FC and Linko, YY 1961 Absorption and deposition of silica by four varieties of sorghum. *Journal of Agricultural and Food Chemistry* 9: 463–465. Si content of 4 var was analyzed at 3-week intervals during growth. Differences in rate and amount correspond to insect and disease resistance. 3 figures (4 photos), 2 tables.

Lanning, FC, Ponnaiya, BWX and Crumpton, CF 1958 The chemical nature of silica in plants. *Plant Physiology* 33: 339–343. Examination of Si from 4 grasses and 2 dicots by XRD and petrographic microscopy indicates the mineral opal. Deposition patterns (location, cell types) are also reported. 12 figures (9 photos), 2 tables.

Laroche, J 1968 Contribution a l'étude de l'*Equisetum arvense* L. III. Recherches sur la nature et la localisation de la silice chez le sporophyte. (Contribution to the study of *Equisetum arvense* L. III. Research on the nature and localization of silica in the sporophyte.) *Revue Generale de Botanique* 75: 65–116. (French, English summary). Si content, composition and distribution in tissues for a horsetail. Si is continuous on the outer surface and decreases to the interior. 8 plates (29 photos), 16 figures, 1 table.

_____1969a Etude des concrétions siliceuses de l'épiderme de l'*Equisetum arvense* L. au microscope à balayage. (Study of silica bodies on the epidermis of *Equisetum arvense* L. by scanning electron microscope.) *Comptes Rendus Hebdomadaires des Seances de l'Academie des Sciences, Paris, Series D, Sciences Naturelles* 268: 2417–2418. (French). Si deposits on horsetail are formed first as vermiculaites concretions. Siliceous cement then forms a homogenous mass. 1 plate (4 figures).

_____1969b État de la silice sur et dans la membrane épidermique des organes aériens stériles d'*Equisetum arvense* L. (State of silica on and in the epidermal membrane of sterile aerial organs of *Equisetum arvense* L.) *Revue Generale de Botanique* 76: 483–489. (French, English summary). Si is deposited as microfibrils in epidermal spicules. 5 figures.

_____1977 La silice et les plantes supérieures. (Silica and higher plants.) *Revue de Cytologie et de Biologie Vegetales* 40: 15–45. (French, English summary). Review of Si formation in plants covers both extracellular and intracellular deposits. 14 photos.

Laroche, J, Guervin, C and Le Coq, C 1975 La silice chez une Equisétacée: *Equisetum maximum* Lamk. I. Distribution dans les organes aériens stériles. (Silica in one Equisetaceae: *Equisetum maximum* Lamk. I. Distribution in sterile aerial organs.) *Le Botaniste* 57: 61–74. (French, English summary). Silicification of the sterile aerial shoot of a horsetail proceeds from the epidermis to the interior. 3 plates (8 photos), 1 figure.

Laroche, J, Guervin, C, Le Coq, C and Dao, VT 1970 Intérêt taxinomique de l'excrétion siliceuse chez les Equisétacées. (Taxonomic interest in silica excretion in the Equisetaceae.) *Comptes Rendus Hebdomadaires des Seances de l'Academie des Sciences, Paris, Series D, Sciences Naturelles* 270: 2958–2960. (French). Epidermal Si concretions in *Equisetum* are constant, characteristic in arrangement and may be of taxonomic significance. 5 photos.

Laroche, J and Robert, D 1976 Mise en évidence, en microscopie électronique, du transit de silice a la faveur des parois dans les feuilles d'*Equisetum arvense* L. (Evidence in electronic microscopy of silica transmitted by means of the wall in the leaves of *Equisetum arvense.*) *Comptes-Rendus Hebdomadaires des Seances de l'Academie des Sciences, Paris, Series D, Sciences Naturelles* 282: 1721–1723. (French). Si is transported to the epidermis through the cell wall. 3 photos.

_____1979 Séquence des événements conduisant à la silicification de certaines ornementations chez la feuille de *Selaginella kraussiana.* (Steps leading to silicification of certain ornamentations in *Selaginella kraussiana* leaf.) *Comptes Rendus Hebdomadaires des Seances de l'Academie des Sciences, Paris, Series D, Sciences Naturelles* 288: 951–953. (French, English abstract). Silicified ornamentations are first callosic in nature. Callose fills empty spaces and then converts to mineral stage. 3 figures.

Laroche, J, Robert, D, Le Coq, C and Guervin, C 1972 Apports de la cytochimie, de la cytophysique, des microscopies électroniques par transmission et à balayage, à l'étude des phénomènes de silicification chez la feuille de *Selaginella kraussiana.* (Reports of the cytochemistry and cytophysiology, by transmission and scanning electron microscope, on the study of the phenomenon of silicification in the leaf of *Selaginella kraussiana.*) *Comptes Rendus Hebdomadaires des Seances de l'Academie des Sciences, Paris, Series D, Sciences Naturelles* 275: 2671–2674. (French). Silicification follows two steps. First, Si substitutes for callose in structures; then the external epidermis becomes silicified. 2 plates (11 photos).

Lau, E, Goldoftas, M, Baldwin, VD, Dayanandan, P, Srinivasan, J and Kaufman, PB 1978 Structure and localization of silica in the leaf and internodal epidermal system of the marsh grass *Phragmites australis.* *Canadian Journal of Botany* 56: 1696–1701. Samples were studied with SEM and x-ray, and wet-ashed for light microscopy. Si was concentrated in guard cells, Si cells and OTW of long cells but not in trichomes. Si body shapes and dimensions are not described. 11 figures (11 photos).

Lawton, JR 1980 Observations on the structure of epidermal cells, particularly the cork and silica cells, from the flowering stem internode of *Lolium temulentum* L. (Gramineae). *Linnean Society of London, Botanical Journal* 80: 161–178. Development of hairs, stomata, cork cells and silica cells is described in *Lolium temulentum.* Transportation and function of Si is not well known. 21 figures (19 photos).

Le Cohu, MC 1973 Examen au microscope électronique à balayage, des cônes de silice chez les Cypéracées. (Scanning electron microscope examination of silica cones in the Cyperaceae.) *Comptes Rendus Hebdomadaires des Seances de l'Academie des Sciences, Paris, Series D, Sciences Naturelles* 277: 1301–1303. (French). Silica cones in sedges develop from a basal socket. The variety correlates somewhat to different growth stages. 1 plate (6 photos), 10 figures.

Lewis, RO 1981 Use of opal phytoliths in paleoenvironmental reconstruction. *Journal of Ethnobiology* 1: 175–181. Phytolith analysis provides paleoenvironmental information in areas lacking pollen preservation. Some examples from the US High Plains are based on the Twiss, Suess and Smith (1969) classification. 4 figures.

Lloyd, FE 1921 The structure of cereal straws. *Pulp and Paper Magazine* 19: 973–976. Description includes limited mention of silicified cells. Epidermal patterns for stem and leaf sheath of barley, oat, rye and wheat are shown. 14 figures (12 photos), 1 table.

Luhan, M 1963 Die epidermis von *Agropyron repens.* (The epidermis of *Agropyron repens.*) *Protoplasma* 56: 645–660. (German). Description of epidermal morphology includes phytoliths. 8 figures.

MacDonald, LL (ms) 1974 Opal phytoliths as indicators of plant succession in northcentral Wyoming. MS thesis, University of Wyoming: 71 p. 11 grasses were dry-ashed and compared to soil samples from 2 sites. The comparative collection generally fit the patterns in Twiss, Suess and Smith (1969). The sites showed domination by grasses with possible changes in composition. 42 figures (36 photos), 5 tables, 6 appendices.

Marumo, Y and Yanai, H 1986 Morphological analysis of opal phytoliths for soil discrimination in forensic science investigation. *Journal of Forensic Sciences* 31: 1039–1049. Soil from different land uses is differentiated by phytolith content even when mineralogical characteristics were similar. Grass and tree morphological types are illustrated. 7 figures, 2 tables.

Matsutani, A 1972 Spodographic analysis of ash from the Kotosh site: A preliminary report. *In* Izum, S and Terada, K, eds, *Andes 4: Excavations at Kotosh, Peru, 1963 and 1966*. Tokyo, University of Tokyo Press: 319–326. Ash from the Kotosh site contained numerous spodograms (silica skeletons) identified to the Gramineae and *Equisetum* sp. The comparative collection included 4 horsetails, 19 grasses and 2 sedges. 12 photos.

Mehra, PN and Sharma, OP 1965 Epidermal silica cells in the Cyperaceae. *Botanical Gazette* 126: 53–58. The cyperaceous phytolith is defined as conical bodies from sedges (60 spp, 13 genera); it is distinct from grass shapes. 18 figures (3 photos), 2 tables.

Metcalfe, CR 1960 *Anatomy of the Monocotyledons. I. Gramineae.* Oxford, Clarendon Press: 731 p. Definitive research on grass anatomy includes phytolith distribution and morphology for numerous genera. Lists of genera containing certain characters are presented. 24 figures.

_____1961 The anatomical approach to systematics: General introduction with special reference to recent work on monocotyledons. In *Recent Advances in Botany*, Ninth International Botanical Congress: 146–150. Toronto, University of Toronto Press. Anatomical data are useful in determining taxonomic relationships. Crystals and Si are present in the monocots; presence and morphology is noted for some families.

_____1971 *Anatomy of the Monocotyledons. V. Cyperaceae.* Oxford, Clarendon Press: 597 p. Definitive description of sedge anatomy includes phytolith distribution and morphology information. Anatomical descriptions of 108 genera are presented. 69 figures.

_____1983 Secreted mineral substances. *In* Metcalfe, CR and Chalk, L, *Anatomy of the Dicotyledons. II. Wood Structure and Conclusion of the General Introduction.* Second edition, Oxford, Clarendon Press: 82–97. General review covers Si and Ca deposits and crystals in dicot families. 3 figures.

Metcalfe, CR and Chalk, L 1950 *Anatomy of the Dicotyledons.* Oxford, Clarendon Press: 1500 p. Comprehensive description of 263 dicot families covers anatomy of leaf, stem, wood and root material. Silica bodies and crystal inclusions are noted; families with crystals are listed (pp 1343–1346). 3 plates (20 photos), 317 figures, 1 appendix.

_____1979 *Anatomy of the Dicotyledons. I. Systematic Anatomy of Leaf and Stem, with a Brief History of the Subject.* Second edition, Oxford, Clarendon Press: 276 p. Scattered references to crystals and cystoliths. 18 plates (119 photos).

_____1983 *Anatomy of the Dicotyledons. II. Wood Structure and Conclusion of the General Introduction.* Second edition, Oxford, Clarendon Press: 197 p. Treatment of Si and Ca bodies. Families listed by type of deposit on pp 215–221. 11 plates (69 photos).

Mika, V 1987 Koncentrace a formy křemíku u srhy říznačky (*Dactylis glomerata* L. [Silicon concentrations and forms in cocksfoot (*Dactylis glomerata* L.).] *Rostlinna Vyroba* (*Plant Production*) 33: 801–806. (Czechoslovakian, English abstract). Si concentration increased with age but soluble Si concentration decreased. 4 tables.

Milanez, FR and Filho, AM 1956 Nota sôbre a ocorrência de sílica no lenho de leguminosas. (Note on an occurrence of silica in legumes.) *Rodrigusia* 18(30): 7–26. (Spanish, English abstract). Si particles occur in secondary wood of some woody legumes. Location and co-occurrence with oxalate crystals are noted. 9 photos.

Milby, TH 1971 The leaf anatomy of buffalo grass, *Buchloe dactyloides* (Nutt.) Engelm. *Botanical Gazette* 132: 308–313. Thin sections and epidermal peels showed saddles and some hairs typical of chloridoid grasses; unicellular hairs and crescentic silica cells on the sheath were unlike chloridoids. 10 figures (10 photos).

Miles, SR (ms) 1973 Vegetative history and boundary stability of park areas in the Medicine Bow National Forest. MS thesis, University of Wyoming: 69 p. Boundaries of two open grass parks were investigated by analysis of soil phytolith content, among other measurements. Analysis of 8 spp for phytolith shapes is reported. 22 figures, 1 table.

Miles, SR and Singleton, PC 1975 Vegetative history of Cinnabar Park in Medicine Bow National Forest, Wyoming. *Soil Science Society of America, Proceedings* 39: 1204–1208. Boundaries of an open grass park within a forested area were examined by various methods, including phytolith content. The park is shifting slowly eastward, as indicated by amounts of opal in the park and surrounding forest. 6 figures (7 photos).

Montgomery, DJ and Parry, DW 1979 The ultrastructure and analytical microscopy of silicon deposition in the intercellular spaces of roots of *Molinia caerulea* (L.) Moench. *Annals of Botany* 44: 79–84. One grass was studied by TEM and Cora analytical microscopy. Si deposits began as spheres and eventually filled the intercellular spaces just external to the endodermis. Taxonomic significance not discussed. 3 plates (12 photos), 1 figure.

Moody, UL (ms) 1972 Phytoliths as an interpretive device of paleoenvironments in archaeological sites. MA thesis, University of Montana: 67 p. Investigation of phytoliths from a bog and an archaeological site for paleoenvironmental reconstruction. Seven plants (grasses, herbs, a pine) analyzed; sketches of plant and sediment phytoliths. 17 figures, 5 tables.

Mulholland, SC 1986a Phytolith studies at Big Hidatsa, North Dakota: Preliminary results. *In* Clambey, GK and Pemble, RH, eds, *The Prairie: Past, Present and Future. Proceedings of the Ninth North American Prairie Conference, July 29 to August 1, 1984, Moorhead, Minnesota*. Tri-College University Center for Environmental Studies, Fargo, North Dakota: 21–24. Examination of Big Hidatsa sediments indicates that cross-shaped phytoliths (a type common in maize) are rare. Types found in cob chaff, as well as many wild grasses, are more abundant. Sediment provenience is a key factor for interpretation of sediment phytolith assemblages. 5 figures (5 photos).

_____1986b Classification of grass silica phytoliths. *In* Rovner, I, ed, Plant opal phytolith analysis in archaeology and paleoecology: Proceedings of the 1984 Phytolith Research Workshop, North Carolina State University, Raleigh, North Carolina, *Occasional Papers of The Phytolitharien* 1: 41–52. Nine major shape categories, based on leaf samples studied with light microscopy and SEM, are defined and illustrated. Shape distributions in 41 spp show some differences from Twiss, Suess and Smith (1969). 8 figures (8 photos), 2 tables.

_____1989 Phytolith shape frequencies in North Dakota grasses: A comparison to general patterns. *Journal of Archaeological Science* 16: 489–511. 19 wet-ashed spp reflected usual patterns, but shapes indicating Pooideae, Panicoideae and Chloridoideae also occurred in other subfamilies in significant amounts. Silica cell phytoliths are clearly described in detail. 1 figure (8 photos), 11 tables.

_____1990 *Arundo donax* phytolith assemblages. *Phytolitharian Newsletter* 6(2): 3–9. Crosses were found that mimic maize in all three criteria reported as corn-specific. Other maize mimics include wild rice (*Zizania aquatica*) and cut-grass (*Leersia oryzoides*). Shape classification by length-width ratio is considered more reliable than by absolute measurements. 4 figures, 2 tables.

Mulholland, SC and Rapp, G, Jr 1985 Grass silica phytoliths. *Society for Archaeological Sciences Newsletter* 8(2): 5–6 (Research Report 5). Brief classification and distribution of silica bodies from grasses. Distribution in four subfamilies is compared to Twiss, Suess and Smith (1969). 1 table.

____1989 Characterization of grass phytoliths for archaeological analysis. *Materials Research Society Bulletin* 14(3): 36–39. Errata, *MRS Bulletin* 14(5): 50. Introductory article summarizing distribution patterns. Provides useful definitions and diagrams of phytolith shapes. See errata notice. 2 figures (8 photos), 1 table.

Mulholland, SC, Rapp, G, Jr and Gifford, JA 1982 Phytoliths. *In* Rapp, G, Jr and Gifford, JA, eds, *Troy: The Archaeological Geology*. Princeton University Press: 117–137. Sediments from Troy indicate *Arundo donax* and other grasses may have been present. Although 35 plants were examined, the study focuses on grasses. The classification of grass silica bodies has since been revised extensively. 8 photos, 3 tables.

Mulholland, SC, Rapp, G, Jr and Ollendorf, AL 1988 Variation in phytoliths from corn leaves. *Canadian Journal of Botany* 66: 2001–2008. Sets of 100 phytoliths from 24 samples of wet-ashed Mandan maize leaves showed ratios of dumbbells to crosses vary consistently with leaf part (base, midsection and tip) and corn variety but not with leaf node, individual plant or geographical location. Three-dimensional morphology and cross size were not considered. 7 figures (3 photos), 6 tables.

Murphy, P 1986 Botanical evidence [Lawson, AJ, The excavation of a ring-ditch at Bowthorpe, Norwich, 1979]. In Lawson, AJ, ed, *Barrow Excavations in Norfolk 1950–82*, East Anglian Archaeology Report No 29: 43–45. Analysis of a stain beneath a skull indicates dumbbell-producing grasses contributed to the sediment. Comparison to modern sediments and a cereal ash were made. 1 figure.

Murthy, LSV 1965 Silica in Sarawak timbers. *Malayan Forester* 28: 27–45. Distribution of Si is indicated for 531 spp in 41 families. Occurrence in fibers, wood parenchyma or ray cells is listed for 98 spp. 3 figures, 2 tables.

Namba, T and Bae, KH 1981 Pharmacognostical studies on the crude drug "zhú-yè" and the bambusaceous plants V. The comparative anatomical studies on the leaves of the genera *Semiarundinaria*, *Sinobambusa*, *Shibataea*, *Tetragonocalamus*, and *Chimonobambusa*. *Shoyakugaku Zasshi* (*Japanese Journal of Pharmacognosy*) 35: 205–216. Anatomical characteristics differentiate leaves of several bamboo genera. Location and size of silica cells is included. 6 figures.

Nanko, H and Cote, WA 1980 Crystals and other inclusions in the phloem. *In* Nanko, H and Cote, WA *Bark Structure of Hardwoods Grown on Southern Pine Sites*. Syracuse, Syracuse University Press: 47–51. Crystals are described for 13 hardwood spp. Calcium oxalate is most common; other crystals and amorphous Si occur. 18 photos.

Netolitzky, F 1900 Mikroskopische Untersuchung gänzlich verkohlter vorge-schichtlicher Nahrungsmittel aus Tirol. (Microscopic investigation of fully carbonized prehistoric foods in the Tirol.) *Zeitschrift fur Untersuchungen Nahrungs-und Genussmittel* 12: 401–407. (German). Discusses cell infillings but not Si specifically.

____1911 Verkieselungen bei den Rubiaceae-Galieae. (Silicification in the Rubiaceae-Galieae.) *Osterreichische Botanische Zeitschrift* 61: 409–412. (German). Discusses silicification in 5 genera--*Rubia*, *Galium*, *Sherardia*, *Crucianella*, *Asperula*. No illustrations.

____1912 Kieselmembranen der Dicotyledonenblätter Mitteleuropas. (Silica membrane of middle Europe Dicotyledon leaves.) *Osterreichische Botanische Zeitschrift* 62: 353–359. (German). Discussion of silicification in 29 dicotyledon families. No illustrations.

____1929 Die Kieselkörper. (Silica bodies.) *Handbuch der Pflanzenanatomie* 3(1a): 1–19, 101–118. (German). Silica phytolith types, chemistry and nature are reviewed. Note that calcium crystals are reviewed in "Die kalksalze as zellinhaltskörper" in the same issue. 26 figures, 2 tables.

Newman, RH 1983 Asbestos-like fibres of biogenic silica in sugar cane. *The Lancet* 2: 857. Sharp fibers from sugarcane are similar to asbestos and may be implicated in carcinogenesis.

Diameters less than 1.5 um and lengths greater than 8 um can penetrate the lungs. 2 figures (1 photo).

Newman, RH and Mackay, AL 1983 Silica spicules in canary grass. *Annals of Botany* 52: 927–929. Hairs on the fruits were studied with SEM and EDX and described in detail, with a discussion of health implications for humans. 1 figure (6 photos).

Norgren, JA (ms) 1973 Distribution, form and significance of plant opal in Oregon soils. PhD dissertation, Oregon State University: 165 p. 15 conifer spp, 5 grasses and 1 sedge were studied by light microscopy. Conifer Si content ranged from trace to 7.9%, varying between locations. Shapes were possibly distinctive in spruce and Douglas fir. The grasses had very high Si content (17–23%), but shapes were not addressed. Several soils had very high quantities of silt- and clay-sized phytoliths. 10 figures (20 photos), 10 tables, 2 appendices.

Norton, BE 1967 Occurrence of silica in *Lepidosperma limicola*. *The Australian Journal of Science* 29: 371–372. Stomatal guard cells and epidermal walls are reported to be silicified, as well as the "cyperaceous" type of hat. In addition, silicified hook-like projections on leaf margins provide a razor-sharp edge. 2 figures (4 photos).

Ohki, K 1932 On the systematic importance of spodograms in the leaves of Japanese Bambusaceae. *Tokyo Daigaku. Rigakubu. Journal of the Faculty of Science, University of Tokyo, Section 3, Botany* 4: 1–129. Spodogram characteristics, including silica cells, are described for 51 spp of 11 genera of bamboos. Shape, distribution and size information are provided for epidermal cells. 43 figures, 6 tables.

_____1939 (Studies on the systematic importance of spodograms in the leaves of Gramineae I.) *Botanical Magazine of Tokyo* 53: 208–212. (Japanese). Illustrates epidermal anatomy (including phytoliths) for *Zizania aquatica* and *Arundo donax*. 10 figures.

Okamoto, Y 1963 (Effects of silicic acid absorbed from the surface of the leaves upon the paddy rice plant growth.) *Nihon Sakumotsa Gakkai Kiji. (Proceedings of the Crop Science Society of Japan)* 32: 61–65. (Japanese, English summary). Si absorbed through leaves (by means of spraying) gives larger and heavier plants (with more phytoliths) than plants lacking Si or with Si in the culture. 10 figures (8 photos), 2 tables.

Ollendorf, AL (ms) 1986 A study of phytoliths from Philistine levels at Tel Miqne (Ekron), Israel. MS thesis, University of Minnesota: 180 p. Separate parts of 9 sedges, 2 rushes and 57 grasses from Israel, and 7 related spp from elsewhere. Sedges had distinctive phytoliths; rushes had little Si. Grass subfamilies showed trends in shape distributions. Archaeological samples showed possible functional differences. Appendices form a manual for procedures. 24 figures (26 photos), 11 tables, 5 appendices.

_____1987 Archaeological implications of a phytolith study at Tel Miqne (Ekron), Israel. *Journal of Field Archaeology* 14: 453–463. Phytolith analysis at Tel Miqne (the Philistine Ekron) focused on identification of occupation surfaces and contributory plants. Analysis of the grass reference collection includes correspondence analysis of the assemblages within subfamilies. 4 figures (5 photos), 4 tables.

Ollendorf, AL, Mulholland, SC and Rapp, G, Jr 1987 Phytoliths from some Israeli sedges. *Israel Journal of Botany* 36: 125–132. 8 spp were wet-ashed and studied with light microscopy and SEM. Round and hexagonal cones were found singly and in groups. Other Si fragments seen were not typical of grasses. 4 figures (16 photos), 1 table.

_____1988 Phytolith analysis as a means of plant identification: *Arundo donax* and *Phragmites communis*. *Annals of Botany* 61: 209–214. 11 wet-ashed leaf samples showed 2 mutually exclusive patterns of Si body shape in these reedgrasses, which are otherwise indistinguishable without inflorescences. 2 figures (7 photos), 3 tables.

O'Neill, CH, Hodges, GM, Riddle, PN, Jordan, PW, Newman, RH, Flood, RJ and Toulson, EC 1980. A fine fibrous silica contaminant of flour in the high oesophageal cancer area of north-east Iran. *International Journal of Cancer* 26: 617–628. Silicified hairs from *Phalaris* inflorescence bracts are implicated in the etiology of esophageal cancer. Sharp, tapering fibers of a certain size range break off during milling, forming a contaminant in flour. 4 figures (13 photos), 4 tables.

O'Neill, C, Jordan, P, Bhatt, T and Newman, R 1986 Silica and oesophageal cancer. *In Silicon Biochemistry*. Ciba Foundation Symposium No 121, Chichester, UK, John Wiley and Sons: 214–243. Narrow fibers 250 um in length stimulate growth of fibroblasts in culture: similar fibers released by *Phalaris* (Iran), millet (China), sugar cane (US, India) and plants in South Africa are correlated to high cancer rates. A preparation of such fibers has been observed to promote cancer in mice. 5 figures.

O'Neill, C, Pan, QQ, Clarke, G, Liu, F-S, Hodges, G, Ge, M, Jordan, P, Chang, Y-M, Newman, R and Toulson, E 1982. Silica fragments from millet bran in mucosa surrounding oesophageal tumors in patients in northern China. *Lancet* 1: 1202–1206. Fragments of millet silicified hairs occur in mucosal samples around tumors. The fibers may provide anchorage for proliferation of cancer cells. 5 figures (5 photos).

Palmer, PG 1976 Grass cuticles: A new paleoecological tool for East African lake sediments. *Canadian Journal of Botany* 54: 1725–1734. Grass epidermal leaf fragments from lake cores can provide paleoecological information. Preservation of many epidermal characters, including silica deposits, allows identification of grass taxa. 15 figures.

Palmer, PG and Gerbeth-Jones, S 1986 A scanning electron microscope survey of the epidermis of East African grasses: IV. *Smithsonian Contributions to Botany*, No 62, Washington, DC, Smithsonian Institution Press: 120 p. Fourth volume in SEM survey covers tribes Arundinelleae, Isachneae and Paniceae. Epidermal anatomy includes phytolith distribution and morphology. 86 plates (344 photos).

_____1988 A scanning electron microscope survey of the epidermis of East African grasses, V, and West African supplement. *Smithsonian Contributions to Botany*, No 67, Washington, DC, Smithsonian Institution Press: 157 p. Fifth volume in SEM survey covers tribes Andropogoneae and Maydeae; West African supplement covers tribes Bambuseae, Agrostideae, Aeluropodeae, Arundinelleae, Paniceae and Andropogoneae. 120 plates (720 photos).

Palmer, PG, Gerbeth-Jones, S and Hutchinson, S 1985 A scanning electron microscope survey of the epidermis of East African grasses: III. *Smithsonian Contributions to Botany*, No 55, Washington, DC, Smithsonian Institution Press: 136 p. Third volume in SEM survey covers tribes Aristideae to Zoysieae. 98 plates (588 photos).

Palmer, PG and Tucker, AE 1981 A scanning electron microscope survey of the epidermis of East African grasses: I. *Smithsonian Contributions to Botany*, No 49, Washington, DC, Smithsonian Institution Press: 84 p. Introductory volume for SEM survey covers tribes Bambuseae to Centosteceae. 48 plates (288 photos).

_____1983 A scanning electron microscope survey of the epidermis of East African grasses: II. *Smithsonian Contributions to Botany*, No 53, Washington, DC, Smithsonian Institution Press: 72 p. Second volume in SEM survey covers tribes Poeae through Stipeae. 52 plates (312 photos).

Parry, DW and Hodson, MJ 1982 Silica distribution in the caryopsis and inflorescence bracts of foxtail millet (*Setaria italica* (L.) Beauv.) and its possible significance in carcinogenesis. *Annals of Botany* 49: 531–540. Si is concentrated in papillae and epidermal cells or the tissues. Sharp fragments can promote esophageal cancer. 3 figures (18 photos).

Parry, DW, Hodson, MJ and Newman, RH 1985 The distribution of silicon deposits in the fronds of *Pteridium aquilinum* L. *Annals of Botany* 55: 77–83. Si deposition in epidermal cell walls and mid-rib laminae was determined by SEM and XRD. Different deposition mechanisms and functions are proposed. 5 figures (13 photos).

Parry, DW, Hodson, MJ and Sangster, AG 1984 Some recent advances in studies of silicon in higher plants. *Philosophical Transactions of the Royal Society of London, Series* B 304:

537–549. Review covers Si deposition patterns and mechanisms but not taxonomic significance. Human health care is discussed. 4 figures (28 photos), 2 tables.

Parry, DW and Kelso, M 1975 The distribution of silicon deposits in the roots of *Molinia caerulea* (L.) Moench. and *Sorghum bicolor* (L.) Moench. *Annals of Botany* 39: 995–1001. Si deposition in endodermal walls and other tissues differs between species. Wall deposits are similar to leaf epidermal deposits while lumina aggregates appear to be a different type. Development and function of Si is briefly discussed. 2 plates (24 photos), 1 figure.

_____1977 The ultrastructure and analytical microscopy of silicon deposits in the roots of *Saccharum officinarum* (L.). *Annals of Botany* 41: 855–862. Si deposition in root endodermis is analyzed by SEM and EPM. Location and development is compared to that of *Sorghum bicolor*. 3 plates (14 photos), 1 figure, 1 table.

Parry, DW, O'Neill, CH and Hodson, MJ 1986 Opaline silica deposits in the leaves of *Bidens pilosa* L. and their possible significance in cancer. *Annals of Botany* 58: 641–647. Si fibers in microhair walls are described from SEM and EPM. The sharply pointed structures have been implicated in esophageal cancer. 2 figures (15 photos), 1 table.

Parry, DW, and Smithson, F 1957 Detection of opaline silica in grass leaves. *Nature* 179: 975–976. Techniques for exposing phytoliths in plants affect the amount and detail of phytoliths observed. Acid baths, epidermal separation and staining are briefly discussed. 5 figures.

_____1958a Silicification of branched cells in the leaves of *Nardus stricta* L. *Nature* 182: 1460–1461. Wet-ashed samples of this grass showed unique branched phytoliths with potential taxonomic significance. 5 figures (5 photos).

_____1958b Silicification of bulliform cells in grasses. *Nature* 181: 1549–1550. Clear description and diagrams of bulliform phytoliths. 7 figures (6 photos).

_____1958c Techniques for studying opaline silica in grass leaves. *Annals of Botany* 22: 543–549. Discussion of preparation techniques focuses on lab and microscopy methods of making phytoliths conspicuous. 2 plates (21 photos).

_____1963 Influence of mechanical damage on opaline silica deposition in *Molinia caerulea* L. *Nature* 199: 925–926. Mechanical damage to apical portions of leaves caused abnormal silicification of bulliform cells in some replicates. Other damage treatments did not cause abnormalities; none of the treatments caused changes in deposition of dumbbell silica bodies. 1 figure (2 photos).

_____1964 Types of opaline silica depositions in the leaves of British grasses. *Annals of Botany* 28: 169–185. Useful introduction to light microscopy of phytoliths with clear descriptions and diagrams of basic Si shapes (80 spp); reiterates the authors' 1958a article. Their goal of providing a key was stymied by further research questions. 4 plates (91 photos), 4 figures, 2 tables.

_____1966 Opaline silica in the inflorescences of some British grasses and cereals. *Annals of Botany* 30: 525–538. Phytoliths from 37 spp are described and classified by cell type. Some types common to leaves are rare in inflorescences. 3 plates (36 photos), 5 figures.

Parry, DW and Soni, SL 1972 Electron-probe microanalysis of silicon in the root of *Oryza sativa* L. *Annals of Botany* 36: 781–783. Si deposition in rice roots occurs in the endodermis. Heaviest deposits occur on the ITW, with some also on radial walls. 1 plate (4 photos).

Parry, DW and Winslow, A 1977 Electron-probe microanalysis of silicon accumulation in the leaves and tendrils of *Pisum sativum* (L.) following root severance. *Annals of Botany* 41: 275–278. Pea seedlings grown without roots in a Si-rich nutrient took up Si, unlike the control plants. Neither group made phytoliths. A mechanism inhibiting Si uptake is inferred at the external surface of the root. 1 plate (12 photos).

Pearsall, DM 1978 Phytolith analysis of archaeological soils: Evidence for maize cultivation in Formative Ecuador. *Science* 199: 177–178. Leaves of 9 maize races and 10 wild grass genera were studied by light microscopy. All but one maize race, and none of the wild grasses, had large crosses. 10 archaeological samples had large crosses, indicating maize cultivation by 2450 BC. 1 table.

_____(ms) 1979 The application of ethnobotanical techniques to the problem of subsistence in the Ecuadorian Formative. PhD dissertation, University of Illinois: 270 p. Separate parts of an unspecified number of species (mostly grasses) were studied with light microscopy. Results were combined with analyses of seeds, charcoal, site catchment and other bioarchaeological data. Conclusions include probable maize cultivation and cultural selection of leguminous wood. 17 figures (25 photos), 13 tables, 3 appendices.

_____1982 Maize phytoliths: A clarification. *Phytolitharien Newsletter* 1(2): 3. Shape and size criteria for maize identification are described in detail.

_____1989 Phytolith analysis. *In* Pearsall, DM, *Paleoethnobotany: A Handbook of Procedures*. New York, Academic Press: 311–438. Comprehensive presentation of phytolith analysis as applied to paleoethnobotany covers topics of phytolith occurrence in plants, archaeobotanical history, status of research, field and lab methods and interpretation. Review of classification is thorough and provides the most recent introduction to the subject. 39 figures (41 photos), 15 tables.

_____1990 Maize imitators: How great a problem? A comment on "*Arundo donax* phytolith assemblages" by S.C. Mulholland. *Phytolitharien Newsletter* 6(2): 10–11. This maize mimic is a problem only in post-Columbian contexts in areas where the plant has naturalized, but is a reminder of the need for extensive local comparative research and detailed study of three-dimensional morphology.

Pearsall, DM and Piperno, DR 1990 Antiquity of maize cultivation in Ecuador: Summary and reevaluation of the evidence. *American Antiquity* 55: 324–337. Clear review of identification procedures for tropical maize. Re-examined samples show that maize was in coastal Ecuador by 5000 BC. 6 figures (1 photo), 2 tables.

Pease, DS (ms) 1967 Opal phytoliths as indicators of paleosols. MS thesis, New Mexico State University: 81 p. Separate parts of 5 grasses, 3 conifers, 2 cactus, yucca, creosote bush and *Ephedra* from the southwest US were studied with XRD and light microscopy. Black grama roots and conifers produced possibly distinctive phytoliths. Phytoliths ranged from trace to 0.001% in 4 paleosols. 45 figures (36 photos), 10 tables, 4 appendices.

Pease, DS and Anderson, JU 1969 Opal phytoliths in *Bouteloua eriopoda* Torr. roots and soils. *Soil Science Society of America, Proceedings* 33: 321–322. Dry-ashed samples were studied with x-ray and light microscopy. Rectangular phytoliths, considered unique among 12 other (unnamed) spp, were recovered from sediment. 2 figures (2 photos).

Perry, CC, Mann, S and Williams, RJP 1984 Structural and analytical studies of the silicified macrohairs from the lemma of the grass *Phalaris canariensis* L. *Proceedings of the Royal Society of London, Series* B 222: 427–438. Silicification of macrohairs follows a sequence of silica particles and inorganic element distribution. Cellular processes are proposed to control the silicification. 10 figures (16 photos).

Perry, CC, Mann, S, Williams, RJP, Watt, F, Grime, GW and Takacs, J 1984 A scanning proton microprobe study of macrohairs from the lemma of the grass *Phalaris canariensis* L. *Proceedings of the Royal Society of London, Series* B 222: 439–445. Distribution of inorganic elements in macrohairs indicates spatial organization varies with silicification stage and low quantities in mature hairs. Silicification may be associated with cellular activity. 3 figures (18 photos), 2 tables.

Perry, CC, Williams, RJP and Fry, SC 1987 Cell wall biosynthesis during silicification of grass hairs. *Journal of Plant Physiology* 126: 437–448. Analysis of hairs on *Phalaris canariensis* lemmas indicates both ultrastructure of silica and organic phase composition changes during development. Changes in polysaccharide synthesis may direct changes of Si ultrastructure during secondary wall formation. 4 figures, 2 tables.

Peters, I 1968 Opalphytolithe, ihre Brauchbarkeit und Verwendungsmoglichkeiten als Pflanzliche Mikrofossilien. (Opal phytoliths, their usefulness and potential as plant microfossils.) *Palaeontographica*, Abt. B. 123: 243–256. (German, English summary). Investigation of chernozem and podzolic soils in central Europe shows some differences in phytolith content. Si content of 15 grasses is described. 4 plates (23 photos).

Peterson, I 1983 Plant stones. *Science News* 124: 88–89, 94. Brief general review article presents an overview of phytolith research in early 1980s. Phytoliths from some grasses and sedges are illustrated. 4 photos.

Piperno, DR (ms) 1979 Phytolith analysis of archaeological soils from central Panama. MA thesis, Temple University: 99 p. Analysis of material from two sites indicates maize in the ceramic levels of one. Phytolith morphology of grasses, herbs, conifers and deciduous trees is discussed. 17 figures, 5 tables.

_____(ms) 1983 The application of phytolith analysis to the reconstruction of plant subsistence and environments in prehistoric Panama. PhD dissertation, Temple University: 459 p. Leaves of 328 spp in 54 families were studied by light microscopy and SEM. Classifies Si by origin in plant and by shape. Much has been published in later articles, but the unpublished plant sampling chapter is helpful for those starting in regions without previous phytolith work. 51 plates (202 photos), 10 figures, 3 appendices.

_____1984 A comparison and differentiation of phytoliths from maize and wild grasses: Use of morphological criteria. *American Antiquity* 49: 361–383. Leaves of 20 maize races, 2 teosinte spp, and 40 wild Panicoid and Bambusoid grasses were studied by light microscopy and SEM. Consistent identification of maize depended on: 1) high percentages of large crosses; 2) three-dimensional morphology of crosses; and 3) low ratios of dumbbells to crosses. Other grasses with potentially distinctive phytoliths were found. Central Panamanian sites had maize phytoliths in levels dating to about 5000 BC. 3 figures (12 photos), 4 tables.

_____1985a Further comments on the maize-diagnostic phytolith. *Phytolitharien Newsletter* 3(2): 3. Work was in progress on research published in Pearsall and Piperno (1990).

_____1985b Phytolith analysis and tropical paleo-ecology: Production and taxonomic significance of siliceous forms in New World plant domesticates and wild species. *Review of Palaeobotany and Palynology* 45: 185–228. Essential reference. 17 crops and 365 wild plants (260 dicots) were studied by SEM and light microscopy. Few crops produced significant phytoliths. Wild monocots produced fewer phytoliths than previously thought, but production in some dicot families was abundant and morphologically diverse. Even phytoliths with little taxonomic value are illustrated and clearly described. 21 plates (84 photos), 3 tables.

_____1985c Phytolith taphonomy and distributions in archaeological sediments from Panama. *Journal of Archaeological Science* 12: 247–267. Phytolith analysis at four Panamanian sites reveals stable deposition and preservation of a wide variety of silicified plant structures. Evidence of maize and squash, as well as other monocots and dicots, illustrates the strengths and weaknesses of this method. 3 figures (14 photos), 5 tables.

_____1986 A survey of phytolith production and taxonomy in non-graminaceous plants: Implications for paleoecological reconstruction. *In* Rovner, I, ed, Plant opal phytolith analysis in archaeology and paleoecology: Proceedings of the 1984 Phytolith Research Workshop, North Carolina State University, Raleigh, North Carolina, *Occasional Papers of The Phytolitharien* 1: 35–40. Expansion of information in Piperno (1985b), based on 500 spp; most published later in Piperno (1988). Errata noted in *Phytolitharien Newsletter* 5(1): 4.

____1988 *Phytolith Analysis: An Archaeological and Geological Perspective.* New York, Academic Press: 280 p. Most recent and comprehensive review of all aspects of phytolith morphology and its applications for prehistory. Includes history of phytolith systematics and keys for New World tropical plants, for North American Central Plains grasses and for Lower Central American grasses. 96 photos, 27 figures, 16 tables.

____1989 The occurrence of phytoliths in the reproductive structures of selected tropical angiosperms and their significance in tropical paleoecology, paleoethnobotany and systematics. *Review of Palaeobotany and Palynology* 61: 147–173. Phytoliths from 254 tropical angiosperms indicated patterns of Si deposition. Three families are discussed in detail and diagnostic forms described. 16 plates (56 photos).

Piperno, DR and Clary, KM 1984 Early plant use and cultivation in the Santa Maria Basin, Panama: Data from phytoliths and pollen. *In* Lange, FW, ed, Recent developments in Isthmian archaeology: Advances in the prehistory of Lower Central America, *British Archaeological Reports, International Series* 212: 85–121. Pollen and phytolith analysis of Panamanian plants and archaeological sediments was applied to prehistoric subsistence. Phytolith distribution in monocots and dicots is discussed, as is identification of maize (compared to 48 wild grasses). Data are similar to other papers by Piperno. 13 figures (12 photos), 4 tables.

Ponnaiya, BWX 1951 Studies in the genus *Sorghum*: II. The cause of resistance in *Sorghum* to the insect pest *Atherigona indica* M. *Madras University Journal, Section B (Science)* 21: 203–217. Epidermal silica bodies in *Sorghum* are correlated to resistance from insect pest. Distribution of dumbbell and irregularly shaped silica bodies is described from spodograms. 6 tables.

____1960 Silica deposition in *Sorghum* roots and its possible roles. *The Madras Agricultural Journal* 47: 31–32. Spodograms of sorghum roots indicate Si deposition on ITW endodermis. These silica knobs may be restricted to Andropogoneae, particularly species resistant to drought.

Postek, MT 1981 The occurrence of silica in the leaves of *Magnolia grandiflora* L. *Botanical Gazette* 142: 124–134. Specimens were studied with SEM, TEM, EDX and light microscopy. Phytoliths were present exterior to epidermal cell walls, in the cuticle, and in tracheids, vein sheath cells, epidermal cells and guard cells. 28 figures (27 photos).

Powers, AH, Gilbertson, DD and Jones, G 1986 *The Prospects for Environmental Discrimination in Coastal Sand Dunes on the Basis of Studies of Phytolith Type and Abundance*, University of Sheffield, Department of Archaeology and Prehistory, Heritage Landscape Programme Contract Report No 1, 133 p, for Central Excavation Unit, Scottish Development Company. Absolute frequencies of phytoliths are established by use of an independent marker (*Lycopodium* spores). Phytolith extraction by ashing is described. 3 tables.

Prat, H 1932 L'épiderme des Graminées. (The epidermis of the Gramineae.) *Annales des Sciences Naturelles: Botanique, Series* 10, 14: 117–324. (French). Epidermal characteristics (including phytoliths) are described for the Gramineae. *Agropyron* and the Hordeae are described in detail. 20 plates, 106 figures.

____1934 Contribution à l'étude systématique et histologique des Chloridées. (Contribution to the systematic and histologic study of the Chlorideae.) *Societe Botanique de France, Bulletin* 81: 475–491. (French). Placement of the Chloridoideae within the Gramineae, as well as subdivision of tribes, is based on epidermal and other characters. Phytoliths are one epidermal characteristic. 6 figures.

____1935 Contribution à l'étude systématique et histologique des Festucées. (Contribution to the systematic and histologic study of the Festuceae.) *Societe Botanique de France, Bulletin* 82: 498–506. (French). Discussion on anatomy, morphology, cytology and histology of the Festuceae shows the heterogeneous nature of the tribe (as defined by Bentham and Hooker). Silica cell information is included. 31 figures.

_____1936 La systématique des Graminées. (Systematics of the Gramineae.) *Annales des Sciences Naturelles, Botanique, Series 10*, 18: 165–258. (French). Grass systematics is reviewed and revised based on histology, anatomy (including phytoliths), cytology, morphology, chemistry, physiology and ecology. A distinction between Festucoid, Panicoid and Bambusoid subfamilies is made. 29 figures.

_____1948 General features of the epidermis in *Zea mays*. *Annals of the Missouri Botanical Garden* 35: 341–351. The background information describes various lab techniques, epidermal cell types (including silica cells) and subfamily distinctions. Distribution of epidermal characters in *Zea mays* is presented. 3 figures.

_____1960 Vers une classification naturelle des Graminées. (Toward a natural classification of the Gramineae.) *Societie Botanique de France, Bulletin* 107: 32–79. (French). Classification of the grasses is based on morphological and non-morphological characteristics. Epidermal description includes silica cell morphology. 6 figures, 5 tables.

Raeside, JD 1970 Some New Zealand plant opals. *New Zealand Journal of Science* 13: 122–132. Description of phytoliths from 3 tussock grasses and 3 sedges identified some types specific to the sedges, but none for the grasses. The families can be distinguished by phytoliths. 48 figures.

Rapp, GR, Jr 1986 Morphological classification of phytoliths. *In* Rovner, I, ed, Plant opal phytolith analysis in archaeology and paleoecology: Proceedings of the 1984 Phytolith Research Workshop, North Carolina State University, Raleigh, North Carolina, *Occasional Papers of The Phytolitharien* 1: 33–34. Short description of data base generation in Minnesota.

Reid, JS 1947 Silica in beech timbers. *The New Zealand Journal of Forestry* 5: 330–331. Si content of 3 *Nothofagus* species compares to that of teak. Silica crystals concentrate in ray cells.

Richter, HG 1980 Occurrence, morphology, and taxonomic implications of crystalline and siliceous inclusions in the secondary xylem of the Lauraceae and related families. *Wood Science and Technology* 14: 35–44. Crystal (calcium oxalate) and silica bodies in Lauraceae wood are described based on about 750 spp in 40 genera. Occurrence in some genera is listed but not detailed distribution in species. 4 figures (12 photos).

Riquer, J 1960 Les phytolithes de certains sols Tropicaux et des podzols. (Phytoliths of certain tropical soils and podzols.) *Seventh International Congress of Soil Science, Transactions* 4: 425–431. (French, English summary). A horizon composed of phytoliths resembles a podzol A_2 but has a different origin. Phytoliths from four plants are present. 2 figures (2 photos), 1 table.

Robert, D, Laroche, J, Guervin, C, Le Coq, C and Sauvanet, A 1973 Mise en place de la silice dans les cellules épidermiques de la feuille d'une Ptéridophyte: *Selaginella kraussiana II.* Étude ultrastructurale. (Study in place of the silica in epidermal cells of the leaf of a pteridophyte: *Selaginella kraussiana*. II. Ultrastructural study.) *Bulletin du Museum National d'Historie Naturelle, Series 3, Botanique* 14: 211–233. (French, English abstract). Initial Si deposition in epidermal ornamentation is related to evolution of the protoplasm and the cell wall. The second phase of silicification is similar to other species. 7 plates (20 photos).

Robinson, RL 1979a Biosilica analysis: Paleoenvironmental reconstruction of 41LL254. *In* Assad, C and Potter, DR, An Intensive Archaeological Survey of Enchanted Rock State Natural Area, Llano and Gillespie Counties, Texas, *Center for Archaeological Research, San Antonio, University of Texas Archaeological Survey Report* 84: 125–144. Paleoenvironment is reconstructed based on 5 samples covering 6000 to 2000 BP. A morphological classification of sediment phytoliths is presented; a comparative collection of 38 spp was used to identify phytolith origin. 3 figures, 3 tables.

_____1979b Biosilica and climatic change at 41GD21 and 41GD21A. *In* Fox, DE, Archaeological Investigations of Two Prehistoric Sites on the Coleto Creek Drainage, Goliad County, Texas, *Center for Archaeological Research, San Antonio, University of Texas Archaeological Survey Report* 69: 102–113. Climatic changes were reconstructed from 6 samples from 2 sites covering 6000 BC to AD 1000. A morphological classification is presented; identification of plant origin is based on 38 spp. 1 figure, 2 tables.

Rosen, AM 1987 Phytolith studies at Shiqmim. *In* Levy, TE, ed, Shiqmim I: Studies concerning Chalcolithic societies in the northern Negev Desert, Israel (1982–1984), *British Archaeological Reports, International Series* 356: 243–249. Phytolith analysis at Shiqmim identified tissue segments of wheat and barley. Irrigation farming is suggested by the multi-celled segments, while proportions of cereals indicate drier conditions than at other sites. 3 photos, 3 tables.

Rovner, I 1971 Potential of opal phytoliths for use in paleoecological reconstruction. *Quaternary Research* 1: 343–359. Modern synthesis and presentation of phytolith analysis as a major paleoecological and archaeological methodology. Phytoliths from 30 taxa are described and taxonomic significance of phytolith shapes is assessed. 2 figures.

_____1975 Plant opal phytolith analysis in Midwestern archaeology. *Michigan Academician* 8: 129–137. Brief review of phytolith analysis describes phytoliths and applications in archaeology. 4 photos.

_____1983 Plant opal phytolith analysis: Major advances in archaeobotanical research. *In* Schiffer, M, ed, *Advances in Archaeological Method and Theory* 6: 225–266. New York, Academic Press. Review of phytolith analysis covers issues of production, taxonomy, taphonomy and applications as well as extraction procedures. Discussion of taxonomy includes Twiss, Suess and Smith (1969), Blackman (1971), Mehra and Sharma (1965), Bertoldi de Pomar (1971), Rovner (1971) and other articles. 3 figures.

Rovner, I, ed 1986 Plant opal phytolith analysis in archaeology and paleoecology: Proceedings of the 1984 Phytolith Research Workshop, North Carolina State University, Raleigh, North Carolina, *Occasional Papers of The Phytolitharien No* 1: 147 p. 17 papers with question/answer sessions and discussant comments. For abstracts, see *Phytolitharien Newsletter* 3(1) (1984). For comments on 7 systematics papers, see individual authors: Bozarth; Brown; Mulholland; Piperno; Rapp; Smith, Kaplan and Davies; Twiss.

Russ, JC and Rovner, I 1989 Stereological identification of opal phytolith populations from wild and cultivated *Zea*. *American Antiquity* 54: 784–792. Identification of phytoliths from wild vs domesticated *Zea* is proposed using a stereological approach. Size parameters were found to be stable. 2 figures, 5 tables.

Salgado-Labouriau, ML 1973 Contribuição à palinologia dos Cerrados. (Contribution to the palynology of the Cerrados.) *Editoria Academia Brasileira de Ciências* 242–247; 261–262. (Portuguese, English summary). Discussion of palynology briefly touches on phytoliths.

Sangster, AG (ms) 1967 Some environmental factors affecting silica deposition in higher plants with special reference to the Gramineae. PhD dissertation, University College of North Wales: 181 p. Deposition patterns of Si in 3 grass subfamilies correlate with age and external Si concentration. A sequence of cell types becomes silicified with time by transport and deposition. 10 plates (15 photos), 28 figures, 20 tables.

_____1968 Studies of opaline silica deposits on the leaf of *Sieglingia decumbens* (L.) Bernh., using the scanning electron microscope. *Annals of Botany* 32: 237–240. Early SEM study of samples describing in detail costal idioblast (dumbbell), long-cell and sclerenchyma-fiber phytoliths. Species has mixed panicoid-festucoid leaf anatomy. 2 plates (12 photos).

_____1970a Intracellular silica deposition in immature leaves in three species of the Gramineae. *Annals of Botany* 34: 245–257. Heath grass, Bermuda grass and rice were moved from minimal to higher Si solutions for different times and wet-ashed. Si was deposited by 2 days and was confined to costal idioblasts of expanding tissue; bulliform cell phytoliths formed later in leaf tips. 4 figures.

____1970b Intracellular silica deposition in mature and senescent leaves of *Sieglingia decumbens* (L.) Bernh. *Annals of Botany* 34: 557–570. Methods were as in Sangster (1970a). Si deposition in heath grass decreased with increasing tissue age for idioblasts (dumbbells) and long cells but increased for bulliform cells. 4 plates (29 photos), 5 figures, 2 tables.

____1977a Characteristics of silica deposition in *Digitaria sanguinalis* (L.) Scop. (crabgrass). *Annals of Botany* 41: 341–350. Methods were as in Sangster (1970a), with SEM and EPM. Si was deposited by 2 days in idioblasts at the leaf base and prickle hairs at the leaf tip, followed by simultaneous waves of deposition in opposite directions. 3 plates (32 photos), 2 figures.

____1977b Electron-probe microassay studies of silicon deposits in the roots of two species of *Andropogon*. *Canadian Journal of Botany* 55: 880–887. Samples were freeze-dried and studied by SEM, EPM and light microscopy. Si was confined to the ITW of endodermal cells, with size and shape similar in both species. 12 figures (8 photos).

____1978a Silicon in the roots of higher plants. *American Journal of Botany* 65: 929–935. 2 freeze-dried grass spp were studied with SEM, TEM, EPM and EDX, and compared with 3 dicots and 12 other grasses. Si distribution depended on phylogeny rather than basic root anatomy. Focus is not on Si shape. 6 figures (6 photos), 1 table.

____1978b The distribution of silicon deposits in the rhizomes of two *Andropogon* species. *Canadian Journal of Botany* 56: 148–156. Methods were as in Sangster (1977b). Si was confined to ITW of cortex just outside the stele and (in *A. gerardii* only) sheath cells around central vascular bundles. 18 figures (14 photos).

____1978c Electron-probe microassays for silicon in the roots of *Sorghastrum nutans* and *Phragmites communis*. *Canadian Journal of Botany* 56: 1074–1080. Methods were as in Sangster (1977b) plus EDX. Si was confined to the endodermis ITW in *S. nutans* nodal roots; none was in *P. communis* roots. 10 figures (8 photos).

____1983a Anatomical features and silica deposition patterns in the rhizomes of the grasses *Sorghastrum nutans* and *Phragmites australis*. *Canadian Journal of Botany* 61: 752–761. Samples were studied with SEM, EPM, EDX and light microscopy. Si was in 4 zones in *S. nutans* but was absent in *P. australis*. 18 figures (13 photos), 1 table.

____1983b Silicon distribution in the nodal roots of the grass *Miscanthus sacchariflorus*. *Canadian Journal of Botany* 61: 1199–1205. Methods were as in Sangster (1983a). Si was restricted to aggregates on the ITW of the endodermis, as in 7 other Andropogoneae and in contrast to roots of other tribes. 10 figures (10 photos).

____1985 Silicon distribution and anatomy of the grass rhizome, with special reference to *Miscanthus sacchariflorus* (Maxim.) Hackel. *Annals of Botany* 15: 621–634. It was studied with SEM, EPM and EDX. Results for 8 other spp are summarized. Si was in epidermis and variable internally. 5 figures (18 photos), 2 tables.

Sangster, AG, Hodson, MJ and Parry, DW 1983 Silicon deposition and anatomical studies in the inflorescence bracts of four *Phalaris* species with their possible relevance to carcinogenesis. *New Phytologist* 93: 105–122. Si depositional patterns and anatomy of inflorescence bracts from 4 spp are reported from SEM and EPM analyses. Hairs and sheet fragments produce sharp fibers that have been implicated in carcinogenesis. 6 figures (47 photos).

Sangster, AG, Hodson, MJ, Parry, DW and Rees, JA 1983 A developmental study of silicification in the trichomes and associated epidermal structures of the inflorescence bracts of the grass, *Phalaris canariensis* L. *Annals of Botany* 52: 171–187. Samples were studied with SEM, EPM and light microscopy. Silicification began before emergence and increased with time, varying among structures of the inflorescence bracts. 4 figures (48 photos), 5 tables.

Sangster, AG and Parry, DW 1969 Some factors in relation to bulliform cell silicification in the grass leaf. *Annals of Botany* 33: 315–323. Heath and Bermuda grass and rice were wet-ashed. Bulliform phytoliths formed by day 2, mostly in leaf tips by day 16, but no more numerous with 500 ppm Si than 50 ppm. 2 plates (10 photos), 2 figures, 1 table.

____1971 Silica deposition in the grass leaf in relation to transpiration and the effect of dinitrophenol. *Annals of Botany* 35: 667–677. Cultured detached leaves of Bermuda and heath grass formed phytoliths, but leaves coated with oil to reduce transpiration did not. Dinitrophenol, added to inhibit Si uptake, shifted Si deposition from mostly abaxial (idioblast) to mostly adaxial (stoma) sites. Germanium dioxide was deposited in sites similar to SiO_2 but more extra- than intracellular. 1 figure, 2 tables.

____1976a Endodermal silicon deposits and their linear distribution in developing roots of *Sorghum bicolor* (L.) Moench. *Annals of Botany* 40: 361–371. Development of Si deposits in sorghum roots (SEM and EPM) indicates a linear gradient from proximal to distal regions. Deposition conditions are discussed. 4 plates (48 photos).

____1976b Endodermal silicification in mature, nodal roots of *Sorghum bicolor* (L.) Moench. *Annals of Botany* 40: 373–379. Si content was analyzed by SEM and EPM. Dome-shaped aggregates occur on the ITW of the endodermis. 1 plate (12 photos).

____1976c The ultrastructure and electron-probe microassay of silicon deposits in the endodermis of the seminal roots of *Sorghum bicolor* (L.) Moench. *Annals of Botany* 40: 447–459. Formative processes and ultrastructure from TEM and EPM indicate Si is involved in the ITW. Transport and deposition mechanisms are proposed. 8 plates (27 photos).

____1981 Ultrastructure of silica deposits in higher plants. *In* Simpson, TL and Volcani, BE, eds, *Silicon and Siliceous Structures in Biological Systems*. New York, Springer-Verlag: 383–407. Structure of Si deposits differs between aerial and subterranean organs. Ultrastructure of aggregates, cell-wall deposits and intercellular deposits is described. 14 figures, 1 table.

Sase, T 1980 (Plant opal analysis of buried soils immediately beneath the Nanbu pumice layer.) *Daiyonki Kenkyu* (*Quaternary Research*) 19: 117–124. (Japanese, English summary). Phytoliths from sediment below a pumice layer indicate grass (Panicoid and Sasaoid) and lesser amounts of trees. 2 plates (61 photos), 4 figures, 2 tables.

____1981 (Analytical study of plant opal in the buried soil immediately beneath the Hachinohe pumice bed.) *Daiyonki Kenkyu* (*Quaternary Research*) 20: 15–20. (Japanese, English summary). Poaceae grasses and conifers are identified by phytolith analysis. A cooler climate than now is indicated. 2 plates (67 photos), 4 figures, 1 table.

____1986 (Opal phytolith analysis of volcanic ash soils from recent Towada ashes.) *Pedorojisuto (Pedologist)* 30: 102–114. (Japanese, English summary). Vegetative history of 3 soils derived from volcanic ash contains Bambusoid, Panicoid, conifer and broad-leaved plant phytoliths. 2 plates (46 photos), 3 figures, 2 tables.

Sase, T, Hosono, M, Utsugawa, T, Kato, S and Komamura, M 1987. Opal phytolith analysis of the Kanto Loam Formation in Musashino Upland, Kanto Plain, Japan. *Daiyonki Kenkyu (Quaternary Research)* 26: 1–11. (Japanese, English summary). Climatic reconstructions of five phytolith zones are correlated to major glacial epochs in Europe. Phytolith classification is illustrated. 1 plate (45 photos), 3 figures, 2 tables.

Sase, T and Kato, Y 1976 (The study on phytogenic particles, especially, on plant opals, in humic horizons of present and buried volcanic ash soils. I. The problem on the source of plant opals.) *Daiyonki Kenkyu (Quaternary Research)* 15: 21–33. (Japanese, English summary). Grass phytoliths are correlated to taxonomic group (Panicoid, Sasaoid, Festucoid, Chloridoid), although some are more widely distributed (Fan, Point, Elongate). Different types occur in different sites. 1 plate (23 photos), 6 figures, 6 tables.

Sase, T, Kato, Y and Makino, S 1985 (Plant opal analysis of volcanic ash soils at the foots of Mt. Fuji and Mt. Amagi.) *Pedorojisuto (Pedologist)* 29: 44–59. (Japanese, English summary).

Phytoliths from volcanic soils were used to reconstruct the vegetation. Phytolith types are illustrated. 2 plates (37 photos), 4 figures, 5 tables.

Sase, T and Kondo, R 1974 (The study of opal phytoliths in the humus horizon of buried volcanic ash soils in Hokkaido.) *Obihiro Chikusan Daigaku Gakujutsu Kenkyo Hokoku Dai-I-Bu, Japan (Research Bulletin of Obihiro Zootechnical University, Series I)* 8: 465–483. (Japanese, English summary). Amount and type of phytoliths in buried volcanic ash soils formed the basis for vegetation and climatic reconstructions. Identification of seven morphological phytolith groups was based on examination of 23 grass spp. 3 figures, 7 tables.

Schall, WM 1943 An investigation of the presence of siliceous rods in the secondary wall of woody tissue. *Madrona* 7: 8–14. No evidence of a "skeleton" of silica rods was found in woody tissue. 1 photo, 1 table.

Schellenberg, HC 1908 The remains of plants from the North Kurgan, Anau. *In* Pumpelly, R, ed, *Explorations in Turkestan: Expedition of 1904* 2: 471–473. Washington, DC, Carnegie Institution. Silica skeletons of wheat and barley were identified in brick samples. Epidermal patterns are described, emphasizing differences between the two species. 1 plate (10 figures).

Schreve-Brinkman, EJ 1978 A palynological study of the Upper Quaternary sequence in the El Abra corridor and rock shelters (Colombia). *Palaeogeography, Palaeoclimatology, Palaeoecology* 25: 1–109. Reconstruction of the climate and vegetation in last Glacial and Holocene periods in the Colombian Eastern Cordillera. A short examination of phytoliths presents shapes from 3 spp. 1 plate (13 photos), 16 figures.

Schuyler, AE 1971 Scanning electron microscopy of achene epidermis in species of *Scirpus* (Cyperaceae) and related genera. *Academy of Naturla Sciences of Philadelphia, Proceedings* 123: 29–52. SEM examination of achenes from 58 sedge spp indicates that cell wall and internal structure are taxonomically useful. 129 figures (129 photos).

Scurfield, G, Anderson, CA and Segnit, ER 1974 Silica in woody stems. *Australian Journal of Botany* 22: 211–229. Two types of silica deposits, aggregate grains and dense silica, were observed in 32 woody perennials. Location in cells and morphology are described. 40 figures (38 photos).

Scurfield, G, Michell, AJ and Silva, SR 1973 Crystals in woody stems. *Botanical Journal of the Linnean Society of London* 66: 277–289. Location, composition and crystal habit for 40 woody perennials are reported from SEM, XRD and other analyses. 4 plates (22 photos), 3 figures.

Scurfield, G, Segnit, ER and Anderson, CA 1974 Silicification of wood. *Proceedings of the Scanning Electron Microscope Conference* 1974: 389–396. Discussion of Australian petrified wood includes short summary and 2 photos of Si deposition in living specimens. 41 photos.

Sendulsky, T and Labouriau, LG 1966 Corpos silicosos de gramineas dos Cerrados. I. (Silica bodies of grasses from "Cerrados." I.) *In* Labouriau, LG, ed, *IIo - Simposio Sobre o Cerrado, Anais Academia Brasileira de Ciencias* 38, *Supplemento*: 159–170. (Portuguese, English summary). From Labouriau 1983: Leaf, stem and inflorescence material from herbarium specimens was dry-ashed. The forest grass *Pseudoechinolaena polystachya* had practically no Si, unlike 132 other Brazilian grasses studied. Camera lucida drawings of silica bodies from 60 grass spp. 60 plates.

Sharma, M and Rao, KR 1970 Investigations on the occurrence of silica in Indian timbers. *Indian Forester* 96: 740–754. Of 134 spp of timber from India, 68 contain silica. Percentage, size, form and distribution data are presented. 21 photos, 1 table.

Sharma, OP 1972 Anatomy of *Scirpus squarrosus*. *Current Science* (Bangalore) 41: 494–497. Anatomical description of this sedge includes silica body morphology and distribution. Calcium oxalate crystals are present in stem and leaf tissues. 14 figures.

Sharp, RB 1962 Plant silica: An abrasive constituent of plant matter? *Journal of Agricultural Engineering Research* 7: 214–220. Introductory examination of plant silica focuses on abrasion and wear potential. Milling does cause damage to silica structures, and appears to be responsible for heavy wear patterns. 7 figures (17 photos), 1 table.

Shaw, RB and Smeins, FE 1979 Epidermal characteristics of the callus in *Eriochloa* (Poaceae). *American Journal of Botany* 66: 907–913. 19 *Eriochloa* and 5 other spp were studied with SEM. Silica bodies are used as a taxonomic character and illustrated but not described at length. 20 figures (20 photos), 1 table.

_____1981 Some anatomical and morphological characteristics of the North American species of *Eriochloa* (Poaceae: Paniceae). *Botanical Gazette* 142: 534–544. 12 spp were studied by SEM and light microscopy. Si dumbbells, nodular shapes and crosses were among the characters supporting inclusion of *Eriochloa* among Panicoids. 25 figures (25 photos), 1 table.

Simmonds, DH, St. Clair, MS and Collins, JD 1970 Surface structure of wheat and its contribution to the dust problem. *Cereal Science Today* 15: 230–234. Examination of wheat and wheat dust indicates presence of brush hairs and pericarp fragments. Variations between wheat varieties are briefly discussed. 5 figures (37 photos).

Singh, V and Pande, PC 1977 A study of leaf epidermal structures in some Scitamineae. *Geobios* (Jodhpur) 4: 18–20. Brief summary of epidermal features from 14 spp focuses on stomatal types. Silica bodies are mentioned. 13 figures, 1 table.

Smith, MB (ms) 1982 Phytolith development and morphology in *Zea mays* L. and *Zea mexicana* (Schrader) Kuntze. MS thesis, University of Massachusetts: 199 p. Developmental processes and deposition patterns of Si in maize and teosinte differ with species, age and leaf position. Controlled conditions included Si-free. 97 figures, 33 tables.

Smith, MB, Kaplan, L and Davis, EA 1986 Biogenic silica: Phytolith formation in maize tissue culture. *In* Rovner, I, ed, Plant opal phytolith analysis in archaeology and paleoecology: Proceedings of the 1984 Phytolith Research Workshop, North Carolina State University, Raleigh, North Carolina, *Occasional Papers of The Phytolitharien* 1: 141–147. Shoots and callus tissue from 2 maize races were cultured in Si-free and Si-rich environments for 9 and 23 days and studied as spodograms. Phytoliths formed only in samples grown in Si-rich media, while Si-free samples had fragile cell walls and collapsed trichomes. 1 figure (2 photos).

Smithson, F 1956a Plant opal in soil. *Nature* 178: 107. Brief report of phytoliths in British soils includes a photomicrograph of *Agrostis tenuis* phytoliths. 3 photos.

_____1958 Grass opal in British soils. *Journal of Soil Science* 9: 148–155. Shape descriptions of phytoliths in grasses, sediments and dung. Phytoliths from *Nardus stricta* are considered distinct; others are identifiable to tribe or a group of tribes. 13 photos.

Soderstrom, TR 1982 *Cryptochloa dressleri* (Poaceae), a new Bambusoid grass from Panama. *Brittonia* 34: 25–29. Description of the species specifically discusses distribution and shape of silica bodies. 1 figure.

Söndahl, MR and Labouriau, LG 1970 Corpos silicosos de gramíneas dos Cerrados. IV. (Silica bodies of grasses from the "Cerrados." IV.) *Pesquisa Agropecuraria Brasileira* 5: 183–207. (Portuguese, English abstract, unpublished translation by Hochberg, V, Hyduke, N and Geis, JW.) From Labouriau 1983: Leaf, stem and inflorescence material from herbarium specimens was combination wet- and dry-ashed. Camera lucida drawings of silica bodies from 19 grass spp. 20 figures.

Soni, SL, Kaufman, PB and Bigelow, WC 1970 Electron microprobe analysis of the distribution of silicon in leaf epidermal cells of the oat plant. *Phytomorphology* 20: 350–363. Distribution of Si in epidermal cell types is given for leaf blade, sheath and ligule. The distribution is not equal through the tissue. 51 photos, 3 tables.

_____1972a Electron probe analysis of silicon and other elements in leaf epidermal cells of the rice plant (*Oryza sativa* L.). *American Journal of Botany* 59: 38–42. Study of silica cells in leaf sheath and blade indicates that other elements are inhibited by presence of Si. The mechanisms of inhibition are discussed. 7 photos, 1 table.

_____1972b Electron microprobe analysis of silicon and other elements in developing silica cells of the leaf and internode of *Cyperus alternifolius*. *Annals of Botany* 36: 611–619. Accumulation of elements changes during development of silica cells. 1 plate (6 photos), 2 tables.

_____1972c Electron microprobe analysis of silica cells in leaf epidermal cells of *Cyperus alternifolius*. *Plant and Soil* 36: 121–128. Freeze-dried peels showed that Si retards accumulation of K and enhances it for Ca, Mg, Na, P, Mn and Fe. Si cell shapes are not discussed. 1 plate (3 photos), 2 tables.

Soni, SL, Kaufman, PB and Jones, RA 1972 Electron microprobe analysis of the distribution of silicon and other elements in rice leaf epidermis. *Botanical Gazette* 133: 66–72. Silica cells, trichomes and epidermal cell walls of leaf blade and sheath were found to be sites of Si deposition. Si affects the accumulation of other elements. 20 photos, 2 tables.

Soni, SL and Parry, DW 1973 Electron probe microanalysis of silicon deposition in the inflorescence bracts of the rice plant (*Oryza sativa*). *American Journal of Botany* 60: 111–116. Si deposition is heavy external to the outer epidermis and sparse within the epidermal cells. Cell walls accumulate Si to the interior when specific accumulation sites are filled. 36 photos.

Sowers, AE and Thurston, EL 1979 Ultrastructural evidence for uptake of silicon-containing silicic acid analogs by *Urtica pilulifera* and incorporation into cell wall silica. *Protoplasma* 101: 11–22. Si deposits in nettle stinging cells require not only Si in groundwater but also a specific molecular structure. Cell wall Si is under control of binding site parameters. 18 photos, 1 table.

Srinivasan, J, Dayanandan, P and Kaufman, PB 1979 Silica distribution in *Equisetum hyemale* var *affine* L. (Engelm) in relation to the negative geotropic response. *New Phytologist* 83: 623–626. Investigation of Si deposition in horsetail indicates a basipetal pattern. Young specimens have a negative geotrophic response, whereas mature specimens with more Si do not respond. 1 plate (5 photos), 1 table.

Srivastava, AK 1978 Study of leaf epidermis in the genus, *Digitaria* Rich, Gramineae. *Journal of the Indian Botanical Society* 57: 155–160. Epidermal patterns of 6 spp show taxonomic differences at the sub-generic level. Morphology and location of silica bodies is described. 32 figures, 2 tables.

Stählin, A and Tirtapradja, H 1971 Über die Aufnahme und Einlagerung von Silizium in Futtergräsern. (Uptake and deposition of silicon in fodder grasses.) *Zeitschrift fur Acker und Pflanzenbau* 134: 295–312. (German, English summary). Si content and phytolith morphology is described for 6 spp. Weather (moisture) and manuring affect Si-uptake. 3 figures, 5 tables.

Stant, MY 1973 Scanning electron microscopy of silica bodies and other epidermal features in *Gibasis (Tradescantia)* leaf. *Linnean Society of London, Botanical Journal* 66: 233–244. SEM analysis of epidermal morphology of 3 species gives detailed information on shape and distribution of cell types. Focus is on hairs, papillae and silica bodies as taxonomic markers. 5 plates (36 photos), 2 tables.

Sterling, C 1947 Sclereid formation in the shoot of *Pseudotsuga taxifolia*. *American Journal of Botany* 34: 45–52. Development of sclereids is described and illustrated. Although mineralization is not discussed, these cells can silicify. 19 figures (10 photos).

_____1967 Crystalline silica in plants. *American Journal of Botany* 54: 840–844. Crystalline and amorphous Si is reported from *Fragaria*, horsetails and tabashir by XRD. Superficial location and/or young ages are associated with amorphous deposits. 4 photos, 1 table.

Stewart, DRM 1965 The epidermal characters of grasses, with special reference to East African plain species. *Botanische Jahrbucher fur Systematik, Pflanzengeschichte, und Pflanzengeographie* 84: 63–116, 117–174. Description of grass epidermal characteristics for 163 spp (57 genera) includes distribution and morphology of silica bodies. The epidermis is taxonomically useful for identification of leaf fragments.

Suess, E (ms) 1966 Opal phytoliths. MS thesis, Kansas State University: 77 p. 7 grass spp from the central plains, US, were dry-ashed. Short- and long-grass prairie yielded, respectively, chloridoid (cuboid) and panicoid (dumbbell-shaped) phytoliths (also in dust and soil samples). Gives details about measurement and statistical treatments. 8 plates (29 photos), 5 figures, 3 tables, 4 appendices.

Sustar, F 1976 Horološka in taksonomska problematika kompleksa *Brachypodium pinnatum* na območju Slovenije. (The chorology [sic] and taxonomy of the complex *Brachypodium pinnatum* in Slovenia.) *Bioloski Vestnik* 24: 1–11. (Yugoslavian, German summary). Description of the *Brachypodium pinnatum*-complex includes epidermal anatomy. 6 figures (4 photos).

Suzuki, H 1937 Studies on the relations between the anatomical characters of the rice plant and its susceptibility to blast disease. *Tokyo Imperial University, College of Agriculture Journal* 14: 181–264. Rice grown under wet conditions has thicker outer cell wall and "silicated" layer than under dry conditions. Other factors also correlating to thicker walls include cell type. 8 plates (39 photos), 47 tables.

Swenson, CF, Le Tourneau, D and Erickson, LC 1964 Silica in Medusahead. *Weeds* 12: 16–18. Si content of *Elymus caput-medusae* is presented by plant part and age. 2 figures, 3 tables.

Sz-Borsos, O 1974 Notes on the leaf anatomy of the *Brachypodium pinnatum* species complex. *Acta Botanica Academiae Scientiarum Hungaricae* 20: 13–21. External and internal morphology differs between *B. pinnatum* and *B. rupestre*. Phytolith morphology is illustrated. 4 figures (5 photos).

Takahashi, E, Tanaka, T and Miyake, Y 1981 Distribution of silica accumulator plants in the plant kingdom (6). Commelinales, Cyperales, and some dicotyledoneae families (Cucurbitaceae and Urticaceae). *Japanese Journal of Soil Science and Plant Nutrition* 52: 511–515. Mineral content of 63 spp indicates silica-type, meso-type, calcium-type and non-silica type plants.

Takeoka, Y 1963 Notes on some grasses. VIII. *Botanical Gazette* 124: 264–270. 56 spp were hand-sectioned. "Oryzoid" silica bodies (dumbbells perpendicular to the blade) usually were found in Oryzeae but never in Erharteae, whose bilobates were parallel with the blade. Disarticulated phytoliths were not addressed. 28 figures, 1 table.

Takeoka, Y, Inoue, S and Kawano, S 1959 Notes on some grasses. IX. Systematic significance of bicellular microhairs of leaf epidermis. *Botanical Gazette* 120: 80–91. 4 measurement ratios are given for 238 spp from 4 subfamilies. 2 groups of index values correspond to panicoid and chloridoid types and distinguish subfamilies. Bambuseae and other genera may have taxonomically significant microhairs. 156 figures, 8 tables.

Takeoka, Y, Kaufman, PB and Matsumura, O 1979 Comparative microscopy of idioblasts in lemma epidermis of some C_3 and C_4 grasses (Poaceae) using sump method. *Phytomorphology* 29: 330–337. 21 spp were studied with light microscopy. C_4 spp had 2.5 times more Si cells per unit area than C_3 spp, but shapes were not described or counted. 29 figures (27 photos), 1 table.

Takeoka, Y, Kondo, K and Kaufman, PB 1983 Leaf surface fine-structures in rice plants cultured under shaded and non-shaded conditions. *Nihon Sakumotsa Gakkai Kiji (Proceedings of the Crop Science Society of Japan or Japanese Journal of Crop Science)* 52: 534–543. Wax on rice cuticles was compared from shaded and non-shaded growth conditions. Amount of wax depended on cell type (including silica cells) as well as shading conditions. 18 photos.

Takeoka, Y, Matsumura, O and Kaufman, PB 1983 Studies on silicification of epidermal tissues of grasses as investigated by soft x-ray image analysis. I. On the method to detect and calculate frequency of silica bodies in bulliform cells. *Nihon Sakumotsa Gakkai Kiji (Proceedings of the Crop Science Society of Japan or Japanese Journal of Crop Science)* 52: 544–550. Area and distribution of silica bodies were analyzed by SIAM. Four cultivars exhibit distributional differences in the leaf blade. 7 figures (3 photos).

Takeoka, Y, Wada, T, Naito, K and Kaufman, PB 1984 Studies on silicification of epidermal tissues of grasses as investigated by soft x-ray image analysis. II. Differences in frequency of silica bodies in bulliform cells at different positions in the leaves of rice plants. *Nihon Sakumotsa Gakkai Kiji (Proceedings of the Crop Science Society of Japan or Japanese Journal of Crop Science)* 53: 197–203. Silica body distribution differs within the leaf blade of rice: tip to base, left to right, costal and intercostal. Differences were correlated to degrees of transpiration. 4 figures (1 photo), 2 tables.

Tatsuhara, A and Fujimaki, Y 1987 (Food habit analysis by using opal phytoliths in *Cervus-nippon-yesoensis*.) *Research Bulletin of Obihiro University, Series I* 15: 85–90. (Japanese, English summary and captions). Examination of 66 spp indicates identifiable phytoliths in 5 spp, 2 genera, 2 subfamilies and 2 families. Phytoliths in deer feces indicate the main dietary components. 2 figures (14 photos), 1 table.

Taugourdeau-Lantz, J, Laroche, J, Lachkar, G and Pons, D 1976 La silice chez les vegetaux: Probleme des phytolithaires. (Silica in plants: Problem of phytoliths.) *Travaux du Laboratoire de Micropaleontologie*, Univ. Pierre et Marie Curie 5: 255–303. (French). Review of physiological and morphological phytolith data distinguishes three types of deposition. A revision of the parataxonomic groups is proposed. 6 plates (68 photos), 4 figures, 6 tables.

Teixeira da Silva, S and Labouriau, LG 1970 Corpos silicosos de gramineas dos Cerrados. III. (Silica bodies of grasses from the "Cerrados." III.) *Pesquisa Agropecuraria Brasileira* 5: 167–182. (Portuguese, English abstract, unpublished translation by Hochberg, V, Geis, JW and Hyduke, N.) From Labouriau 1983: Leaf, stem, and inflorescence material from herbarium specimens was combination wet- and dry-ashed. Camera lucida drawings of silica bodies from 10 grass spp. 11 figures.

Terrell, EE and Wergin, WP 1979 Scanning electron microscopy and energy dispersive x-ray analysis of leaf epidermis in *Zizania* (Gramineae). *Scanning Electron Microscopy* 3: 81–88. Leaf epidermal morphology is described from SEM and x-ray analysis of 4 species. Si is higher in silica-bodies than in the papillae. 18 photos.

_____1981 Epidermal features and silica deposition in lemmas and awns of *Zizania* (Gramineae). *American Journal of Botany* 68: 697–707. 4 spp were studied with SEM and EDX. Differences among pistillate lemma phytoliths of different species were judged taxonomically important, but even phytoliths with little taxonomic value are illustrated and clearly described. 22 figures (22 photos), 1 table.

Ter Welle, BJH 1976a On the occurrence of silica grains in the secondary xylem of the Chrysobalanaceae. *International Association of Wood Anatomists, Bulletin* 1976 (2): 19–29. Silica grains occur in ray cells of 62 spp (12 genera); three distribution patterns are described. Diameter of the globular grains is influenced by distance from the cambium and environmental conditions. 6 figures (5 photos), 5 tables.

_____1976b Silica grains in woody plants of the neotropics, especially Surinam. *In* Baas, P, Bolton, AJ and Catling, DM, eds, *Wood Structure in Biological and Technological Research*. Leiden Botanical Series 3, Leiden, University Press: 107–142. Silica grains were identified in secondary xylem of 300 genera. Shape, size and surface texture are examined. 2 plates (11 photos), 2 tables.

Theobald, WL, Krahulik, JL and Rollins, RC 1979 Trichome description and classification. In Metcalfe, CR and Chalk, L, eds, *Anatomy of the Dicotyledons*. Second edition, London, Clarendon Press: 40–53. Detailed description of trichome morphology. Mineralization is not discussed but trichomes are often deposition sites. 6 figures, 1 table.

Thomasson, JR 1978a Epidermal patterns of the lemma in some fossil and living grasses and their phylogenetic significance. *Science* 199: 975–977. Epidermal cell patterns of fossil grass floral bracts indicate taxonomic distinctions. Comparison to modern genera illustrates phylogenetic relations. 2 figures (7 photos).

_____1978b Clearing, cuticle removal, and staining for the fertile bracts (lemmas and paleas) of grass anthoecia. *Stain Technology* 53: 233–236. Preparation of grass inflorescence parts is described. Epidermal cell patterns are taxonomically significant. 8 photos.

_____1980a *Archaeoleersia nebraskensis* gen. et sp. Nov. (Gramineae-Oryzeae), a new fossil grass from the Late Tertiary of Nebraska. *American Journal of Botany* 67: 876–882. Description of a new grass genus and species based on paleas and lemmas includes silica cells. Comparisons to modern taxa indicate similarities to *Leersia ligularis*. 11 photos, 1 table.

_____1980b *Paleoriocoma* (Gramineae, Stipeae) from the Miocene of Nebraska: Taxonomic and phylogenetic significance. *Systematic Botany* 5: 233–240. Epidermal patterns of fossil grass paleas and lemmas are similar to those of the modern genus *Oryzopsis*. Evolutionary relations of these genera, *Stipa* and *Nassella* are discussed. 11 photos.

_____1981 Micromorphology of the lemma in *Stipa robusta* and *Stipa viridula* (Gramineae; Stipeae): Taxonomic significance. *The Southwest Naturalist* 26: 211–214. Epidermal cells of the lemma are described in 2 spp of *Stipa*. The presence/absence and distribution pattern of silica cells are taxonomically useful. 4 figures (2 photos).

_____1983 *Carex graceii* sp. n., and *Cyperocarpus eliasii* sp. n., *Cyperocarpus terrestris* sp. n., and *Cyperocarpus pulcherrima* sp. n. (Cyperaceae) from the Miocene of Nebraska. *American Journal of Botany* 70: 435–449. 4 new Miocene sedges are described from fossilized achene and perigynia material. Epidermal anatomy includes silica cells. 38 photos, 1 table.

Thomasson, JR, Nelson, ME and Zakrzewski, RJ 1986 A fossil grass (Gramineae: Chloridoideae) from the Miocene with Kranz anatomy. *Science* 233: 876–878. Leaf epidermal characters of a fossil are comparable to modern chloridoid grasses. Dumbbell silica bodies are noted, indicating C_4 photosynthesis. 10 photos.

Thomson, MC (ms) 1982 Phytolith studies: Techniques and their application to the examination of phytoliths from plants and sediments collected at Tel Michal, Israel. MS Plan B paper, University of Minnesota: 175 p. Preparation and examination techniques for plants and sediments from a study of material from Israel. Phytoliths from 4 plants are reported, focusing on trapezoid types. 10 plates (66 photos), 10 figures, 11 tables.

Thomson, M and Rapp, G, Jr 1989 Paleobotany from phytoliths. *In* Herzog, Z, Rapp, G, Jr and Negbi, O, eds, *Excavations at Tel Michal, Israel*. Minneapolis, University of Minnesota Press: 223–225. Brief application of phytolith analysis to paleobotany at Tel Michal presents analysis of 4 grasses and 2 sediments. Overall phytolith shape and size ranges of short trapezoids are discussed for 3 spp.

Thurston, EL 1974 Morphology, fine structure, and ontogeny of the stinging emergence of *Urtica dioica*. *American Journal of Botany* 61: 809–817. The stinging structure of nettles, studied by SEM, TEM, EDX and light microscopy, was found to have an elongate stinging cell and a multicellular pedestal. The stinging celi's apical wall is composed of silica bodies that decrease basipetally in concentration; these are measured but not described as to shape. 30 figures (28 photos).

Toivonen, H and Timonen, T 1976 Perigynium and achene epidermis in some species of *Carex*, subg. Vignea (Cyperaceae), studied by scanning electron microscopy. *Annales Botanici*

Fennici 13: 49–59. Epidermal characters of 23 spp are described by SEM, including silica bodies. Classification of the species is discussed. 48 photos (including good illustrations of cell shapes), 3 tables.

Tomlinson, PB 1961 *Anatomy of the Monocotyledons. II. Palmae.* Oxford, Clarendon Press: 453 p. Description of palm anatomy and morphology includes silica body and stegmata description and morphology. A key for identification of major groups is based in part on silica bodies (p 71). 9 plates (52 photos), 45 figures.

_____1969 *Anatomy of the Monocotyledons. III. Commelinales-Zingiberales.* Oxford, Clarendon Press: 446 p. Anatomical description of 16 monocotyledon families including silica body shape and location. Distribution in the families is summarized (pp 410–412). 86 figures.

Twiss, PC 1983 Dust deposition and opal phytoliths in the Great Plains. *In* Caldwell, WH, Schultz, CB and Stout, TM, eds, *Man and the Changing Environments in the Great Plains, Transactions of the Nebraska Academy of Sciences* 11: 73–82. Patterns and rates of dust deposition and grass phytolith analysis yield information on the development of the Great Plains. Distribution of phytolith types in 22 grasses. 2 figures, 10 tables.

_____1986 Morphology of opal phytoliths in C3 and C4 grasses. *In* Rovner, I, ed, Plant opal phytolith analysis in archaeology and paleoecology: Proceedings of the 1984 Phytolith Research Workshop, North Carolina State University, Raleigh, North Carolina, *Occasional Papers of The Phytolitharien* 1: 4–11. Festucoid class phytoliths (circular, elliptical, crescent, rectangular or oblong shapes) occur only in C_3 spp, generally reflecting high latitudes or high altitudes. Chloridoid shapes (saddles or barrels and double axes) occur only in C_4 spp (eg, short grass prairie) and indicate warm, arid regions. Panicoid shapes (crosses, dumbbells) are problematic since the subfamily Panicoideae has C_3 and C_4 grasses even in the same genera. Assumes that Festucoid, Chloridoid, and Panicoid class phytoliths do not occur in other subfamilies (but see Mulholland 1986, 1989).

_____1987 Grass-opal phytoliths as climatic indicators of the Great Plains Pleistocene. *In* Johnson, WC, ed, *Quaternary Environments of Kansas.* Kansas Geological Survey Guidebook Series 5: 179–188. Grass silica bodies are correlated to three subfamilies and C_3/C_4 pathways. Important paleoclimatic information is produced by phytoliths, especially when used in conjunction with palynomorphs. 11 figures (25 photos).

Twiss, PC, Suess, E and Smith, RM 1969 Morphological classification of grass phytoliths. *Soil Science Society of America, Proceedings* 33: 109–115. Widely cited article. Leaves of 17 spp of the Central Plains, US, were dry-ashed as spodograms and in disarticulation. Si bodies were classified as distinctive Festucoid, Chloridoid and Panicoid as well as redundant Elongate types. The distinctive types were recovered from soils and paleosols but not Pleistocene sediments. Useful review of phytolith systematics. 4 figures, 4 tables.

Uhl NW 1976a Developmental studies in *Ptychosperma* (Palmae). I. The inflorescence and the flower cluster. *American Journal of Botany* 63: 82–96. The inflorescences of 2 spp of arecoid palms are described in detail. Raphides and spherical silica bodies are present in the structures. 33 figures, 2 tables.

_____1976b Developmental studies in *Ptychosperma* (Palmae). II. The staminate and pistillate flowers. *American Journal of Botany* 63: 97–109. Structure and development of flowers in 2 spp of arecoid palms is described in detail. Raphides and spherical silica bodies are reported in the structures. 46 figures.

Umemoto, K 1973 Mode of morphological forms of crystalline inorganic component in plants: Silicon bodies in wheat. *Chemical and Pharmaceutical Bulletin* 21: 1391–1393. Shape and location of silica bodies in flag leaves of 4 wheat spp is inherited by descendants (2 amphidiploids). 2 photos.

Umlauf, M 1989 Occurrence of unusual short cell types in tropical grasses. *Phytolith Basic Research Project, Technical Paper No 1*, Columbia, University of Missouri. The morphology of unusual short cell phytoliths is described; distribution in 30 grasses is reported. 1 figure, 3 tables.

Uphof, JCT 1962 *Plant Hairs*. Berlin, Gebruder Borntraeger: 292 p. Comprehensive treatment of plant hairs includes information on Si and Ca incrustations. Description and classification of types of hairs are helpful since many phytoliths originate in hairs.

Velain, C 1878 Étude microscopique des verres résultant de la fusion des cendres de graminées. (Microscopic study of glasses resulting in the fusion of grass ashes.) *Bulletin de la Societe Mineralogique de France* 1: 113–124. (French). Vitreous masses of glass form by fusion of Si in grasses and other plants during fires. Glass from 3 fires described. 2 figures.

Vignal, C 1969 Contribution à l'étude épidermique de quelques Cypéracées: *Carex, Cyperus, Scirpus* et *Kobresia*. (Contribution to the epidermal study of several sedges: *Carex, Cyperus, Scirpus* and *Kobresia*.) *Annales de la Faculte des Sciences de Marseilles* 42: 313-321. (French). Epidermal characteristics (including phytoliths) can aid in taxonomic determinations. 7 sedge spp are described. 2 plates (13 figures).

Von Aderkas, P, Rogerson, A and de Freitas, ASW 1986 Silicon accumulation in fronds of the ostrich fern, *Matteuccia struthiopteris*. *Canadian Journal of Botany* 64: 696–699. Si was detected in mature petioles by EM and XRD. Deposition patterns are random and restricted to cell walls; no aggregates were observed. 6 figures, 1 table.

Watanabe, N 1968 Spodographic evidence of rice from prehistoric Japan. *Journal of the Faculty of Science* 3: 217–235. Spodograms, epidermal ash or silica skeletons, contain features that are characteristic of grass taxa. Rice can be identified by spodograms recovered from sediments. 3 plates (14 photos).

_____1970 A spodographic analysis of millet from prehistoric Japan. *Journal of the Faculty of Science* 3: 357–384. Rice and millet (*Setaria italica*) were identified by silica cells and other epidermal features. Comparison was made to spodograms of 28 varieties of millet (including *Panicum miliaceum* and *Echinochloa crus-galli*). 5 plates (62 photos).

Waterkeyn, L and Bienfait, A 1967 Les émergences callosiques et silicifiées des feuilles de Sélaginelles. (The callosic and silicified emergences of leaves of Selaginelles.) *Comptes-Rendus Hebdomadaires des Seances de l'Academie des Sciences, Paris, Series D, Sciences Naturelles* 264: 1608-1611. (French) Callose deposits on the leaf margin of *Selaginella kraussiana* become filled with Si. These sharp structures may help control transpiration. 4 figures.

Waterkeyn, L, Bienfait, A and Peeters, A 1982 Callose et silice épidermiques: Rapports avec la transpiration cuticulaire. (Epidermal callose and silica: Connection with cuticular transpiration.) *Cellule* 73: 267–287. (French, English summary). Silicification is preceded by callose deposits in several species. Morphology of the epidermal appendages is illustrated for *Selaginella, Urtica* and *Blumenbachia*. 5 figures.

Waterkeyn, L and Dupont, C 1982 L'observation des dépôts pariétaux de silice au microscope électronique à balayage. (Scanning electron microscopy of silica deposits isolated from plant cell walls.) *Bulletin Societe Royale de Botanique de Belgique* 115: 156–160. (French, English summary). Development of phytoliths in *Selaginella, Urtica* and *Cannabis* is related to presence of callose. Silicified hairs are described. 14 figures (11 photos).

Werner, O 1928 Blatt-Aschenbilder heimischer Wiesengräser als Mittel ihrer Verwandtschafts- und Wertbestimmung. (Leaf-ash picture of native meadow grass as means of its relationship and disposition.) *Biologia Generalis* 4: 403–446. (German). Spodograms of 47 grasses are described, including phytolith data. 4 plates (53 drawings), 7 figures.

Wilding, LP and Drees, LR 1971 Biogenic opal in Ohio soils. *Soil Science Society of America, Proceedings* 35: 1004–1010. Total amount and depth-distribution of 50–20 um opal is similar between Alfisols and Mollisols. Tree opal is typically smaller than 20 um but appears morphologically distinct from grass opal. 5 figures (29 photos).

_____1972 Biogenic opal in Ohio soils: Errata. *Soil Science Society of America, Proceedings* 36: 383–385. Reprint of photographs in more legible form from Wilding and Drees 1971. 3 figures (29 photos).

_____1973 Scanning electron microscopy of opaque opaline forms isolated from forest soils in Ohio. *Soil Science Society of America, Proceedings* 37: 647–650. SEM of opaque opal from soils shows great similarity to opal from deciduous trees. Some shapes are considered diagnostic of forest vegetation. 8 figures (9 photos), 1 table.

_____1974 Contributions of forest opal and associated crystalline phases to fine silt and clay fractions of soils. *Clays and Clay Minerals (Clay Mineral Society, Proceedings of the Conference)* 22: 295–306. Opal phytolith fraction of 6 deciduous trees includes crystalline phases, most in the 2–5 um-fraction. Tree opal is more soluble than grass opal. 11 figures (26 photos).

Wilding, LP, Smeck, NE and Drees, LR 1977 Silica in soils: Quartz, cristobalite, tridymite, and opal. *In* Dixon, JB and Weed, SB, eds, *Minerals in Soil Environments.* Madison, Wisconsin, Soil Science Society of America: 471–552. Comprehensive treatment of four minerals includes structural properties, quantification, formation and occurrence data. Mineralogic and chemical data on opal phytoliths is presented. 27 figures.

Wilson, SM 1985 Phytolith analysis at Kuk, an early agricultural site in Papua New Guinea. *Archaeology in Oceania* 20: 90–96. Phytoliths provide a means to study early plant domestication. A distinctive phytolith is in *Musa* sp (banana). 4 figures, 3 tables.

Witty, JE (ms) 1962 Grass opal in some chestnut and forested soils of Wasco County, Oregon. MS thesis, Oregon State University: 47 p. 11 grass spp and 8 other spp were studied. No distinction among grasses, sedges and rushes was made. Ponderosa pine, *Eriogonum* (buckwheat) and *Balsamorhiza* (balsamroot) might have distinctive shapes, but pine also had confusers for grass rods. *Chrysothamnus nauseosus* (rabbitbrush) and *Libocedrus decurrens* (incense cedar) had little or no Si. Grass/monocot Si was concluded to be a useful index mineral. 7 figures (6 photos), 4 tables, 1 appendix.

Witty, JE and Knox, EG 1964 Grass opal in some chestnut and forested soils in North Central Oregon. *Soil Science Society of America, Proceedings* 28: 685–688. Summary of Witty's (1962) thesis, without full list of species. Forest soils showed little or no recent history of grass cover, and segments of forest/grass boundary had persisted for several thousand years. 2 figures (4 photos), 1 table.

Yoshida, S, Ohnishi, Y and Kitagishi, K 1962a Histochemistry of silica in rice plant II. Localization of silicon within rice tissues. *Soil Science and Plant Nutrition* 8: 36–41. Location of Si within plant parts is determined. The epidermis and vascular bundles have the greatest amount; Si deposition is thought to be related to transpiration. 11 figures (10 photos).

_____1962b Histochemistry of silica in rice plant III. The presence of cuticle-silica double layer in the epidermal tissue. *Soil Science and Plant Nutrition* 8: 1–5. Examination of Si deposition indicates a layer is present beneath the cuticle. The "silica-cuticle double layer" is thought to provide protection against pests and reduce water evaporation. 11 figures (8 photos), 1 table.

PLANT NAME INDEX

GENERAL INDEX

ABOUT THE EDITORS

George (Rip) Rapp, Jr., is Professor of Geology and Archaeology and Director of the Archaeometry Laboratory, University of Minnesota, Duluth; Professor of Ancient Studies, University of Minnesota, Twin Cities; and Research Professor of Archaeology, Boston University. He has been involved as a geoarchaeologist in major projects in Greece, Cyprus, Israel, Turkey, Tunisia, China, and Italy as well as in North America.

Susan C. Mulholland is Associate Director of the Archaeometry Laboratory and Adjunct Assistant Professor of Anthropology, University of Minnesota, Duluth as well as Adjunct Assistant Professor of Ancient Studies, University of Minnesota, Twin Cities. She has been involved in phytolith analysis for 14 years, focusing on archaeology and paleoecology of the Upper Midwest and Northern Great Plains.

CORRESPONDENCE

Steven R. Bozarth
Department of Geography
University of Kansas
Lawrence, KS 66045

Vaughn M. Bryant, Jr.
Palynology Laboratory
Texas A&M University
College Station, TX 77843

Linda Scott Cummings
Paleoresearch Laboratories
15485 West 44th Avenue, Suite A
Golden, CO 80403

Elizabeth H. Dinan
Department of Anthropology
University of Missouri
103 Swallow Hall
Columbia, MO 65211

Martin J. Hodson
School of Biological Sciences
University of Birmingham
P.O. Box 363
Birmingham B15 2TT
United Kingdom

John G. Jones
Smithsonian Tropical
Research Institute
APO Miami, 34002

Lawrence Kaplan
Department of Biology
University of Massachusetts
Boston, MA 02125

Elizabeth J. Lawlor
Department of Anthropology
University of California, Riverside
Riverside, CA 92521

Susan C. Mulholland
Archaeometry Laboratory
University of Minnesota, Duluth
10 University Drive
Duluth, MN 55812-2496

Amy L. Ollendorf
3915 Elliot Avenue, South
Minneapolis, MN 55455

Deborah M. Pearsall
Department of Anthropology
American Archeology Division
University of Missouri, Columbia
103 Swallow Hall
Columbia, MO 65211

Alix H. Powers
Department of Archaeology
University of Cambridge
Downing Street
Cambridge CB2 3DZ
United Kingdom

George Rapp, Jr.
Archaeometry Laboratory
University of Minnesota, Duluth
10 University Drive
Duluth, MN 55812-2496

Arlene Miller Rosen
Archaeology Division
Ben Gurion University of the Negev
P.O. Box 653
Beer Sheva 84 105, Israel

Irwin Rovner
Department of Sociology
and Anthropology
North Carolina State University
Raleigh, NC 27650

John C. Russ
College of Materials Science
and Engineering
North Carolina State University
Raleigh, NC 27650

Allan G. Sangster
Division of Natural Sciences
Glendon College, York University
2275 Bayview Avenue
Toronto, Ontario M4N 3M6, Canada

Mary B. Smith
Department of Biology
University of Massachusetts
Boston, MA 02125

Lesley A. Sneddon
Department of Biology
University of Massachusetts
Boston, MA 02125

Page C. Twiss
Department of Geology
Kansas State University
Manhattan, KS 66506

GPSR Compliance

The European Union's (EU) General Product Safety Regulation (GPSR) is a set of rules that requires consumer products to be safe and our obligations to ensure this.

If you have any concerns about our products, you can contact us on ProductSafety@springernature.com

In case Publisher is established outside the EU, the EU authorized representative is:

Springer Nature Customer Service Center GmbH
Europaplatz 3
69115 Heidelberg, Germany

Batch number: 10761599

Printed by Printforce, the Netherlands